INTERNATIONAL CENTRE FOR MECHANICAL SCIENCES

COURSES AND LECTURES - No. 240

THIN SHELL THEORY NEW TRENDS AND APPLICATIONS

EDITED BY

W. OLSZAK

POLISH ACADEMY OF SCIENCES
INTERNATIONAL CENTRE FOR MECHANICAL SCIENCES
UDINE

Springer-Verlag Wien GmbH

ISBN 978-3-211-81602-8 ISBN 978-3-7091-2442-0 (eBook)
DOI 10.1007/978-3-7091-2442-0

Originally published by CISM, Udine in 1980.

PREFACE

The design of thin shell structures constitutes one of the most challenging fields in modern civil engineering, architecture, and aeronautics. With ever increasing spans and the shapes always more fanciful, the shell thickness is continually decreasing. Research and design face difficult and responsible problems. The texts included in this volume are intended to present some characteristic trends of the present-day developments.

In the last two decades, several international meetings of various kinds have been organized in order to formulate and discuss some particular questions of the shell theory. One of them, held in 1963 in Warsaw under the auspices of the International Association for Shell and Spatial Structures (IASS), was focussed on "Nonclassical Shell Problems" being mainly interested in some new trends of this theory and the ensuing engineering applications.(*)

Somewhat on similar lines much effort has been made in the last years to further deepen our understanding of the response of thin shell structures under condictions which previously were not taken into consideration and to find answers to virtually open questions.

The present volume offers seven contributions delivered during the CISM Huber Session on the above topics; they reflect the state-of-the-art in some "nonclassical" fields of the shell theory and/or present new results of recent research work in this challenging domain.

The contribution by W.B. Krätzig *offers a short introduction into the variables and governing equations of the general linear shell theory. Starting with a review of differential geometry of surfaces, the variables of the dynamic field are defined and the corresponding equations of motion and dynamic boundary conditions are formulated. Thereafter the kinematic field is treated in the same way. Constitutive equations for hyperelastic shells are then derived from a rate of energy equation as a first order approximation. Finally, two consistent formulations of Kirchhoff-Love-type shell theories and of those including shear deformation are formulated by virtue of the corresponding variational principles, their primal operators are derived and presented in a matrix notation.*

For a long time the diverse and often precocious buckling behaviour of shell structures seemed to defy rational analysis. Great efforts, theoretical and experimental, have been made to correctly seize these phenomena. W.T. Koiter's *contributions to the development of stability theory opened a new approach to the analysis of buckling phenomena and of nonlinear post-buckling behaviour of structures. He has also examined their sensitivity to initial imperfections and provided useful quantitative estimates for the magnitude of this sensitivity.*

This is well reflected in his lectures where firstly, the foundations and justification of shell theory, both in the linear and nonlinear range, are discussed. Furher on, the basic principles of elastic stability are treated. Finally, buckling of cylindrical shells under axial compression and external pressure are examined.

In a series of lectures delivered by M. Dikmen *a general linear theory of vibrations of thin elastic shells of arbitrary shape is presented and a number of particular cases of shells with simple geometry and various boundary conditions are treated as applications. After some information on the history of shell dynamics, the general form of the basic differential equations is discussed, whereupon the particular differential operator used throughout the lectures is given in detail.*

The following topics are covered and discussed: free vibrations, the use of asymptotic approach in obtaining a classification of this type of vibrations and the corresponding simplified equations, edge effects and the possibility of defining vibration zones, nodal lines and symmetries, spectral theorems, wave propagation, as well as a brief discussion of the inverse problem. The text is accompanied by a comprehensive list of references.

In the next contribution by W. Pietraszkiewicz *the deformation of the shell under Kirchhoff-Love constraints is decomposed into a translation, a pure stretch along principal directions of strains followed by a finite rotation of the principal directions. General formulae for the finite rotation are given in terms of displacements. The concept of a total finite rotation of the boundary element allows to construct the geometrical boundary conditions in terms of displacements as well as in terms of the finite rotations on the strains alone. The relations of the first-approximation geometrically nonlinear theory of elastic shells are given and their consistent simplification for small, moderate, and large rotations are discussed.*

Finally, an exact theory of finite rotations in shells is outlined and the results valid under the assumption of small strains are given. This allows for a refinement of some results of the first approximation shell-theory.

With ever increasing spans of modern shell structures and with continuously growing working and loading programmes, inelastic effects are not to be avoided. In the next contribution, W. Olszak *and* A. Sawczuk *present a concise survey on what has been and is being done in the analysis of inelastic response of thin shells. The general form of constitutive equations, linear and nonlinear visco-elastic response, creep and relaxation phenomena and elastic-plastic behaviour are briefly characterised and treated. Limit analysis and limit design, cyclic and transitory effects, geometric nonlinearity and dynamic phenomena are concisely characterized. Finally, the evolution of the notion of yield phenomena is presented in a concise way.*

Quite a class of structural materials can be regarded to exhibit a linear viscoelastic response of a general type. Convolutional and variational principles and their Laplace transforms can then successfully be applied in the formulation and solution of the involved time dependent problems.

Introducing the generalized potential energy, J. Brilla *derives in his contribution the general and convolutional variational principles in view of their applications to the analysis of quasistatic and dynamic phenomena occuring in visco-elastic shallow shells. An analogous approach is proved to be valid for a generalization of the finite element method for time dependent problems.*

It is also shown that a similar approach is possible in the analysis of stability problems, with applications to buckling and post-buckling phenomena of visco-elastic shallow shells.

There exists in Mechanics, similarly for the rest as in other fields of Sciences, certain formal analogies which may considerably facilitate the understanding and solution of the investigated phenomena. C.R. Calladine's *contribution explains the basis of the "static-geometric analog" in thin shell theory. It is related to the formal correspondence between the equilibrium equations on the one hand and the geometric compatibility equations on the other. Discussing the conditions prevailing in shallow shells, an explanation for this analogy is given. Paper also indicates the various limitations of the analogy and discusses its practical implications.*

We hope the present volume will constitute a useful source of information on some present-day problems of the shell theory.

W. Olszak

(*) cf. W. Olszak and A. Sawczuk (Editors), "Nonclassical Shell Problems", Proc. IASS International Congress 1963, North Holland Publ. Comp. Amsterdam and PWN Polish Scient. Publishers, Warsaw 1964, p. 1708.

LIST OF CONTRIBUTORS

J. BRILLA:	Professor, Comenius University, Bratislava, Czechoslovakia.
C.R. CALLADINE:	Professor, Cambridge University, Cambridge, U.K.
M. DIKMEN:	Professor, Technical University, Istanbul, Turkey.
W.T. KOITER:	Professor, University of Technology, Delft, The Netherlands.
W.B. KRÄTZIG:	Professor, Ruhr University, Bochum, Federal Republic of Germany.
W. OLSZAK:	Professor, Polish Academy of Sciences, Warsaw, Poland; International Centre for Mechanical Sciences (C.I.S.M), Udine, Italy.
W. PIETRASZKIEWICZ:	Assoc. Professor, Institute of Fluid-Flow Machinery of the Polish Academy of Sciences, Gdańsk, Poland.
A. SAWCZUK:	Professor, Polish Academy of Sciences, Warsaw, Poland.

CONTENTS

Introduction to General Shell Theory

by W.B. Krätzig

Page

1. Introductory Remarks 3
2. Shell Space and Middle Surface 6
   2.1. Geometry of the Middle Surface 6
   2.2. The Shell Space 11
3. The Dynamic Field 14
   3.1. External Dynamic Variables 14
   3.2. Internal Dynamic Variables 17
   3.3. Conditions of Equilibrium in Vector Form 21
   3.4. Conditions of Equilibrium in Component Form 23
   3.5. Dynamic Boundary Conditions 25
4. The Kinematic Field 28
   4.1. General Assumptions 28
   4.2. External Kinematic Variables 30
   4.3. Internal Kinematic Variables 32
   4.4. Kinematic Relations 36
5. Constitutive Equations 39
   5.1. The Elastic Potential for Hyperelastic Shells 39
   5.2. First-Order Approximation of Constitutive Equations 43
6. Linear Shell Theory Including Shear Deformations 45
   6.1. Review of Variables and Fundamental Equations 45
   6.2. The Principal of Virtual Displacement 47
   6.3. A Consistent Formulation 49
7. Linear Kirchhoff-Love-Type Shell Theory 52
   7.1. Derivation of the Basic Equations 52
   7.2. Consistent Formulation 55
8. Concluding Remarks 59

References 59

General Theory of Shell Stability
by W.T. Koiter

1. Basic Principles of Elastic Stability 65
   1.1. Energy Criteria for Conservative Systems 65
   1.2. The Energy Functional for Elastic Structures 67
   1.3. Neutral Equilibrium and Buckling Modes 69
   1.4. Stability as Critical Load 70
   1.5. Post-buckling Behaviour 71
   1.6. Influence of Geometric Imperfections 73
2. General Theory of Shell Buckling 76
3. Buckling of Cylindrical Shells under Axial Compression and External Pressure 77
   3.1. Neutral Equilibrium 77
   3.2. Evaluation for Simply Supported Edges 79
   3.3. Special Cases 80
   3.4. Nonlinear Theory of Shallow Buckling Modes 83
   3.5. Influence of Imperfection 86

Vibrations of Thin Elastic Shells
by M. Dikmen

Introduction 91
   1. Brief Historical Remarks 91
   2. Scope of the Lectures 92
   3. Description of the Shell 93
   4. General Form of the Differential Equations 95
   5. Choice of a Particular Differential Operator 95
Chapter I. Free Vibrations 100
   1. Standing Waves 100
   2. Ideas Underlying the Asymptotic Approach to Shell Problems 102
   3. Definitions and Preliminaries 103
   4. The Different Cases of Free Vibrations 109
Chapter II. The Central Set of Equations 113
   1. The Central Set of Equations 113
   2. The Frequency Equation 114
Chapter III. Edge Effect 117
   1. Assumptions Concerning Geometric Quantities 117

2. The Frequency Equation with Constant Coefficients . . . 117
3. Governing Equation for the Edge Effect . . . 119
4. Classification of the Edge Effects . . . 119
5. Vibration Zones . . . 123
Chapter IV. Nodal Lines . . . 126
1. Density of Nodal Lines . . . 126
2. Symmetry Properties of the Nodal Pattern and Group Theory . . . 127
Chapter V. Spectral Theorems . . . 129
1. The Eigenvalue Problem and Further Properties of the Differential Operator . . 129
2. Distribution of Eigenvalues . . . 130
3. Relation to the Spectrum of the Membrane Operator . . . 131
4. Ellipticity of the operator L− λI . . . 132
5. Density of Distribution of the Eigenfrequencies . . . 132
Chapter VI. The Inverse Problem . . . 134
1. General Formulation . . . 134
2. Perturbation and Moment Problems . . . 136
Chapter VII. Wave Propagation . . . 138
1. Some Definitions . . . 138
2. Monochromatic Wave Solution . . . 138
3. Propagation of Discontinuities . . . 140
Complements . . . 142
a. Loaded Shells . . . 142
b. Thick Shells . . . 142
c. Additional References . . . 142
References . . . 143

Finite Rotations of the Nonlinear Theory of Thin Shells
by W. Pietraszkiewicz

Abstract . . . 153
1. Introduction . . . 155
2. Strains and Rotations in Thin Shells . . . 157
2.1. Notations and Preliminary Relations . . . 157
2.2. Deformation of Thin Shells under Kirchhoff-Love Constraints . . . 160
2.3. Finite Rotation Tensor and Vector . . . 164
3. Deformation of Shell Boundary . . . 169
3.1. Total Rotation of a Boundary . . . 169

3.2. Differentiation Along the Boundary . . . 172
3.3. Geometric Boundary Conditions . . . 175
4. Basic Shell Equations . . . 178
4.1. Equilibrium Equations and Static Boundary Conditions . . . 178
4.2. Modified Static Boundary Conditions . . . 184
4.3. Simplified Shell Relations Under Small Elastic Strains . . . 187
4.4. Canonized Intrinsic Shell Equations . . . 193
5. Geometrically Nonlinear Theory of Shells . . . 196
5.1. Classification of Rotations . . . 196
5.2. Moderate Rotation Theory of Shells . . . 198
5.3. Large Rotation Theory of Shells . . . 201
References . . . 205

Inelastic Response of Thin Shells. Basic Problems and Applications
by W. Olszak and A. Sawczuk

1. Introduction . . . 211
2. Constitutive Equations . . . 212
3. Assumptions of the Shell Theory . . . 214
4. Linear Visco-elasticity . . . 215
5. Steady Creep . . . 216
6. Elastic-plastic Deformations . . . 217
7. Limit Analysis and Limit Design. Fundamentals and Applications . . . 218
8. Cyclic and Transitory Effects . . . 219
9. Nonlinearity; Dynamic Effects . . . 221
10. Evolution of Yield Criteria . . . 221
11. Concluding Remarks . . . 222
References . . . 224

Variational Principles and Methods for Viscoelastic Shallow Shells
by J. Brilla

Abstract . . . 245
1. Introduction . . . 246
2. Constitutive Equations . . . 247
3. Generalized Variational Theorems and Methods . . . 248
4. Generalized Variational Methods . . . 254

5. Dynamic Problems of Viscoelastic Shells 260
6. Convolutional Variational Principles and Methods 263
7. Convolutional Variational Principles of Stability of Viscoelastic Shallow Shells . . 267
References 277

The Static-geometric Analogy in the Equations of Thin Shell Structures
by C.R. Calladine

Abstract 281
Introduction 281
Gaussian Curvature and Change of Gaussian Curvature 283
A Two-Surface Model of a Shell 285
Equations of the Problem 288
Explanation of the Static-Geometric Analogy 290
Boundary Conditions 295
General Surface Tractions and the Need for Numerical Procedures 296
Practical Applications 298
References 300

INTRODUCTION TO GENERAL SHELL THEORY

by

W.B. Krätzig

Ruhr University, Bochum

INTRODUCTION TO GENERAL SHELL THEORY

W.B. Krätzig

1. INTRODUCTORY REMARKS

As we all know from daily experience we are living in a geometrically three-dimensional world. The aim of any shell theory is to describe the mechanical behaviour of thin, three-dimensional bodies in a two-dimensional manner, namely by only two spatial coordinates. Because any unique mapping from a three- to a two-dimensional space is incompatible with our experience, this goal obviously can only be achieved in an *approximative* sense.

There are two conceptually different ways to derive general shell equations. The first one is the derivation and approximation from the point of view of classical continuum mechanics. After defining a reference surface F all basic mechanical equations of a three-dimensional space are transformed such that the mechanical variables remain only functions of the co-ordinates of F. If we then restrict the deformations of the body to surface-like ones, we end up with two-dimensional shell equations.

This roughly sketched way is full of various approximations of the shell space and the states of stresses and deformations. Indeed, only recently systematic procedures have been developed, which allow a critical judgement of the effect of such approximations [7, 8, 15, 16] and an estimate of the errors involved [9, 10]. Still, a lot of questions remains unanswered regarding the evaluation of error bounds of this method.

The second way to derive general shell equations starts off with the definition of a mass-covered reference surface F, which has a load-bearing capacity. To every point of F a vector quantity, called a *director*, is assigned, which is able to portray its kinematic behaviour. The definition of internal and external mechanical variables in terms of corresponding quantities finally leads to a two-dimensional set of basic equations of shell theory [2, 12, 16, 18].

The problems of this alternative way can be succinctly summarized in the following questions: How can the two-dimensional *Cosserat Model* be related to our three-dimensional world? How can a two-dimensional internal energy or two-dimensional constitutive relations portray the known behaviour of our three-dimensional bodies?

Apparently both ways have their advantages and their disadvantages. The first one carries the burden of a whole pile of geometrical and mechanical approximations and thus may lose much of its clearness. The second one overcomes all these difficulties by the *a-priori-definition* of the *Cosserat Surface*, but it leaves all questions of the relationship between the model and the reality to the user.

For an introduction into general shell theory we favour a mixed derivation, which is oriented towards the concept of *Cosserat Surfaces*, but emphasizes the close connections to three-dimensional continuum mechanics. For all those readers, who thereafter look forward to a deeper understanding of this exciting part of mechanics the monograph [16] may be recommended.

Finally the question may be raised, why shell theory is used after all. In the first place, because of our limited mathematical tools: two-dimensional problems can be treated much easier in practical applications than three-dimensional ones. Secondly, because many *thin* structures behave in a surface-like manner and in the light of our limited capacity of experience and insight a two-dimensional description then is more convenient. Thirdly, because well-developed three-dimensional computational tools may lose two-dimensional phenomena in a haystack of (mostly irrelevant) computer output.

2. SHELL SPACE AND MIDDLE SURFACE

2.1 Geometry of the Middle Surface

Our reader is assumed to have a thorough knowledge of differential geometry, as it is expounded in many mathematical textbooks and monographs. In this section we repeat the basic formulae and some selected results mainly in order to introduce our notation.

According to Fig. 1 we consider a surface F, to be later denoted by *middle surface*, imbedded in the three-dimensional *Euclidean* space. This E3 is described by the system x^i $(i = 1,2,3)$ of normal coordinates and the ortho-normal vector base $\underset{\sim}{i}_i$. The surface F is also expressed as a two-dimensional manifold θ^α $(\alpha = 1,2)$ of *Gaussian* parameters.

The position vector $\underset{\sim}{r}$ of any arbitrary point $P(x^i)$ of F can be represented in the following way:

$$\underset{\sim}{r} = \underset{\sim}{r}\ (x^i) = \underset{\sim}{r}\ (\theta^1,\ \theta^2) = \underset{\sim}{r}\ (\theta^\alpha) \quad \text{with:} \quad x^i = x^i\ (\theta^\alpha) \tag{2.1}$$

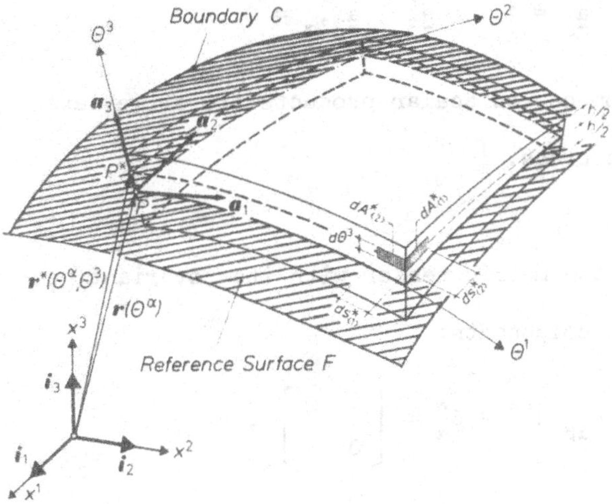

Fig. 1: Shell Space, Euclidean Space and Middle Surface

Herein and further on the tilde as a subscript denotes *vector* and *tensor quantities*, which are printed in bold types in the figures. Provided $\underset{\sim}{r}$ and $\underset{\sim}{r}_{,\alpha}$ are continuous we derive from (2.1) the covariant base vectors of point P

$$\underset{\sim}{a}_\alpha = \frac{\partial \underset{\sim}{r}}{\partial \theta^\alpha} = \underset{\sim}{r}_{,\alpha}\ , \tag{2.2}$$

where partial differentiation is denoted by a comma. The corresponding unit normal vector

$$\underset{\sim}{a}_3 = \underset{\sim}{a}^3 = \frac{\underset{\sim}{a}_1 \times \underset{\sim}{a}_2}{|\underset{\sim}{a}_1 \times \underset{\sim}{a}_2|} = \underset{\sim}{a}_1 \times \underset{\sim}{a}_2 \Big/ \sqrt{a} \tag{2.3}$$

is related to $\underset{\sim}{a}_\alpha$ by:

$$\underset{\sim}{a}_\alpha \cdot \underset{\sim}{a}_3 = 0, \qquad \underset{\sim}{a}_3 \cdot \underset{\sim}{a}_3 = 1, \qquad \underset{\sim}{a}_3 \cdot \underset{\sim}{a}_{3,\alpha} = 0 \,. \tag{2.4}$$

In both equations and from now on scalar products are abbreviated by a dot, vector products by a cross.

Now we proceed to define the metric tensor with its covariant, contravariant and mixed variant components:

$$a_{\alpha\beta} = \underset{\sim}{a}_\alpha \cdot \underset{\sim}{a}_\beta \,, \qquad a_{\alpha\beta}\, a^{\beta\lambda} = \delta^\lambda_\alpha = \begin{bmatrix} 1 & 0 \\ 0 & 1 \end{bmatrix},$$

$$\underset{\sim}{a}_\alpha = a_{\alpha\beta}\, \underset{\sim}{a}^\beta \,, \qquad \underset{\sim}{a}^\alpha = a^{\alpha\beta}\, \underset{\sim}{a}_\beta \,, \tag{2.5}$$

$$a = \det a_{\alpha\beta} \,, \qquad a^{-1} = \det a^{\alpha\beta}.$$

The metric tensor enables us to raise and lower tensor indices and to compute the length of the base vectors:

$$|\underset{\sim}{a}_\alpha| = \sqrt{a_{\alpha\alpha}} \,, \qquad |\underset{\sim}{a}^\alpha| = \sqrt{a^{\alpha\alpha}} \,. \tag{2.6}$$

In (2.5) we have already introduced *Einstein's* summation convention, according to which two equal indices, one as a sub-, the other as a superscript, require a summation; e.g.:

$$\underset{\sim}{a}_\alpha = a_{\alpha\beta}\, \underset{\sim}{a}^\beta = a_{\alpha 1}\, \underset{\sim}{a}^1 + a_{\alpha 2}\, \underset{\sim}{a}^2 \,, \qquad \delta^\alpha_\alpha = \delta^1_1 + \delta^2_2 = 2 \,. \tag{2.7}$$

δ^λ_α used herein and in (2.5) denotes the *Kronecker Symbol*; in all previous and future formulae *latin (greek) indices* represent the numbers 1, 2, 3 (1, 2). Finally we recall the permutation tensor $\underset{\sim}{\varepsilon}$ with its components

$$\varepsilon_{\alpha\beta} = \begin{bmatrix} 0 & \sqrt{a} \\ -\sqrt{a} & 0 \end{bmatrix} , \quad \varepsilon^{\alpha\beta} = \begin{bmatrix} 0 & 1/\sqrt{a} \\ -1/\sqrt{a} & 0 \end{bmatrix} \tag{2.8}$$

and their part they are playing in the evaluation of the vector products:

$$\begin{aligned} \underset{\sim}{a}_\alpha \times \underset{\sim}{a}_\beta &= \varepsilon_{\alpha\beta}\, \underset{\sim}{a}^3 , & \underset{\sim}{a}^\alpha \times \underset{\sim}{a}^\beta &= \varepsilon^{\alpha\beta} \underset{\sim}{a}_3 , \\ \underset{\sim}{a}_3 \times \underset{\sim}{a}_\beta &= \varepsilon_{\beta\rho}\, \underset{\sim}{a}^\rho , & \underset{\sim}{a}^3 \times \underset{\sim}{a}^\beta &= \varepsilon^{\beta\rho} \underset{\sim}{a}_\rho . \end{aligned} \tag{2.9}$$

We now turn to the curvature tensor $\underset{\sim}{b}$ of the reference surface F ; its covariant components are defined by:

$$b_{\alpha\beta} = -\underset{\sim}{a}_\alpha \cdot \underset{\sim}{a}_{3,\beta} = -\underset{\sim}{a}_\beta \cdot \underset{\sim}{a}_{3,\alpha} = \underset{\sim}{a}_3 \cdot \underset{\sim}{a}_{\alpha,\beta} = \underset{\sim}{a}_3 \cdot \underset{\sim}{a}_{\beta,\alpha} . \tag{2.10}$$

Mixed and contravariant components can be evaluated by introducing (2.5):

$$b^\alpha_\beta = a^{\alpha\lambda}\, b_{\lambda\beta} , \qquad b^{\alpha\beta} = a^{\alpha\lambda}\, a^{\beta\nu}\, b_{\lambda\nu} , \tag{2.11}$$

and for the mean curvature H and the *Gaussian* curvature K we find:

$$H = \frac{1}{2}\, b^\alpha_\alpha = \frac{1}{2}\,(b^1_1 + b^2_2) , \qquad K = \frac{|b_{\alpha\beta}|}{a} = b^1_1\, b^2_2 - b^1_2\, b^2_1 . \tag{2.12}$$

The partial differentiations of the tangential base vectors $\underset{\sim}{a}_\alpha$, $\underset{\sim}{a}^\alpha$ and the unit normal vector $\underset{\sim}{a}_3$ are described by the equations of

Weingarten and *Gauss:*

$$\underset{\sim}{a}_{\alpha},_{\beta} = \Gamma^{\lambda}_{\alpha\beta}\, \underset{\sim}{a}_{\lambda} + b_{\alpha\beta}\, \underset{\sim}{a}_{3} , \qquad \underset{\sim}{a}^{\alpha},_{\beta} = -\Gamma^{\alpha}_{\beta\lambda}\, \underset{\sim}{a}^{\lambda} + b^{\alpha}_{\beta}\, \underset{\sim}{a}_{3} ,$$
$$\underset{\sim}{a}_{3},_{\alpha} = -b^{\lambda}_{\alpha}\, \underset{\sim}{a}_{\lambda} . \tag{2.13}$$

At the same time these relations define the covariant differentiation of the base vectors, denoted by a vertical bar:

$$\underset{\sim}{a}_{\alpha},_{\beta} - \underset{\sim}{a}_{\lambda}\Gamma^{\lambda}_{\alpha\beta} = \underset{\sim}{a}_{\alpha}\big|_{\beta} = b_{\alpha\beta}\underset{\sim}{a}^{3}, \quad \underset{\sim}{a}^{\alpha},_{\beta} + \underset{\sim}{a}^{\lambda}\Gamma^{\alpha}_{\lambda\beta} = \underset{\sim}{a}^{\alpha}\big|_{\beta} = b^{\alpha}_{\beta}\, \underset{\sim}{a}^{3}, \tag{2.14}$$

and the *Christoffel Symbols*, for which the following relations hold:

$$\Gamma_{\beta\gamma\alpha} = \frac{1}{2}\,(a_{\alpha\beta},_{\gamma} + a_{\alpha\gamma},_{\beta} - a_{\beta\gamma},_{\alpha}) ,$$
$$\Gamma^{\alpha}_{\beta\gamma} = a^{\alpha\lambda}\,\Gamma_{\beta\gamma\lambda} , \quad \Gamma^{\lambda}_{\lambda\alpha}\sqrt{a} = \sqrt{a},_{\alpha} . \tag{2.15}$$

The *Gauss-Codazzi* equation and the *Ricci-Lemma* conclude our review of some formulae of differential geometry:

$$b_{\alpha 1}\big|_{2} = b_{\alpha 2}\big|_{1} , \quad a_{\alpha\beta}\big|_{\lambda} = a^{\alpha\beta}\big|_{\lambda} = \varepsilon_{\alpha\beta}\big|_{\lambda} = \varepsilon^{\alpha\beta}\big|_{\lambda} = 0 . \tag{2.16}$$

The base vectors at point P of F may be used advantageously to decompose vectors or tensors defined at this point:

$$\underset{\sim}{v} = v_{i}\underset{\sim}{a}^{i} = v_{\alpha}\underset{\sim}{a}^{\alpha} + v_{3}\underset{\sim}{a}^{3} = v^{i}\underset{\sim}{a}_{i} = v^{\alpha}\underset{\sim}{a}_{\alpha} + v^{3}\underset{\sim}{a}_{3} ,$$
$$\underset{\sim}{t} = t_{ij}\underset{\sim}{a}^{i}\underset{\sim}{a}^{j} = t_{\alpha\beta}\underset{\sim}{a}^{\alpha}\underset{\sim}{a}^{\beta} + t_{\alpha 3}\underset{\sim}{a}^{\alpha}\underset{\sim}{a}^{3} + t_{3\beta}\underset{\sim}{a}^{3}\underset{\sim}{a}^{\beta} + t_{33}\underset{\sim}{a}^{3}\underset{\sim}{a}^{3} . \tag{2.17}$$

In this decomposition the normal components v^3 and v_3 are equal because of $\underset{\sim}{a}^3 = \underset{\sim}{a}_3$. The second order tensor $\underset{\sim}{t}$ of course can also be decomposed with respect to $\underset{\sim}{a}_i\underset{\sim}{a}_j$, $\underset{\sim}{a}^i\underset{\sim}{a}_j$ or $\underset{\sim}{a}_i\underset{\sim}{a}^j$. Proceeding in the same manner, we find for surface vectors and surface tensors:

$$\underset{\sim}{v} = v_\alpha \underset{\sim}{a}^\alpha = v^\alpha \underset{\sim}{a}_\alpha \ ,$$

$$\underset{\sim}{t} = t_{\alpha\beta}\underset{\sim}{a}^\alpha \underset{\sim}{a}^\beta = t^{\alpha\beta}\underset{\sim}{a}_\alpha \underset{\sim}{a}_\beta = t_{\alpha\cdot}^{\ \beta}\, a^\alpha\, a_\beta = t^{\alpha}_{\cdot\beta}\, a_\alpha a^\beta \ . \tag{2.18}$$

Physical vector and tensor components can be defined by means of (2.6), (2.17) through:

$$\underset{\sim}{v} = v_\alpha \underset{\sim}{a}^\alpha + v_3 \underset{\sim}{a}^3 = v_\alpha \sqrt{a^{\alpha\alpha}}\, \underset{\sim}{a}^{\langle\alpha\rangle} + v_3 \underset{\sim}{a}^{\langle 3\rangle} = v_{\langle\alpha\rangle} \underset{\sim}{a}^{\langle\alpha\rangle} + v_{\langle 3\rangle} \underset{\sim}{a}^{\langle 3\rangle} ,$$

$$= v^\alpha \underset{\sim}{a}_\alpha + v^3 \underset{\sim}{a}_3 = v^\alpha \sqrt{a_{\alpha\alpha}}\, \underset{\sim}{a}_{\langle\alpha\rangle} + v^3 \underset{\sim}{a}_{\langle 3\rangle} = v^{\langle\alpha\rangle} \underset{\sim}{a}_{\langle\alpha\rangle} + v^{\langle 3\rangle} \underset{\sim}{a}_{\langle 3\rangle} ,$$

$$v_{\langle\alpha\rangle} = v_\alpha \sqrt{a^{\alpha\alpha}} , \quad v^{\langle\alpha\rangle} = v^\alpha \sqrt{a_{\alpha\alpha}} , \quad v_{\langle 3\rangle} = v^{\langle 3\rangle} = v_3 \ . \tag{2.19}$$

If we extend the definition of a covariant differentiation (2.14) to the components of a surface vector we find:

$$v_\lambda\big|_\alpha = v_{\lambda,\alpha} - v_\mu \Gamma^{\mu}_{\lambda\alpha} , \qquad v^\lambda\big|_\alpha = v^\lambda{}_{,\alpha} + v^\mu\, \Gamma^{\lambda}_{\mu\lambda} \ . \tag{2.20}$$

2.2 The Shell Space

The shell space shall be separated from the remaining *Euclidean* space by the *outer* and *inner face* and by the *boundary* of the shell. The middle surface F bisects the distance h - the shell thickness - between both faces. Any arbitrary point P^* of the shell space will be represented by:

$$\underset{\sim}{r}^* (\theta^i) = \underset{\sim}{r} (\theta^1, \theta^2) + \theta^3 \underset{\sim}{a}_3 (\theta^1, \theta^2) = \underset{\sim}{r} + \theta^3 \underset{\sim}{a}_3 \quad , \qquad (2.21)$$

where $\underset{\sim}{r}$ abbreviates the position vector, θ^α the curvilinear *Gaussian* parameters of the middle surface F and θ^3 a straight coordinate line in the direction of $\underset{\sim}{a}_3$:

$$-h/2 \leq \theta^3 \leq +h/2 \quad . \qquad (2.22)$$

This representation (2.21) enables us together with (2.2), to define at any point P^* a triple of base vectors $\underset{\sim}{a}^*_i$

$$\begin{aligned} \underset{\sim}{a}^*_\alpha &= \underset{\sim}{r}^*{}_{,\alpha} = \underset{\sim}{r}{}_{,\alpha} + \theta^3 \underset{\sim}{a}_{3,\alpha} = (\delta^\beta_\alpha - \theta^3 b^\beta_\alpha)\, \underset{\sim}{a}_\beta = \mu^\beta_\alpha \underset{\sim}{a}_\beta \; , \\ \underset{\sim}{a}^*_3 &= \underset{\sim}{r}^*{}_{,3} = \underset{\sim}{a}_3 \end{aligned} \qquad (2.23)$$

as well as the covariant and contravariant metric tensors:

$$\begin{aligned} a^*_{\alpha\beta} &= \underset{\sim}{a}^*_\alpha \cdot \underset{\sim}{a}^*_\beta = a_{\alpha\beta} - 2\theta^3 b_{\alpha\beta} + (\theta^3)^2 b^\lambda_\alpha b_{\lambda\beta} \; , \\ a^*_{\alpha\beta}\, a^{*\beta\lambda} &= \delta^\lambda_\alpha \; , \\ a^{*\alpha\beta} &= a^{\alpha\beta} + 2\theta^3 b^{\alpha\beta} + 3(\theta^3)^2 b^\alpha_\lambda b^{\lambda\beta} + \ldots \; . \end{aligned} \qquad (2.24)$$

The tensor function μ^β_α in (2.23) shifts the base vectors $\underset{\sim}{a}_\beta$ of the middle surface to those of point P^*, and it is accordingly called *shifter-tensor*. We would like to call the reader's attention to several restrictions imposed upon μ_{α}^{β} (see [17], p. 17).

The definition of $\overset{*}{\underset{\sim}{a}}_i$ enables us to relate all geometrical quantities of the shell space to those of the middle surface F , e.g.:

$$\sqrt{\overset{*}{a}} = (\overset{*}{\underset{\sim}{a}}_1 \times \overset{*}{\underset{\sim}{a}}_2) \cdot \overset{*}{\underset{\sim}{a}}_3$$

$$= \sqrt{a}\ (1 - 2\theta^3 H + (\theta^3)^2 K) \ , \tag{2.25}$$

$$\mu = \det \mu_{\alpha}^{\beta} = \sqrt{\frac{\overset{*}{a}}{a}} = 1 - 2\theta^3 H + (\theta^3)^2 K \ .$$

All starred quantities of the shell space obviously transform into their corresponding unstarred quantities of F , when θ^3 approaches 0 . Finally we derive the elements of the coordinate lines

$$d\overset{*}{s}_{\langle 1\rangle} = \sqrt{\overset{*}{a}_{11}}\ d\theta^1 = \sqrt{\overset{*}{a}\,\overset{*}{a}{}^{22}}\ d\theta^1 \ : \tag{2.26}$$

$$d\overset{*}{s}_{\langle\alpha\rangle} = \sqrt{\overset{*}{a}_{\alpha\alpha}}\ d\theta^{\alpha} = \sqrt{\overset{*}{a}\,\overset{*}{a}{}^{\beta\beta}}\ d\theta^{\alpha}, \quad (\alpha \neq \beta)$$

and the surface elements of the cross-section in Fig. 1:

$$d\overset{*}{A}_{\langle 1\rangle} = d\overset{*}{s}_{\langle 2\rangle}\, d\theta^3 = \sqrt{\overset{*}{a}\,\overset{*}{a}{}^{11}}\ d\theta^2 d\theta^3 \ : \tag{2.27}$$

$$d\overset{*}{A}_{\langle\alpha\rangle} = d\overset{*}{s}_{\langle\beta\rangle}\, d\theta^3 = \sqrt{\overset{*}{a}\,\overset{*}{a}{}^{\alpha\alpha}}\ d\theta^{\beta} d\theta^3 \ , \ (\alpha \neq \beta) \ .$$

Both elements are physical (real) quantities; both equations use solution of

$$a^*_{\alpha\beta}\, a^{*\beta\lambda} = \delta^\lambda_\alpha \; : \quad a^{*\mu\mu} = a^* a^*_{\nu\nu} \;, \qquad (\mu \neq \nu) \tag{2.28}$$

and are valid - without a star - also for the middle surface.

3. THE DYNAMIC FIELD

3.1 External Dynamic Variables

We now consider a loaded and deformed shell structure, which will be represented by its middle surface F. Because we restrict our derivation to infinitesimally small displacements, any distinction between the deformed and the undeformed configuration for all equilibrium questions is immaterial. In this and the next section we follow the more detailed treatment in [11].

Let us consider in Fig. 2 an element dF of the middle surface, which is loaded by a *force vector* $\underset{\sim}{p}$ and a *moment vector* $\underset{\sim}{c}$. Both *external dynamic variables* are related to the real area of F , such that the resulting external force and moment acting on dF is given by:

$$\underset{\sim}{p}dF = \underset{\sim}{p}\sqrt{a}\, d\theta^1 d\theta^2 \;, \quad \underset{\sim}{c}dF = \sqrt{a}\, d\theta^1 d\theta^2 \;. \tag{3.1}$$

The force load vector $\underset{\sim}{p}$ will be decomposed with respect to the unit vectors $\underset{\sim}{a}_{\langle i\rangle}$ and also with respect to $\underset{\sim}{a}_i$:

$$\begin{aligned} \underset{\sim}{p} &= p^{\langle 1\rangle}\underset{\sim}{a}_{\langle 1\rangle} + p^{\langle 2\rangle}\underset{\sim}{a}_{\langle 2\rangle} + p^{\langle 3\rangle}\underset{\sim}{a}_{\langle 3\rangle} = p^{\langle\alpha\rangle}\underset{\sim}{a}_{\langle\alpha\rangle} + p^{\langle 3\rangle}\underset{\sim}{a}_{\langle 3\rangle} \\ &= p^{\langle 1\rangle}\frac{\underset{\sim}{a}_1}{\sqrt{a_{11}}} + p^{\langle 2\rangle}\frac{\underset{\sim}{a}_2}{\sqrt{a_{22}}} + p^{\langle 3\rangle}\underset{\sim}{a}_3 \\ &= p^1\underset{\sim}{a}_1 + p^2\underset{\sim}{a}_2 + p^3\underset{\sim}{a}_3 = p^\alpha\underset{\sim}{a}_\alpha + p^3\underset{\sim}{a}_3 . \end{aligned} \tag{3.2}$$

(3.2) leads to the following relations between the tensorial (p^α, p^3) and the physical $(p^{\langle\alpha\rangle}, p^{\langle 3\rangle})$ components

$$p^{\langle\alpha\rangle} = p^\alpha\sqrt{a_{\alpha\alpha}} , \quad p^{\langle 3\rangle} = p^3 , \tag{3.3}$$

which are of great importance for the transformation of the acting physical loads into the tensor space.

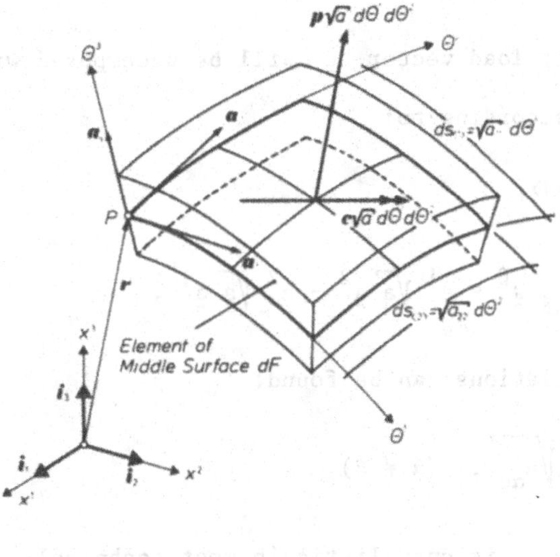

Fig. 2: Loaded element dF

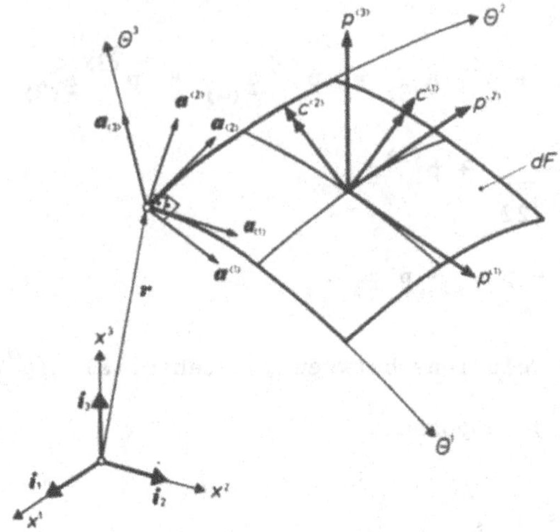

Fig. 3: Positive directions of all physical load variables with unit vector base

On the other hand the moment load vector $\underset{\sim}{c}$ will be decomposed with respect to $\underset{\sim}{a}^{\langle\beta\rangle}$ and a^{β} according to:

$$\begin{aligned} \underset{\sim}{c} &= c_{\langle 1\rangle} \underset{\sim}{a}^{\langle 2\rangle} - c_{\langle 2\rangle} \underset{\sim}{a}^{\langle 1\rangle} \\ &= c^{\rho} \underset{\sim}{a}_3 \times \underset{\sim}{a}_{\rho} = c^{\rho} \varepsilon_{\rho\beta} \underset{\sim}{a}^{\beta} = c^1 \sqrt{a}\, \underset{\sim}{a}^2 - c^2 \sqrt{a}\, \underset{\sim}{a}^1 , \end{aligned} \tag{3.4}$$

from which the following relations can be found:

$$c_{\langle\alpha\rangle} = c^{\alpha} \sqrt{a\, a^{\beta\beta}} = c^{\alpha} \sqrt{a_{\alpha\alpha}} , \quad (\alpha \neq \beta) . \tag{3.5}$$

Because the existence of $c_{\langle 3\rangle}$ is unrealistic in most technical applications, we have deleted this component completely. Finally the previously defined positive directions of all physical components (3.3, 3.5) are shown in Fig. 3.

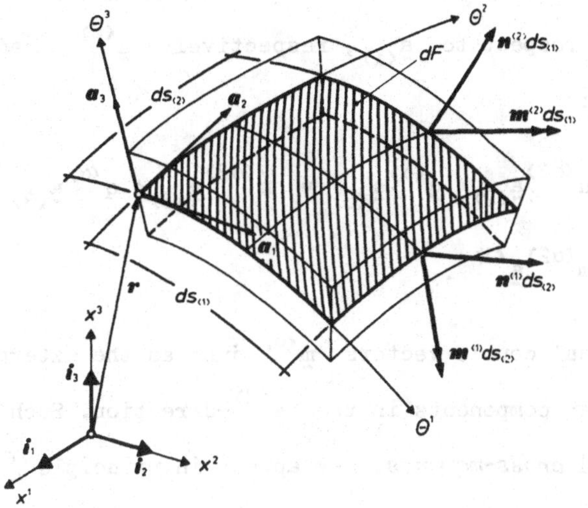

Fig. 4: Element of middle surface with vectorial internal dynamic variables

3.2 Internal Dynamic Variables

Let us now consider a loaded element dF of the middle surface in a state of equilibrium. In order to balance the external forces and couples the following internal dynamic variables must act along the section lines of dF (see Fig. 4):

$$\begin{aligned} &\text{along } \theta^1 = \text{const.}: \quad \underset{\sim}{n}^{\langle 1\rangle}\sqrt{a_{22}}\; d\theta^2, \quad \underset{\sim}{m}^{\langle 1\rangle}\sqrt{a_{22}}\; d\theta^2, \\ &\text{along } \theta^2 = \text{const.}: \quad \underset{\sim}{n}^{\langle 2\rangle}\sqrt{a_{11}}\; d\theta^1, \quad \underset{\sim}{m}^{\langle 2\rangle}\sqrt{a_{11}}\; d\theta^1. \end{aligned} \tag{3.6}$$

The *internal force vectors* $\underset{\sim}{n}^{\langle\alpha\rangle}$ and the *internal couple vectors* $\underset{\sim}{m}^{\langle\alpha\rangle}$ are defined as physical (real) quantities and are related to the real lengths (2.26) of the parameter lines θ^α. Their decomposition into physical components with respect to $\underset{\sim}{a}_{\langle i\rangle}$, respectively $\underset{\sim}{a}^{\langle\beta\rangle}$, leads to the following results:

$$\underset{\sim}{n}^{\alpha} = n^{\langle\alpha 1\rangle}\underset{\sim}{a}_{\langle 1\rangle} + n^{\langle\alpha 2\rangle}\underset{\sim}{a}_{\langle 2\rangle} + q^{\langle\alpha\rangle}\underset{\sim}{a}_{\langle 3\rangle} = n^{\langle\alpha\beta\rangle}\underset{\sim}{a}_{\langle\beta\rangle} + q^{\langle\alpha\rangle}\underset{\sim}{a}_{\langle 3\rangle} ,$$

$$\underset{\sim}{m}^{\langle\alpha\rangle} = m^{\langle\alpha 1\rangle}\underset{\sim}{a}^{\langle 2\rangle} - m^{\langle\alpha 2\rangle}\underset{\sim}{a}^{\langle 1\rangle} . \tag{3.7}$$

We assume that the internal couple vectors $\underset{\sim}{m}^{\langle\alpha\rangle}$ just as the external couple $\underset{\sim}{c}$ do not have any components in the $\underset{\sim}{a}^{\langle 3\rangle}$-direction. Such components, the so-called *cross-moments*, may appear in principle [2, 8, 15]. However, they are deleted in all engineering theories:

$$m^{\langle\alpha 3\rangle} \equiv 0 . \tag{3.8}$$

For orthogonal parameter lines $a_{12} = 0$ we find from (3.7) all well-known stress-resultants and couples of the classical theory of plates and shells, namely

normal stress resultants:	$n^{\langle 11\rangle}$, $n^{\langle 22\rangle}$,
in-plane shear stress resultants:	$n^{\langle 12\rangle}$, $n^{\langle 21\rangle}$,
transverse shear resultants:	$q^{\langle 1\rangle}$, $q^{\langle 2\rangle}$,
bending moments:	$m^{\langle 11\rangle}$, $m^{\langle 22\rangle}$,
twisting moments:	$m^{\langle 12\rangle}$, $m^{\langle 21\rangle}$.

Fig. 5 shows these physical components in their positive directions, yet again for an arbitrarily skew and curvilinear coordinate system.

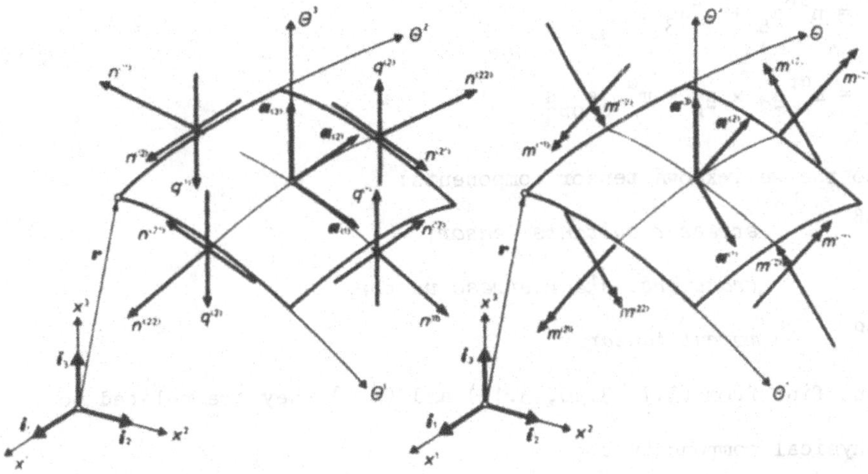

Fig. 5: Positive physical components of internal dynamic variables

We now introduce the stress resultant vectors $\underset{\sim}{n}^{\alpha}$ and the couple vectors $\underset{\sim}{m}^{\alpha}$ by the following transformations:

$$\underset{\sim}{n}^{\langle\alpha\rangle}\sqrt{a_{\beta\beta}}\, d\theta^{\beta} = \underset{\sim}{n}^{\langle\alpha\rangle}\sqrt{a\, a^{\alpha\alpha}}\, d\theta^{\beta} = \underset{\sim}{n}^{\alpha}\sqrt{a}\, d\theta^{\beta} ,$$
$$\underset{\sim}{m}^{\langle\alpha\rangle}\sqrt{a_{\beta\beta}}\, d\theta = \underset{\sim}{m}^{\langle\alpha\rangle}\sqrt{a\, a^{\alpha\alpha}}\, d\theta^{\alpha} = \underset{\sim}{m}^{\alpha}\sqrt{a}\, d\theta^{\beta} . \tag{3.9}$$

Both variables are tensors of rank 1 [11] and connected with their physical components by:

$$\underset{\sim}{n}^{\alpha} = \underset{\sim}{n}^{\langle\alpha\rangle}\sqrt{a^{\alpha\alpha}} , \quad \underset{\sim}{m}^{\alpha} = \underset{\sim}{m}^{\langle\alpha\rangle}\sqrt{a^{\alpha\alpha}} . \tag{3.10}$$

Decomposition with respect to the vector base of point P of the middle surface F

$$\underset{\sim}{n}^{\alpha} = n^{\alpha\beta}\underset{\sim}{a}_{\beta} + q^{\alpha}\underset{\sim}{a}_{3} ,$$
$$\underset{\sim}{m}^{\alpha} = m^{\alpha\rho}\underset{\sim}{a}_{3} \times \underset{\sim}{a}_{\rho} = m^{\alpha\rho}\, \varepsilon_{\rho\beta}\underset{\sim}{a}^{\beta} \qquad (3.11)$$

leads to the well-known tensor components:

$n^{\alpha\beta}$ *stress resultants tensor,*

q^{α} *transverse shear stress vector,*

$m^{\alpha\rho}$ *moment tensor.*

As we can find from (3.7, 3.10, 3.11) and (2.6) they are related to their physical components by:

$$n^{\langle\alpha\beta\rangle} = \sqrt{\frac{a_{\beta\beta}}{a^{\alpha\alpha}}}\; n^{\alpha\beta} \;, \quad q^{\langle\alpha\rangle} = \frac{1}{\sqrt{a^{\alpha\alpha}}}\; q^{\alpha} \;,$$
$$m^{\langle\alpha\beta\rangle} = \sqrt{\frac{a_{\beta\beta}}{a^{\alpha\alpha}}}\; m^{\alpha\beta} \;. \qquad (3.12)$$

In our previous derivation the *stress resultant tensor*, the *transverse shear stress vector* and the *moment* or *couple tensor* were defined as independent variables. Starting from the context of a classical three-dimensional continuum, these variables of course can be evaluated by integrating the stresses over the thickness of the shell. To the interested reader we recommend the detailed evaluation and discussion in [11, 16, 17].

3.3 Conditions of Equilibrium in Vector Form

We now consider in Fig. 6 again an element

$$dF = \sqrt{a}\, d\theta^1\, d\theta^2$$

of the middle surface, loaded by the force vector $\underset{\sim}{p}dF$ and the couple vector $\underset{\sim}{c}dF$. Then, for equilibrium, equivalent internal forces and moments

$$-\underset{\sim}{n}^{\alpha} \sqrt{a}\, d\theta^{\beta}, \qquad -\underset{\sim}{m}^{\alpha} \sqrt{d}\, d\theta^{\beta}, \qquad (\alpha \neq \beta) \tag{3.13}$$

must act along the negative section lines θ^{α}=const.; in addition to them we find along the positive section lines $(\theta^{\alpha} + d\theta^{\alpha})$=const. the differentials

$$(\underset{\sim}{n}^{\alpha} \sqrt{a}),_{\lambda} d\theta^{\lambda} d\theta^{\beta}, \qquad (\underset{\sim}{m}^{\alpha} \sqrt{a}),_{\lambda} d\theta^{\lambda} d\theta^{\beta}, \qquad (\alpha \neq \beta)\ . \tag{3.14}$$

The equilibrium condition of vanishing forces furnishes on the basis of Fig. 6, apart from higher order differentials:

$$\begin{aligned} &-\underset{\sim}{n}^1 \sqrt{a}\, d\theta^2 + \underset{\sim}{n}^1 \sqrt{a}\, d\theta^2 + (\underset{\sim}{n}^1 \sqrt{a}),_1 d\theta^1 d\theta^2 \\ &-\underset{\sim}{n}^2 \sqrt{a}\, d\theta^1 + \underset{\sim}{n}^2 \sqrt{a}\, d\theta^1 + (\underset{\sim}{n}^2 \sqrt{a}),_2 d\theta^2 d\theta^1 + \underset{\sim}{p} \sqrt{a}\, d\theta^1 d\theta^2 = 0\ , \end{aligned} \tag{3.15}$$

which can be simplified by use of Einstein's summation convention to:

$$(\underset{\sim}{n}^{\alpha} \sqrt{a}),_{\alpha} + \underset{\sim}{p} \sqrt{a} = 0\ . \tag{3.16}$$

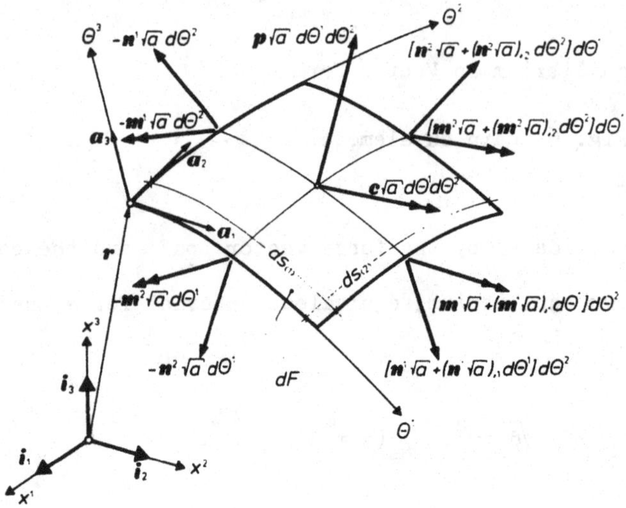

Fig. 6: Element of middle surface in equilibrium state

In a similar manner and after similar simplifications we find the equilibrium condition for vanishing moments:

$$(\underset{\sim}{m}^{\alpha} \sqrt{a})_{,\alpha} + (\underset{\sim}{a}_{\alpha} \times \underset{\sim}{n}^{\alpha} \sqrt{a}) + \underset{\sim}{c} \sqrt{a} = 0 . \tag{3.17}$$

If we finally introduce (2.15, 2.20) into these conditions we obtain their invariant vector form

$$\begin{aligned} &\underset{\sim}{n}^{\alpha}|_{\alpha} + \underset{\sim}{p} = 0 , \\ &\underset{\sim}{m}^{\alpha}|_{\alpha} + \underset{\sim}{a}_{\alpha} \times \underset{\sim}{n}^{\alpha} + \underset{\sim}{c} = 0 \end{aligned} \tag{3.18}$$

with the abbreviations

$$\underset{\sim}{n}^{\alpha}\big|_{\alpha} = \underset{\sim}{n}^{\alpha}{}_{,\alpha} + \underset{\sim}{n}^{\lambda}\,\Gamma^{\alpha}_{\alpha\lambda} \,,$$
$$\underset{\sim}{m}^{\alpha}\big|_{\alpha} = \underset{\sim}{m}^{\alpha}{}_{,\alpha} + \underset{\sim}{m}^{\lambda}\,\Gamma^{\alpha}_{\alpha\lambda} \qquad (3.19)$$

according to (2.20). As mentioned previously, these equations are valid for infinitesimally small displacements, and thus all appearing tensors and covariant differentiations are related to the undeformed configuration. For geometrically non-linear shell theories the equilibrium conditions remain formally unchanged, if $\underset{\sim}{n}^{\alpha}$, $\underset{\sim}{m}^{\alpha}$, $\underset{\sim}{p}$, $\underset{\sim}{c}$, the covariant differentiations and the base vectors $\underset{\sim}{a}_{\alpha}$ are related to the deformed middle surface [12, 15].

3.4 Conditions of Equilibrium in Component Form

For all computational purposes it is unavoidable to decompose the conditions of equilibrium. Introducing $\underset{\sim}{n}^{\alpha}$, $\underset{\sim}{m}^{\alpha}$ (3.11), $\underset{\sim}{p}$ (3.2) and $\underset{\sim}{c}$ (3.4) into (3.16, 3.17), we obtain from the first condition

$$n^{\alpha\beta}\big|_{\alpha}\underset{\sim}{a}_{\beta} + n^{\alpha\beta}\underset{\sim}{a}_{\beta}\big|_{\alpha} + q^{\alpha}\big|_{\alpha}\underset{\sim}{a}_{3} + q^{\alpha}\underset{\sim}{a}_{3}\big|_{\alpha} + p^{\beta}\underset{\sim}{a}_{\beta} + p^{3}\underset{\sim}{a}_{3} = 0 \,, \qquad (3.20)$$

which can be transformed by use of (2.13, 2.14) into:

$$(n^{\alpha\beta}\big|_{\alpha} - q^{\alpha}b^{\beta}_{\alpha} + p^{\beta})\,\underset{\sim}{a}_{\beta} + (n^{\alpha\beta}b_{\alpha\beta} + q^{\alpha}\big|_{\alpha} + p^{3})\,\underset{\sim}{a}_{3} = 0 \,. \qquad (3.21)$$

Analogously we find from the second condition

$$(m^{\alpha\beta}\big|_{\alpha} - q^{\beta} + c^{\beta})\,\underset{\sim}{a}_{3}\times\underset{\sim}{a}_{\beta} + m^{\alpha\beta}(\varepsilon_{\beta\rho}\underset{\sim}{a}^{\rho})\big|_{\alpha} + n^{\alpha\beta}\varepsilon_{\alpha\beta}\underset{\sim}{a}^{3} = 0, \qquad (3.22)$$

which will be converted with the help of (2.13, 2.16) into:

$$(m^{\alpha\beta}\big|_{\alpha} - q^{\beta} + c^{\beta})\, \underset{\sim}{a}_3 \times \underset{\sim}{a}_{\beta} + \varepsilon_{\alpha\beta}\, (n^{\alpha\beta} + m^{\alpha\rho} b^{\beta}_{\rho})\, \underset{\sim}{a}^3 = 0 \, . \tag{3.23}$$

Scalar multiplication of the final equations (3.21, 3.23), firstly by $\underset{\sim}{a}^{\lambda}$, thereafter by $\underset{\sim}{a}^3$, leads under observation of (2.4) to the following *six conditions of equilibrium* in the form of vector components:

$$\begin{aligned} &n^{\alpha\beta}\big|_{\alpha} - q^{\alpha} b^{\beta}_{\alpha} + p^{\beta} = 0 \, , \\ &n^{\alpha\beta} b_{\alpha\beta} + q^{\alpha}\big|_{\alpha} + p^3 = 0, \\ &m^{\alpha\beta}\big|_{\alpha} - q^{\beta} + c^{\beta} = 0 \, , \end{aligned} \tag{3.24}$$

$$\varepsilon_{\alpha\beta}\, (n^{\alpha\beta} + m^{\alpha\rho} b^{\beta}_{\rho}) = 0 \, , \tag{3.25}$$

$$\tilde{n}^{(\alpha\beta)} = n^{\alpha\beta} + m^{\alpha\rho} b^{\beta}_{\rho} \, , \tag{3.26}$$

in which the abbreviations

$$\begin{aligned} &n^{\alpha\beta}\big|_{\alpha} = n^{\alpha\beta}{}_{,\alpha} + n^{\rho\beta}\, \Gamma^{\alpha}_{\rho\alpha} + n^{\alpha\rho}\, \Gamma^{\beta}_{\rho\alpha} \, , \\ &q^{\alpha}\big|_{\alpha} = q^{\alpha}{}_{,\alpha} + q^{\rho}\, \Gamma^{\alpha}_{\rho\alpha} \, , \\ &m^{\alpha\beta}\big|_{\alpha} = m^{\alpha\beta}{}_{,\alpha} + m^{\rho\beta}\, \Gamma^{\alpha}_{\rho\alpha} + m^{\alpha\rho}\, \Gamma^{\beta}_{\rho\alpha} \, , \end{aligned} \tag{3.27}$$

are used (2.20). The first two lines of (3.24) describe the equilibrium conditions of the forces in the tangential and normal directions. The equilibrium of the moments in the directions of the tangent base vectors $\underset{\sim}{a}^\rho$ is formulated by the last line of (3.24), while (3.25) ensures the equilibrium of vanishing moments in the normal direction $\underset{\sim}{a}^3$. We interpret this condition as the definition (3.26) of the new *symmetric pseudo-stress tensor* $\tilde{n}^{(\alpha\beta)}$ [21], which for that reason carries round superscript brackets.

3.5 Dynamic Boundary Conditions

Let us now assume an arbitrary curve C in the middle surface F

$$C = C(\theta^\alpha) \; ; \tag{3.28}$$

the real arc length of which is measured by the coordinate s. The in-plane tangent vector $\underset{\sim}{t}$ and the normal vector $\underset{\sim}{u}$ of C, both unit vectors, obey the conditions

$$\underset{\sim}{t} = t^\alpha \underset{\sim}{a}_\alpha = t_\alpha \underset{\sim}{a}^\alpha , \qquad \underset{\sim}{u} = \underset{\sim}{t} \times \underset{\sim}{a}_3 = u^\alpha \underset{\sim}{a}_\alpha = u_\alpha \underset{\sim}{a}^\alpha \tag{3.29}$$

with

$$t^\alpha = \frac{d\theta^\alpha}{ds}, \quad u_\beta = \varepsilon_{\beta\alpha} \frac{d\theta^\alpha}{ds} = \varepsilon_{\beta\alpha} t^\alpha, \quad t^\alpha = \varepsilon^{\rho\alpha} u_\rho . \tag{3.30}$$

$(\underset{\sim}{u}, \underset{\sim}{t}, \underset{\sim}{a}_3)$ serve in this order as the right-handed vector triple of C.

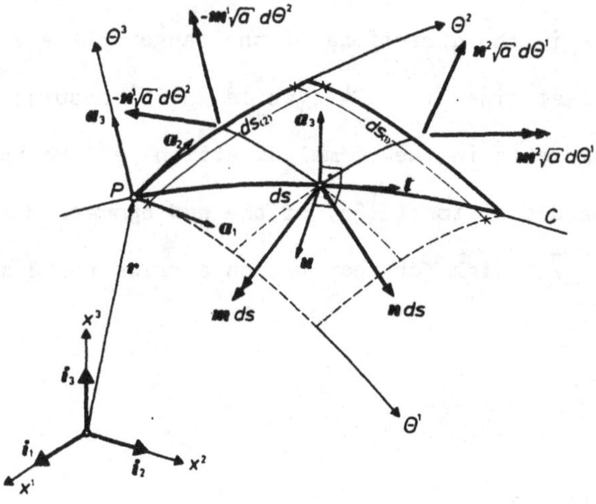

Fig. 7: Triangular element of middle surface

Fig. 7 shows a triangular element of the middle surface, bounded by two parameter lines and the curve C , in a state of equilibrium. The vector triple respectively the coordinate s is oriented in such a way, that $\underset{\sim}{u}$ always is an *outside normal*. According to Fig. 6 we find

$$-\underset{\sim}{n}^1 \sqrt{a}\, d\theta^2 \quad \text{and} \quad -\underset{\sim}{m}^1 \sqrt{a}\, d\theta^2$$

acting along $ds_{\langle 2\rangle}$, and

$$\underset{\sim}{n}^2 \sqrt{a}\, d\theta^1 \quad \text{and} \quad \underset{\sim}{m}^2 \sqrt{a}\, d\theta^1$$

acting along $ds_{\langle 1\rangle}$. If these internal dynamic variables are balanced by the force $\underset{\sim}{n}ds$ and moment $\underset{\sim}{m}ds$, acting along the section line of C , the equilibrium conditions can be broken down as:

$$-\underset{\sim}{n}^{1} \sqrt{a}\, d\theta^{2} + \underset{\sim}{n}^{2} \sqrt{a}\, d\theta^{1} + \underset{\sim}{n} ds = 0 \;,$$
$$-\underset{\sim}{m}^{1} \sqrt{a}\, d\theta^{2} + \underset{\sim}{m}^{2} \sqrt{a}\, d\theta^{1} + \underset{\sim}{m} ds = 0 \;. \tag{3.31}$$

Introducing (3.30) we find with *Einstein's* summation convention:

$$\underset{\sim}{n} = \underset{\sim}{n}^{\alpha} u_{\alpha}, \quad \underset{\sim}{m} = \underset{\sim}{m}^{\alpha} u_{\alpha} \;. \tag{3.32}$$

We now use herein the decomposition (3.11) and select for $\underset{\sim}{n}$ ($\underset{\sim}{m}$) a decomposition with respect to $\underset{\sim}{u}$, $\underset{\sim}{t}$, $\underset{\sim}{a}_3$ ($\underset{\sim}{u}$, $\underset{\sim}{t}$). This finally leads to:

$$\underset{\sim}{n} = n_t \underset{\sim}{t} + n_u \underset{\sim}{u} + n_3 \underset{\sim}{a}_3 \;, \qquad \underset{\sim}{m} = m_t \underset{\sim}{t} + m_u \underset{\sim}{u} \tag{3.33}$$

with

$$n_t = n^{\alpha\beta} u_\alpha t_\beta \;, \qquad n_u = n^{\alpha\beta} u_\alpha u_\beta, \qquad n_3 = q^{\alpha} u_\alpha \;,$$
$$m_t = m^{\alpha\beta} u_\alpha u_\beta \;, \qquad m_u = -m^{\alpha\beta} u_\alpha t_\beta \;. \tag{3.34}$$

However, if the boundary of the shell is formed by the curve C, force and moment boundary conditions can be formulated for all components (3.34), which must correspond to the prescribed external forces and couples:

$$\underset{\sim}{n} = \underset{\sim}{n}^{o} \;, \qquad \underset{\sim}{m} = \underset{\sim}{m}^{o} \;. \tag{3.35}$$

4. THE KINEMATIC FIELD

4.1 General Assumptions

According to Fig. 8 we now study the deformation of an arbitrary point P of the middle surface and its corresponding point $\overset{*}{P}$ of the shell space, situated at a distance θ^3 along the normal $\underset{\sim}{a}_3$ of P. The position vector of the undeformed location of $\overset{*}{P}$ can be written as a sum of two contributions:

$$\overset{*}{\underset{\sim}{r}} = \underset{\sim}{r} + \theta^3 \underset{\sim}{a}_3 . \qquad (4.1)$$

In the same manner the deformed position of P and $\overset{*}{P}$, denoted by a bar, may be described by:

$$\overset{*}{\bar{\underset{\sim}{r}}} = \bar{\underset{\sim}{r}} + \theta^3 \bar{\underset{\sim}{d}}_3 , \qquad (4.2)$$

where the undeformed unit normal vector $\underset{\sim}{a}_3$ is replaced by the deformed director $\bar{\underset{\sim}{d}}_3$.

We now postulate three important assumptions for the further treatment:

a) All points within the shell space are described by a system of *convected coordinates* θ^i, which means that the undeformed point $\overset{*}{P}(P)$ is characterized by the same triple of numbers θ^α, θ^3 (θ^α, $\theta^3 = 0$) as the deformed point $\overset{*}{\bar{P}}(\bar{P})$.

b) The deformation of the shell continuum shall be described by a constant term $\underset{\sim}{v}$ and a term $\theta^3 \underset{\sim}{w}$ varying linearly with θ^3 over the thickness.

c) There shall be no elongation in the direction of the shell thickness during deformation:

$$|\underset{\sim}{a}_3| = |\bar{\underset{\sim}{a}}_3| = |\bar{\underset{\sim}{d}}_3| = 1 \,. \qquad (4.3)$$

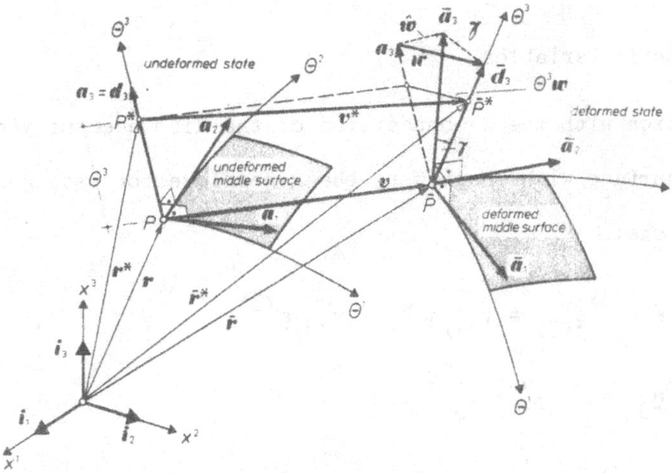

Fig. 8: The kinematics of the undeformed and deformed shell space

Hence we are able to define the *displacement vector of the shell space*, the *displacement vector of the middle surface* and the *displacement vector of the director* from the difference of the position vectors (4.1, 4.2) by:

$$\underbrace{\bar{\underset{\sim}{r}}^* - \underset{\sim}{r}^*}_{\underset{\sim}{v}^*} = \underbrace{\bar{\underset{\sim}{r}} - \underset{\sim}{r}}_{\underset{\sim}{v}} + \theta^3 \underbrace{(\bar{\underset{\sim}{d}}_3 - \underset{\sim}{a}_3)}_{\underset{\sim}{w}}$$

$$\underset{\sim}{v}^* = \underset{\sim}{v} + \theta^3 \underset{\sim}{w}$$

- $\underset{\sim}{w}$: displacement vector of the director
- $\underset{\sim}{v}$: displacement vector of the middle surface F
- $\underset{\sim}{v}^*$: displacement vector of the shell space

For one and the same point P all vector quantities in (4.4) are functions of the same pair θ^α of convected coordinates.

4.2 External Kinematic Variables

We start this section with the decomposition of the displacement vector $\underset{\sim}{v}$ of the middle surface with respect to the general vector base and to the normalized one:

$$\begin{aligned}\underset{\sim}{v} &= v^{\langle\alpha\rangle} \underset{\sim}{a}_{\langle\alpha\rangle} + v^{\langle 3\rangle} \underset{\sim}{a}_{\langle 3\rangle} = v_{\langle\alpha\rangle} \underset{\sim}{a}^{\langle\alpha\rangle} + v_{\langle 3\rangle} \underset{\sim}{a}^{\langle 3\rangle} \\ &= v^\alpha \underset{\sim}{a}_\alpha + v^3 \underset{\sim}{a}_3 = v_\alpha \underset{\sim}{a}^\alpha + v_3 \underset{\sim}{a}^3 .\end{aligned} \tag{4.5}$$

As consequence of (2.6), physical and tensorial components of $\underset{\sim}{v}$ are connected by:

$$v_{\langle\alpha\rangle} = v_\alpha \sqrt{a^{\alpha\alpha}}, \quad v^{\langle\alpha\rangle} = v^\alpha \sqrt{a_{\alpha\alpha}}, \quad v_{\langle 3\rangle} = v^{\langle 3\rangle} = v_3 . \tag{4.6}$$

Because we have excluded any elongation of $\bar{\underset{\sim}{a}}_3$, $\bar{\underset{\sim}{d}}_3$, compared to the original length $|\underset{\sim}{a}_3| = 1$, and because of the assumption of infinitesimally small rotations, the decomposition of $\underset{\sim}{w}$ cannot contain the component w_3 :

$$\underset{\sim}{w} = w^{\langle\alpha\rangle} \underset{\sim}{a}_{\langle\alpha\rangle} = w_{\langle\alpha\rangle} \underset{\sim}{a}^{\langle\alpha\rangle} = w^\alpha \underset{\sim}{a}_\alpha = w_\alpha \underset{\sim}{a}^\alpha . \tag{4.7}$$

Physical and tensorial components of this vector are connected by:

$$w_{\langle\alpha\rangle} = [illegible] \sqrt{a^{\alpha\alpha}}, \quad w^{\langle\alpha\rangle} = w^\alpha \sqrt{[illegible]} \tag{4.8}$$

For later use we partition the displacement vector $\underset{\sim}{w}$ of the director into the rotation $(\bar{\underset{\sim}{a}}_3 - \underset{\sim}{a}_3)$ of the unit normal vector and into the contribution of the shear distorsion $(\bar{\underset{\sim}{d}}_3 - \bar{\underset{\sim}{a}}_3)$:

$$\underset{\sim}{w} = \bar{\underset{\sim}{d}}_3 - \underset{\sim}{a}_3 = (\bar{\underset{\sim}{d}}_3 - \bar{\underset{\sim}{a}}_3) + (\bar{\underset{\sim}{a}}_3 - \underset{\sim}{a}_3) = \underset{\sim}{\gamma} + \hat{\underset{\sim}{w}} \ . \tag{4.9}$$

Both vectors can be decomposed in a similar way to (4.7):

$$\begin{aligned} \underset{\sim}{\gamma} &= \gamma_\alpha \underset{\sim}{a}^\alpha = \gamma^\alpha \underset{\sim}{a}_\alpha = \gamma_{\langle\alpha\rangle} \underset{\sim}{a}^{\langle\alpha\rangle} = \gamma^{\langle\alpha\rangle} \underset{\sim}{a}_{\langle\alpha\rangle} , \qquad |\underset{\sim}{\gamma}| = \gamma , \\ \hat{\underset{\sim}{w}} &= \hat{w}_\alpha \underset{\sim}{a}^\alpha = \hat{w}^\alpha \underset{\sim}{a}_\alpha = \hat{w}_{\langle\alpha\rangle} \underset{\sim}{a}^{\langle\alpha\rangle} = \hat{w}^{\langle\alpha\rangle} \underset{\sim}{a}_{\langle\alpha\rangle} \end{aligned} \tag{4.10}$$

Because of (4.3) the length of the vector $\underset{\sim}{\gamma}$ is equal to the shear angle γ in Fig. 8.

Applying the deformed base vectors $\underset{\sim}{a}_\alpha$ and the deformed unit normal vector $\underset{\sim}{a}_3$:

$$\begin{aligned} &\bar{\underset{\sim}{r}} = \underset{\sim}{r} + \underset{\sim}{v} , \qquad \bar{\underset{\sim}{r}},_\alpha = \bar{\underset{\sim}{a}}_\alpha = \underset{\sim}{r},_\alpha + \underset{\sim}{v},_\alpha = \underset{\sim}{a}_\alpha + \underset{\sim}{v},_\alpha , \\ &\bar{\underset{\sim}{a}}_3 = \underset{\sim}{a}_3 + \hat{\underset{\sim}{w}} \end{aligned} \tag{4.11}$$

the orthogonality condition

$$\begin{aligned} \bar{\underset{\sim}{a}}_\alpha \cdot \bar{\underset{\sim}{a}}_3 &= 0 = (\underset{\sim}{a}_\alpha + \underset{\sim}{v},_\alpha) \cdot (\underset{\sim}{a}_3 + \hat{\underset{\sim}{w}}) \\ &= \underbrace{\underset{\sim}{a}_\alpha \cdot \underset{\sim}{a}_3}_{=0} + \underset{\sim}{v},_\alpha \cdot \underset{\sim}{a}_3 + \underset{\sim}{a}_\alpha \cdot \hat{\underset{\sim}{w}} + \underbrace{\underset{\sim}{v},_\alpha \cdot \hat{\underset{\sim}{w}}}_{\approx 0} \end{aligned} \tag{4.12}$$

leads with the help of

$$\mathbf{v}_{,\alpha} = (v_{\rho|\alpha} - v_3 b_{\rho\alpha})\, \mathbf{a}^\rho + (v_{3,\alpha} + v^\lambda b_{\lambda\alpha})\, \mathbf{a}^3 = \varphi_{\alpha\rho}\mathbf{a}^\rho + \varphi_{\alpha 3}\mathbf{a}^3 \qquad (4.13)$$

to an important relation between the components of $\hat{\mathbf{w}}$ and $\mathbf{v}$:

$$\hat{w}_\alpha = -\varphi_{\alpha 3} = -(v_{3,\alpha} + v_\lambda b^\lambda_\alpha) \; . \qquad (4.14)$$

4.3 Internal Kinematic Variables

The surface metric tensor of an arbitrary surface $\overset{*}{F}$, "parallel" to the middle surface F in a certain distance θ^3, can be expressed in the undeformed and the deformed state by:

$$\overset{*}{a}_{\alpha\beta} = \overset{*}{\mathbf{a}}_\alpha \cdot \overset{*}{\mathbf{a}}_\beta \,, \quad \bar{\overset{*}{a}}_{\alpha\beta} = \bar{\overset{*}{\mathbf{a}}}_\alpha \cdot \bar{\overset{*}{\mathbf{a}}}_\beta \,, \quad (\alpha,\beta = 1,2) \qquad (4.15)$$

Analogously the spatial metric tensor of the shell space can be formulated by:

$$\overset{*}{a}_{ij} = \overset{*}{\mathbf{a}}_i \cdot \overset{*}{\mathbf{a}}_j \,, \quad \bar{\overset{*}{a}}_{ij} = \bar{\overset{*}{\mathbf{a}}}_i \cdot \bar{\overset{*}{\mathbf{a}}}_j \,, \quad (i,j = 1,2,3) \; . \qquad (4.16)$$

The base vectors $\overset{*}{\mathbf{a}}_i$ of the undeformed shell space, which appear in (4.16), can be taken from (2.23), those of the deformed one can be evaluated by using (4.2):

$$\overset{*}{\mathbf{r}} = \mathbf{r} + \theta^3\, \mathbf{a}_3 : \quad \overset{*}{\mathbf{r}}_{,\alpha} = \overset{*}{\mathbf{a}}_\alpha = \mathbf{a}_\alpha + \theta^3 \mathbf{a}_{3,\alpha} \,,$$

$$\overset{*}{\mathbf{r}}_{,3} = \mathbf{a}_3 \; ;$$

$$\bar{\underset{\sim}{r}}^{*} = \bar{\underset{\sim}{r}} + \theta^3 \bar{\underset{\sim}{d}}_3 : \qquad \bar{\underset{\sim}{r}},\overset{*}{_\alpha} = \bar{\underset{\sim}{a}}\overset{*}{_\alpha} = \bar{\underset{\sim}{a}}_\alpha + \theta^3 \bar{\underset{\sim}{d}}_{3},_\alpha ,$$

$$\bar{\underset{\sim}{r}},\overset{*}{_3} = \bar{\underset{\sim}{a}}\overset{*}{_3} = \bar{\underset{\sim}{d}}_3 . \qquad (4.17)$$

The basis of our further derivations will be the classical *Cauchy-Green strain tensor*, formed by the difference of the spatial metric tensors. His well-known components

$$\gamma_{ij} = \gamma_{ji} = \gamma_{(ij)} = \frac{1}{2} (\bar{a}\overset{*}{_{ij}} - \overset{*}{a}_{ij}) = \frac{1}{2} (\bar{\underset{\sim}{a}}\overset{*}{_i} \cdot \bar{\underset{\sim}{a}}\overset{*}{_j} - \overset{*}{\underset{\sim}{a}}_i \cdot \overset{*}{\underset{\sim}{a}}_j) \qquad (4.18)$$

can be transformed by virtue of

$$\bar{\underset{\sim}{r}}^{*} = \overset{*}{\underset{\sim}{r}} + \overset{*}{\underset{\sim}{v}} ,$$

$$\bar{\underset{\sim}{r}},\overset{*}{_i} = \bar{\underset{\sim}{a}}\overset{*}{_i} = \underset{\sim}{r},\overset{*}{_i} + \underset{\sim}{v},\overset{*}{_i} = \overset{*}{\underset{\sim}{a}}_i + \underset{\sim}{v},\overset{*}{_i} \qquad (4.19)$$

into

$$\gamma_{(ij)} = \frac{1}{2} \left[(\overset{*}{\underset{\sim}{a}}_i + \underset{\sim}{v},\overset{*}{_i}) \cdot (\overset{*}{\underset{\sim}{a}}_j + v,\overset{*}{_j}) - \overset{*}{\underset{\sim}{a}}_i \cdot \overset{*}{\underset{\sim}{a}}_j \right] \qquad (4.20)$$

and further into their linearized form for small displacements $(\underset{\sim}{v},\overset{*}{_i} \cdot \underset{\sim}{v},\overset{*}{_j} \approx 0)$:

$$\gamma_{(ij)} = \frac{1}{2} \left[\overset{*}{\underset{\sim}{a}}_i \cdot \underset{\sim}{v},\overset{*}{_j} + \overset{*}{\underset{\sim}{a}}_j \cdot \underset{\sim}{v},\overset{*}{_i} \right]. \qquad (4.21)$$

For the evaluation of the components (4.18) of the *Cauchy-Green strain tensor* we start with the tangential ones:

$$\gamma_{(\alpha\beta)} = \frac{1}{2}\,(\bar{\underset{\sim}{a}}^*_\alpha \cdot \bar{\underset{\sim}{a}}^*_\beta - \underset{\sim}{a}^*_\alpha \cdot \underset{\sim}{a}^*_\beta)$$

$$= \frac{1}{2}\Big[(\bar{\underset{\sim}{a}}_\alpha + \theta^3 \bar{\underset{\sim}{d}}_{3,\alpha}) \cdot (\bar{\underset{\sim}{a}}_\beta + \theta^3 \bar{\underset{\sim}{d}}_{3,\beta})$$

$$- (\underset{\sim}{a}_\alpha + \theta^3 \underset{\sim}{a}_{3,\alpha}) \cdot (\underset{\sim}{a}_\beta + \theta^3 \underset{\sim}{a}_{3,\beta})\Big]$$

$$= \frac{1}{2}\,(\bar{\underset{\sim}{a}}_\alpha \cdot \bar{\underset{\sim}{a}}_\beta - \underset{\sim}{a}_\alpha \cdot \underset{\sim}{a}_\beta)$$

$$+ \frac{\theta^3}{2}\Big[(\bar{\underset{\sim}{a}}_\alpha \cdot \bar{\underset{\sim}{d}}_{3,\beta} + \bar{\underset{\sim}{a}}_\beta \cdot \bar{\underset{\sim}{d}}_{3,\alpha}) - (\underset{\sim}{a}_\alpha \cdot \underset{\sim}{a}_{3,\beta} + \underset{\sim}{a}_\beta \cdot a_{3,\alpha})\Big]$$

$$+ \frac{(\theta^3)^2}{2}\Big[\bar{\underset{\sim}{d}}_{3,\alpha} \cdot \bar{\underset{\sim}{d}}_{3,\beta} - \underset{\sim}{a}_{3,\alpha} \cdot \underset{\sim}{a}_{3,\beta}\Big] . \qquad (4.22)$$

As we find out the assumption of a linear displacement field over the shell thickness (4.4) leads to quadratic tangential strains. However, there doesn't exist a corresponding dynamic variable for this last term of (4.22). Thus quadratic tangential strains must be considered as immaterial in shell theory and can therefore be deleted. Using further (4.9)

$$\bar{\underset{\sim}{d}}_3 = \bar{\underset{\sim}{a}}_3 + \underset{\sim}{\gamma} , \qquad (4.23)$$

we finally find as a linear approximation of the tangential strain:

$$\gamma_{(\alpha\beta)} = \frac{1}{2}\,(\bar{a}_{\alpha\beta} - a_{\alpha\beta})$$

$$+ \theta^3\Big[b_{\alpha\beta} - \bar{b}_{\alpha\beta} + \frac{1}{2}\,(\bar{a}_\alpha \cdot \gamma,_\beta + \bar{a}_\beta \cdot \gamma,_\alpha)\Big] \qquad (4.24)$$

$$\gamma_{(\alpha\beta)} = \alpha_{(\alpha\beta)} + \theta^3 \beta_{(\alpha\beta)} ,$$

in which

$$\alpha_{(\alpha\beta)} = \frac{1}{2}\,(\bar{a}_{\alpha\beta} - a_{\alpha\beta})\ ,$$

$$\beta_{(\alpha\beta)} = b_{\alpha\beta} - \bar{b}_{\alpha\beta} + \frac{1}{2}\,(\bar{\underset{\sim}{a}}_{\alpha} \cdot \underset{\sim}{\gamma},_{\beta} + \bar{\underset{\sim}{a}}_{\beta} \cdot \gamma,_{\alpha}) \tag{4.25}$$

abbreviate the *first* and *second strain tensor of the middle* surface F. Both tensors are symmetric with respect to α,β.

For the component $\gamma_{\alpha 3}$ of the *Cauchy-Green strain tensor* we find from (4.18)

$$\gamma_{\alpha 3} = \frac{1}{2}\,(\bar{\underset{\sim}{a}}^{*}_{\alpha} \cdot \bar{\underset{\sim}{a}}^{*}_{3} - \underset{\sim}{a}^{*}_{\alpha} \cdot \underset{\sim}{a}^{*}_{3})$$

$$= \frac{1}{2}\left[(\bar{\underset{\sim}{a}}_{\alpha} + \theta^{3}\bar{\underset{\sim}{d}}_{3},_{\alpha}) \cdot \bar{\underset{\sim}{d}}_{3} - (\underset{\sim}{a}_{\alpha} + \theta^{3}\underset{\sim}{a}_{3},_{\alpha}) \cdot a_{3}\right]$$

$$= \frac{1}{2}\Big[\bar{\underset{\sim}{a}}_{\alpha} \cdot \bar{\underset{\sim}{d}}_{3} - \underbrace{\underset{\sim}{a}_{\alpha} \cdot \underset{\sim}{a}_{3}}_{=0} + \theta^{3}\,(\underbrace{\bar{\underset{\sim}{d}}_{3},_{\alpha} \cdot \bar{\underset{\sim}{d}}_{3}}_{=0} - \underbrace{\underset{\sim}{a}_{3},_{\alpha} \cdot \underset{\sim}{a}_{3}}_{=0})\Big]\ , \tag{4.26}$$

$$(\bar{\underset{\sim}{d}}_{3} \cdot \bar{\underset{\sim}{d}}_{3}),_{\alpha} = 1,_{\alpha} = 0 = 2\,\bar{\underset{\sim}{d}}_{3},_{\alpha} \cdot \underset{\sim}{d}_{3}$$

which transforms by introduction of (4.23) into:

$$\gamma_{\alpha 3} = \frac{1}{2}\left[\bar{\underset{\sim}{a}}_{\alpha} \cdot (\bar{\underset{\sim}{a}}_{3} + \underset{\sim}{\gamma})\right] = \frac{1}{2}\left[\bar{\underset{\sim}{a}}_{\alpha} \cdot \underset{\sim}{a}_{3} + \bar{\underset{\sim}{a}}_{\alpha} \cdot \underset{\sim}{\gamma}\right] = \frac{1}{2}\,\bar{\underset{\sim}{a}}_{\alpha} \cdot \underset{\sim}{\gamma}\ . \tag{4.27}$$

Finally we find for the component γ_{33} of (4.18):

$$\gamma_{33} = \frac{1}{2}\,(\bar{\mathbf{a}}_3^{*} \cdot \bar{\mathbf{a}}_3^{*} - \mathbf{a}_3^{*} \cdot \mathbf{a}_3^{*})$$

$$= \frac{1}{2}\,(\bar{\mathbf{d}}_3 \cdot \bar{\mathbf{d}}_3 - \mathbf{a}_3 \cdot \mathbf{a}_3) = \frac{1}{2}\,(1-1) = 0 \,. \qquad (4.28)$$

As expected there exists no strain component in the θ^3-direction. Hence we find the complete *Cauchy-Green strain tensor* in the context of shell theory approximated by:

$$\gamma_{(ij)} = \left[\begin{array}{c|c} \alpha_{(\alpha\beta)} + \theta^3 \beta_{(\alpha\beta)} & \gamma_{\alpha 3} \\ \hline \gamma_{3\alpha} & 0 \end{array}\right] \qquad (4.29)$$

4.4 Kinematic Relations

In order to evaluate the kinematic relations we repeat the following formulae

$$\mathbf{a}_\alpha^{*} = \mathbf{a}_\alpha - \theta^3 b_\alpha^\lambda \mathbf{a}_\lambda \,,$$

$$\mathbf{v},_\alpha = (v_\lambda|_\alpha - b_{\lambda\alpha})\,\mathbf{a}^\lambda + (v_3,_\alpha + b_\alpha^\lambda v_\lambda)\,\mathbf{a}^3 = \varphi_{\alpha\lambda}\mathbf{a}^\lambda + \varphi_{\alpha 3}\mathbf{a}^3 \,,$$

$$\mathbf{w},_\alpha = w_\lambda|_\alpha\,\mathbf{a}^\lambda + b_\alpha^\lambda w_\lambda \mathbf{a}^3 \,, \qquad (4.30)$$

$$\boldsymbol{\gamma} = \gamma_\beta \mathbf{a}^\beta \qquad (4.31)$$

from (2.23, 4.13, 4.7, 4.10). Introducing now the equations (4.4) and (4.30) into (4.21)

$$\gamma_{(\alpha\beta)} = \frac{1}{2} (\overset{*}{\underset{\sim}{a}}_{\alpha} \cdot \overset{*}{\underset{\sim}{v}}_{,\beta} + \overset{*}{\underset{\sim}{a}}_{\alpha} \cdot \overset{*}{\underset{\sim}{v}}_{,\alpha})$$

$$= \frac{1}{2} \Big[(v_{\alpha|\beta} + v_{\beta|\alpha} - 2b_{\alpha\beta}v_3) \tag{4.32}$$

$$+ \theta^3 (w_{\alpha|\beta} + w_{\beta|\alpha} - b^{\lambda}_{\alpha}v_{\lambda|\beta} - b^{\lambda}_{\beta}v_{\lambda|\alpha} + 2b^{\lambda}_{\alpha}b_{\lambda\beta}v_3) + \dots \Big] ,$$

we find for the first and second strain tensor:

$$\alpha_{(\alpha\beta)} = \frac{1}{2} (v_{\alpha|\beta} + v_{\beta|\alpha} - 2b_{\alpha\beta}v_3) = \frac{1}{2} (\varphi_{\alpha\beta} + \varphi_{\beta\alpha}) ,$$

$$\beta_{(\alpha\beta)} = \frac{1}{2} (w_{\alpha|\beta} + w_{\beta|\alpha} - b^{\lambda}_{\alpha}v_{\lambda|\alpha} - b^{\lambda}_{\beta}v_{\lambda|\alpha} + 2b^{\lambda}_{\alpha}b_{\lambda\beta}v_3) \tag{4.33}$$

$$= \frac{1}{2} (w_{\alpha|\beta} + w_{\beta|\alpha} - b^{\lambda}_{\alpha}\varphi_{\beta\lambda} - b^{\lambda}_{\beta}\varphi_{\alpha\lambda}) .$$

The transverse shear component $\gamma_{\alpha 3}$ (4.27) may be simplified by use of (4.31):

$$\gamma_{\alpha 3} = \frac{1}{2} \bar{\underset{\sim}{a}}_{\alpha} \cdot \underset{\sim}{\gamma} = \frac{1}{2} \underset{\sim}{a}_{\alpha} \cdot \gamma_{\beta}\underset{\sim}{a}^{\beta} = \frac{1}{2} \gamma_{\alpha} . \tag{4.34}$$

Introducing (4.9, 4.14) herein we obtain:

$$\gamma_{\alpha 3} = \frac{1}{2} (w_{\alpha} + v_{3,\alpha} + b^{\lambda}_{\alpha}v_{\lambda}) ,$$

$$\gamma_{\alpha} = w_{\alpha} + v_{3,\alpha} + b^{\lambda}_{\alpha}v_{\lambda} = w_{\alpha} + \varphi_{\alpha 3} . \tag{4.35}$$

4.5 Kinematic Boundary Conditions

Again we consider a triangular element of the middle surface F, bounded by two parameter lines θ^{α} and by an arbitrary curve C (Fig.

7). The components of the displacement vector $\underset{\sim}{v}$ with respect to the vector triple $(\underset{\sim}{u}, \underset{\sim}{t}, \underset{\sim}{a}_3)$ can be evaluated by use of (3.29) from (4.5):

$$\underset{\sim}{v} = v^\alpha \underset{\sim}{a}_\alpha + v^3 \underset{\sim}{a}_3 = v_t \underset{\sim}{t} + v_u \underset{\sim}{u} + v_3 \underset{\sim}{a}_3 \ ,$$
$$v_t = v^\alpha t_\alpha, \quad v_u = v^\alpha u_\alpha, \quad v_3 = v^3, \quad v^\alpha = v_t t^\alpha + v_u u^\alpha . \tag{4.36}$$

For the components of the displacement vector $\underset{\sim}{w}$ we find analogously:

$$\underset{\sim}{w} = w^\alpha \underset{\sim}{a}_\alpha = w_t \underset{\sim}{t} + w_u \underset{\sim}{u} \ ,$$
$$w_t = w^\alpha t_\alpha, \quad w_u = w^\alpha u_\alpha, \quad w^\alpha = w_t t^\alpha + w_u u^\alpha . \tag{4.37}$$

We now complete the previously defined external kinematic variables by the *rotation vector* $\underset{\sim}{\omega}$ *of the director* in the following way:

$$\underset{\sim}{w} = \underset{\sim}{\omega} \times \underset{\sim}{a}_3, \qquad \underset{\sim}{\omega} = \underset{\sim}{a}_3 \times \underset{\sim}{w} \ , \tag{4.38}$$

the components of which

$$\underset{\sim}{\omega} = \omega^\alpha \underset{\sim}{a}_\alpha = \omega_\alpha \underset{\sim}{a}^\alpha = \omega^{\langle\alpha\rangle} \underset{\sim}{a}_{\langle\alpha\rangle} = \omega_{\langle\alpha\rangle} \underset{\sim}{a}^{\langle\alpha\rangle} \ ,$$
$$\omega_{\langle\alpha\rangle} = \omega_\alpha \sqrt{a^{\alpha\alpha}}\ , \qquad \omega^{\langle\alpha\rangle} = \omega^\alpha \sqrt{a_{\alpha\alpha}} \tag{4.39}$$

are related to $\underset{\sim}{w}$ by

$$w_\beta = \omega^\alpha \varepsilon_{\beta\alpha}\ , \qquad \omega^\rho = w_\beta \varepsilon^{\beta\rho} . \tag{4.40}$$

A decomposition of $\underset{\sim}{\omega}$ with respect to $\underset{\sim}{u}, \underset{\sim}{t}$ leads to:

$$\underset{\sim}{\omega} = \omega_t \underset{\sim}{t} + \omega_u \underset{\sim}{u} \,, \qquad (4.41)$$

$$\omega_t = \omega^\alpha t_\alpha, \quad \omega_u = \omega^\alpha u_\alpha, \quad \omega^\alpha = \omega_t t^\alpha + \omega_u u^\alpha \,.$$

If the curve C is assumed to form the shell boundary, kinematic boundary conditions can be prescribed advantageously by the displacement vector $\underset{\sim}{v}$ and by the rotation vector $\underset{\sim}{\omega}$:

$$\underset{\sim}{v} = v_t \underset{\sim}{t} + v_u \underset{\sim}{u} + v_3 \underset{\sim}{a}_3 = \underset{\sim}{v}^{\circ}, \qquad (4.42)$$

$$\underset{\sim}{\omega} = \omega_t \underset{\sim}{t} + \omega_u \underset{\sim}{u} = \underset{\sim}{\omega}^{\circ}.$$

5. CONSTITUTIVE EQUATIONS

5.1 The Elastic Potential for Hyperelastic Shells

We start our survey of the constitutive equations of hyperelastic shells with the *residual energy equation* (see e.g. [2, 8, 15]). This relation originates in the first law of thermodynamics - formulated in rates of energy - for isothermal conditions, when additionally invariance against rigid body motions is required:

$$\dot{u} - (\tilde{n}^{(\alpha\beta)} \dot{\alpha}_{(\alpha\beta)} + m^{(\alpha\beta)} \dot{\beta}_{(\alpha\beta)} + q^{\alpha\cdot} \dot{\gamma}_\alpha) = 0 \,. \qquad (5.1)$$

The dot in this rate of energy equation denotes a material time derivation

$$(\dot{..}) = \frac{D...}{D\,t} \tag{5.2}$$

or, for the case of equilibrium, any increment. u abbreviates the *internal energy density*, which, for the isothermal case corresponds with the *elastic potential* or *strain energy density*.

We now define the constitutive behaviour of hyperelastic shells in such a way, that the strain energy density u shall only depend on the internal kinematic variables $\alpha_{(\alpha\beta)}$, $\beta_{(\alpha\beta)}$, γ_α (and on some suitable parameters of the shell space, e.g. $a_{\alpha\beta}$, $b_{\alpha\beta}$, h):

$$u = u\;(\alpha_{(\alpha\beta)}, \beta_{(\alpha\beta)}, \gamma_\alpha)\;. \tag{5.3}$$

Forming the time derivative of u

$$\dot{u} = \frac{1}{2}\left(\frac{\partial u}{\partial \alpha_{\alpha\beta}} + \frac{\partial u}{\partial \alpha_{\beta\alpha}}\right)\dot{\alpha}_{\alpha\beta} + \frac{1}{2}\left(\frac{\partial u}{\partial \beta_{\alpha\beta}} + \frac{\partial u}{\partial \beta_{\beta\alpha}}\right)\dot{\beta}_{\alpha\beta} + \frac{\partial u}{\partial \gamma_\alpha}\,\dot{\gamma}_\alpha\;, \tag{5.4}$$

we obtain by comparison with (5.1) the following general constitutive equations:

$$\begin{aligned} \tilde{n}^{(\alpha\beta)} &= \frac{1}{2}\left(\frac{\partial u}{\partial \alpha_{\alpha\beta}} + \frac{\partial u}{\partial \alpha_{\beta\alpha}}\right), \\ m^{(\alpha\beta)} &= \frac{1}{2}\left(\frac{\partial u}{\partial \beta_{\alpha\beta}} + \frac{\partial u}{\partial \beta_{\beta\alpha}}\right), \\ q^{\alpha} &= \frac{\partial u}{\partial \gamma_\alpha}\;. \end{aligned} \tag{5.5}$$

The further treatment starts with the elastic potential of a three-dimensional continuum

$$u^* = \frac{1}{2} E^{ijrs} \overset{*}{\gamma}_{ij} \overset{*}{\gamma}_{rs} \tag{5.6}$$

$$= \frac{G}{2}\left[a^{*ir} a^{*js} + a^{*is} a^{*jr} + \frac{2\nu}{1-2\nu} a^{*ij} a^{*rs} \right] \overset{*}{\gamma}_{ij} \overset{*}{\gamma}_{rs} ,$$

in which the following abbreviations have been used:

$G = \frac{E}{2(1+\nu)}$, E Young's modulus, ν Poisson's ratio.

If we now introduce (2.24, 2.25) and (4.29) together with (4.34) into (5.6) and integrate over the thickness h of the shell

$$u = \frac{1}{2} \int_{-h/2}^{+h/2} \sqrt{\frac{\overset{*}{a}}{a}} E^{ijrs} \gamma_{ij} \gamma_{rs} \, d\theta^3 , \tag{5.7}$$

we end up after several transformations with an approximately 50-terms expression, if terms of the relative order of magnitude $\leq \lambda/80$ are cancelled. λ abbreviates the *shell parameter*

$$\lambda = \frac{h}{\min\{L, R_{min}\}} \tag{5.8}$$

with L smallest length of the shell metric,

R_{min} smallest principal radius of curvature.

If all terms in (5.7), with the exception of the most important ones for each component $\alpha_{(\alpha\beta)}$, $\beta_{(\alpha\beta)}$, γ_α, are cancelled too, we obtain the *first-order approximation* of the strain energy as the simpliest expression for u :

$$u = \frac{1}{2}\left[D\,H^{\alpha\beta\rho\lambda}\,\alpha_{(\alpha\beta)}\,\alpha_{(\rho\lambda)} + B\,H^{\alpha\beta\rho\lambda}\,\beta_{(\alpha\beta)}\,\beta_{(\rho\lambda)} + Gha^{\alpha\beta}\,\gamma_\alpha\,\gamma_\beta\right]. \qquad (5.9)$$

We have used herein the following abbreviations:

$$D = \frac{Eh}{1-\nu^2} \quad \text{membrane stiffness,}$$

$$B = \frac{Eh^3}{12(1-\nu^2)} \quad \text{bending stiffness,}$$

$$Gh = \frac{Eh}{2(1+\nu)} \quad \text{transverse shear stiffness,} \qquad (5.10)$$

$$H^{\alpha\beta\rho\lambda} = \frac{1-\nu}{2}\left[a^{\alpha\rho}\,a^{\beta\lambda} + a^{\alpha\lambda}\,a^{\beta\rho} + \frac{2\nu}{1-\nu}\,a^{\alpha\beta}\,a^{\rho\lambda}\right]$$

$$= \frac{1}{2}\left[a^{\alpha\rho}\,a^{\beta\lambda} + a^{\alpha\lambda}\,a^{\beta\rho} + \nu\,(\varepsilon^{\alpha\rho}\,\varepsilon^{\beta\lambda} + \varepsilon^{\alpha\lambda}\,\varepsilon^{\beta\rho})\right],$$

$$H^{\alpha\beta\rho\lambda} = H^{(\alpha\beta)(\rho\lambda)} = H^{(\rho\lambda)(\alpha\beta)} \quad . \qquad (5.11)$$

It is obvious that the first-order approximation (5.9) contains many geometrical, mechanical and formal approximations, which imply a certain inexactness of the shell equations. The errors involved can be estimated by global error bounds [4, 9, 10].

5.2 First-Order Approximation of Constitutive Equations

Applying the general relations (5.5) to the strain energy function (5.9), we obtain the explicit constitutive equations for homogeneous, isotropic shells of (approximately) constant thickness:

$$\begin{aligned} \tilde{n}^{(\alpha\beta)} &= D\, H^{\alpha\beta\rho\lambda}\, \alpha_{(\rho\lambda)} \,, \\ m^{(\alpha\beta)} &= B\, H^{\alpha\beta\rho\lambda}\, \beta_{(\rho\lambda)} \,, \\ q^{\alpha} &= Gh a^{\alpha\beta}\, \gamma_{\beta} \,. \end{aligned} \tag{5.12}$$

Equivalent equations exist for anisotropic shells [11].

If the above mentioned approximations, which result from the transformations of (5.6) into (5.11), are investigated carefully [13, 22], it will be found out, that the errors will be minimized in the following limit cases:

$\rho = L/R_{min} = 0$ plane problems,

$\alpha_{(\alpha\beta)} = \gamma_\alpha = 0$ inextensional deformations,

$\beta_{(\alpha\beta)} = \gamma_\alpha = 0$ membrane theory.

These limit cases are represented by the thick boundary lines of the domain of Fig. 9. If a special shell problem deviates from the boundary, which will usually be the case, the errors in (5.9), respectively in (5.12), will increase. Because of that, there were many attempts to construct higher order approximations for the elastic potential or the constitutive equations, which intend to produce smaller error

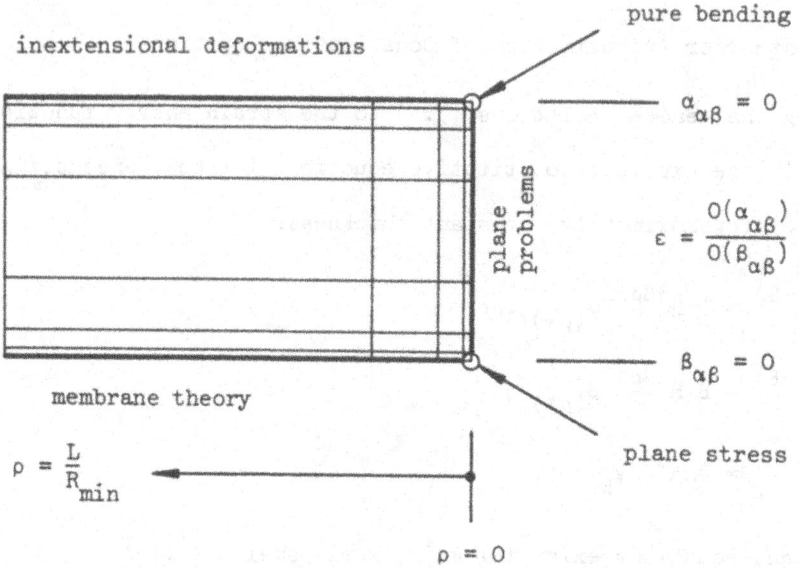

Fig. 9: On error margins in shell theory

bounds. As a typical example we give the following equations:

$$\tilde{n}^{(\alpha\beta)} = D\left[\left[H^{\alpha\beta\rho\lambda} + \frac{h^2}{12}\left(\overset{2}{H}{}^{\alpha\beta\rho\lambda} - 2H\,\overset{1}{H}{}^{\alpha\beta\rho\lambda} + K\,H^{\alpha\beta\rho\lambda}\right)\right]\alpha_{(\rho\lambda)} + \frac{h^2}{12}\left[(1.00 \div 1.75)\,\overset{1}{H}{}^{\alpha\beta\rho\lambda} - 2H\,H^{\alpha\beta\rho\lambda}\right]\beta_{(\rho\lambda)}\right],$$

$$m^{(\alpha\beta)} = B\left[H^{\alpha\beta\rho\lambda}\,\beta_{(\rho\lambda)} + \left[(1.00 \div 1.75)\,\overset{1}{H}{}^{\alpha\beta\rho\lambda} - 2H\,H^{\alpha\beta\rho\lambda}\right]\alpha_{(\rho\lambda)}\right], \tag{5.13}$$

in which $\overset{1}{H}{}^{\alpha\beta\rho\lambda}$, $\overset{2}{H}{}^{\alpha\beta\rho\lambda}$ denote higher order elasticity tensors [13, 14]. In our belief, supported by noting the variability range $1.00 \leq \varkappa \leq 1.75$ in the above equations, any higher approximation with the purpose of decreasing the general error bound of (5.9, 5.12) leads at the same time to further ambiguities in the constitutive equations. At the present state of research no simple method to bypass this problem is known.

6. LINEAR SHELL THEORY INCLUDING SHEAR DEFORMATIONS

6.1 Review of Variables and Fundamental Equations

We now summarize all previously defined or derived variables and equations. For this purpose we start with the external *surface loads* $\underset{\sim}{p}$ (3.2), $\underset{\sim}{c}$ (3.4) and the corresponding *displacement vectors* $\underset{\sim}{v}$ (4.5), $\underset{\sim}{w}$ (4.8):

$$\underset{\sim}{p} = p^{\alpha}\underset{\sim}{a}_{\alpha} + p^{3}\underset{\sim}{a}_{3} , \qquad \underset{\sim}{v} = v_{\alpha}\underset{\sim}{a}^{\alpha} + v_{3}\underset{\sim}{a}^{3} ,$$
$$\underset{\sim}{c} = c^{\rho}\varepsilon_{\rho\beta}\underset{\sim}{a}^{\beta} , \qquad \underset{\sim}{w} = w_{\alpha}\underset{\sim}{a}^{\alpha} . \qquad (6.1)$$

As *internal forces* the symmetric pseudo stress tensor $\tilde{n}^{(\alpha\beta)}$ (3.27), the transverse shear stress vector q^{α} and the moment tensor $m^{(\alpha\beta)}$ (3.11) have been introduced, as *strain variables* the first - $\alpha_{(\alpha\beta)}$ - and second - $\beta_{(\alpha\beta)}$ - strain tensor of the middle surface (4.25) and the shear angles γ_{α} (4.10):

$$\tilde{n}^{(\alpha\beta)} = n^{\alpha\beta} + m^{(\alpha\rho)}b_{\rho}^{\beta} , \qquad \alpha_{(\alpha\beta)} ,$$
$$m^{(\alpha\beta)} , \qquad \beta_{(\alpha\beta)}$$

$$q^{\alpha} \, , \qquad \gamma_{\alpha} \, . \tag{6.2}$$

External and internal dynamic variables are connected by the *equations of equilibrium* (3.24):

$$\begin{aligned} -p^{\beta} &= n^{\alpha\beta}|_{\alpha} - q^{\alpha} b^{\beta}_{\alpha} \, , \\ -p^{3} &= n^{\alpha\beta} b_{\alpha\beta} + q^{\alpha}|_{\alpha} \, , \\ -c^{\beta} &= m^{\alpha\beta}|_{\alpha} - q^{\beta} \, ; \end{aligned} \tag{6.3}$$

external and internal kinematic variables by the *kinematic relations* (4.33, 4.35):

$$\begin{aligned} \alpha_{(\alpha\beta)} &= \frac{1}{2} (v_{\alpha}|_{\beta} + v_{\beta}|_{\alpha} - 2b_{\alpha\beta} v_{3}) = \frac{1}{2} (\varphi_{\alpha\beta} + \varphi_{\beta\alpha}) \, , \\ \beta_{(\alpha\beta)} &= \frac{1}{2} (w_{\alpha}|_{\beta} + w_{\beta}|_{\alpha} - b^{\lambda}_{\alpha} \varphi_{\beta\lambda} - b^{\lambda}_{\beta} \varphi_{\alpha\lambda}) \, , \\ \gamma_{\alpha} &= w_{\alpha} + v_{3,\alpha} + b^{\lambda}_{\alpha} v_{\lambda} \, . \end{aligned} \tag{6.4}$$

The *constitutive equations* can be taken from (5.12):

$$\begin{aligned} \tilde{n}^{(\alpha\beta)} &= D \, H^{\alpha\beta\rho\sigma} \, \alpha_{(\rho\sigma)} \, , \quad m^{(\alpha\beta)} = B \, H^{\alpha\beta\rho\sigma} \, \beta_{(\rho\sigma)} \, , \\ q^{\alpha} &= Gha^{\alpha\beta} \gamma_{\beta} \, . \end{aligned} \tag{6.5}$$

To condude this review, we also mention the possible *dynamic* (3.35) and *kinematic* (4.42) *boundary conditions:*

$$
\begin{aligned}
\underset{\sim}{n}^{o} &= n^{\alpha\beta}u_{\alpha}t_{\beta}\underset{\sim}{t} + n^{\alpha\beta}u_{\alpha}u_{\beta}\underset{\sim}{u} + q^{\alpha}u_{\alpha}\underset{\sim}{a}_{3} , \\
\underset{\sim}{m}^{o} &= m^{\alpha\beta}u_{\alpha}u_{\beta}\underset{\sim}{t} - m^{\alpha\beta}u_{\alpha}t_{\beta}\underset{\sim}{u} , \\
\underset{\sim}{v}^{o} &= v^{\alpha}t_{\alpha}\underset{\sim}{t} + v^{\alpha}u_{\alpha}\underset{\sim}{u} + v_{3}\underset{\sim}{a}_{3} \\
\underset{\sim}{\omega}^{o} &= \omega^{\alpha}t_{\alpha}\underset{\sim}{t} + \omega^{\alpha}u_{\alpha}\underset{\sim}{u} .
\end{aligned}
\tag{6.6}
$$

6.2 The Principle of Virtual Displacement

As a next step we will formulate the *principle of virtual displacements*

$$
\overset{*}{\delta} A_{ext} + \overset{*}{\delta} A_{int} = 0 \tag{6.7}
$$

in terms of shell variables. We start with the contribution of the external variables, in which the virtual work of the forces and moments acting on the middle surface F , and of those along the boundary C must be distinguished:

$$
\begin{aligned}
\overset{*}{\delta} A_{ext} = &\iint_F (\underset{\sim}{p} \cdot \delta\underset{\sim}{v} + \underset{\sim}{c} \cdot \delta\underset{\sim}{\omega})\, dF \\
&+ \oint_C (\underset{\sim}{n} \cdot \delta\underset{\sim}{v} + \underset{\sim}{m} \cdot \delta\underset{\sim}{\omega})\, ds .
\end{aligned}
\tag{6.8}
$$

We now eliminate the load variables $\underset{\sim}{p}, \underset{\sim}{c}$ by the vectorial equations of equilibrium (3.18) and thus obtain:

$$\delta^* A_{ext} = \iint_F \{ \underset{\sim}{n}^\alpha|_\alpha \cdot \delta \underset{\sim}{v} + \underset{\sim}{m}^\alpha|_\alpha \cdot \delta\underset{\sim}{\omega} + [\underset{\sim}{a}_\alpha, \underset{\sim}{n}^\alpha, \delta\underset{\sim}{\omega}] \} \, dF + \oint_C (\underset{\sim}{n} \cdot \delta\underset{\sim}{v} + \underset{\sim}{m} \cdot \delta\underset{\sim}{\omega}) \, ds \; , \tag{6.9}$$

where sharp brackets denote a triple product. After several suitable transformations, in which all remaining external variables are transformed into internal ones, we find:

$$\begin{aligned} \delta^* A_{ext} &= \iint_F \left[n^{\alpha\beta} \delta\varphi_{\alpha\beta} + m^{(\alpha\beta)} \delta\varkappa_{(\alpha\beta)} + q^\alpha \delta\gamma_\alpha \;\; dF \right] \\ &= \iint_F \left[n^{(\alpha\beta)} \delta\alpha_{(\alpha\beta)} + m^{(\alpha\beta)} \delta\beta_{(\alpha\beta)} + q^\alpha \delta\gamma_\alpha \;\; dF \right] \end{aligned} \tag{6.10}$$

with the set of strain variables $\alpha_{(\alpha\beta)}, \beta_{(\alpha\beta)}, \gamma_\alpha$ from (6.4) and the alternative set of quantities:

$$\begin{aligned} \varphi_{\alpha\beta} &= v_{\beta|\alpha} - b_{\alpha\beta} \, v_3 \; , \\ \gamma_\alpha &= w_\alpha + v_{3},_\alpha + b_\alpha^\lambda v_\lambda \; , \\ \varkappa_{(\alpha\beta)} &= \frac{1}{2} (w_{\alpha|\beta} + w_{\beta|\alpha}) \; . \end{aligned} \tag{6.11}$$

By comparison with (6.7) we then find two equivalent forms of the principle of virtual work:

$$\begin{aligned} &\iint_F (\underset{\sim}{p} \cdot \delta\underset{\sim}{v} + \underset{\sim}{c} \cdot \delta\underset{\sim}{\omega}) \, dF + \oint_C (\underset{\sim}{n} \cdot \delta\underset{\sim}{v} + \underset{\sim}{m} \cdot \delta\underset{\sim}{\omega}) \, ds \\ &\qquad - \iint_F \left[\tilde{n}^{(\alpha\beta)} \delta\alpha_{(\alpha\beta)} + m^{(\alpha\beta)} \delta\beta_{(\alpha\beta)} + q^\alpha \delta\gamma_\alpha \right] dF = 0 \; , \end{aligned} \tag{6.12}$$

$$\iint_F (\underset{\sim}{p} \cdot \delta \underset{\sim}{v} + \underset{\sim}{c} \cdot \delta \underset{\sim}{\omega}) \, dF + \oint_C (\underset{\sim}{n} \cdot \delta_v + \underset{\sim}{m} \cdot \delta \underset{\sim}{\omega}) \, ds$$
$$- \iint_F \left[n^{\alpha\beta} \delta\varphi_{\alpha\beta} + m^{(\alpha\beta)} \delta\varkappa_{(\alpha\beta)} + q^{\alpha} \delta\gamma_{\alpha} \right] dF = 0 , \tag{6.13}$$

which differ in the internal variables. All variables, which contribute to the work in one and the same product, are called *corresponding variables*. A formulation of all shell equations in terms of corresponding variables is called a *consistent formulation* [1, 11]. For any type of shell theory there exists an arbitrary large number of consistent formulations with different internal (or external) variables [11].

6.3 A Consistent Formulation

We intend to show the advantage of consistency by deriving such a formulation, in which $n^{\alpha\beta}$, $m^{(\alpha\beta)}$ and q^{α} act as internal kinematic variables. As we find by inspection of (6.3), the equations of equilibrium are already formulated in the desired consistent way, if the modified kinematic relations (6.11) are used. The corresponding constitutive equations have to be derived from (6.5) with the help of (6.11):

$$n^{\alpha\beta} = \tilde{n}^{(\alpha\beta)} - m^{(\alpha\sigma)} b^{\beta}_{\sigma}$$

$$= D\, H^{\alpha\beta\sigma\lambda} \alpha_{(\sigma\lambda)} - B\, H^{\alpha\sigma\rho\lambda} b^{\beta}_{\sigma} \beta_{(\rho\lambda)}$$

$$= D\, H^{\alpha\beta\rho\lambda} \varphi_{\rho\lambda} - B\, H^{\alpha\sigma\rho\lambda} b^{\beta}_{\sigma} \underbrace{(\varkappa_{(\rho\lambda)} - \varphi_{\rho\varepsilon} b^{\varepsilon}_{\lambda})}_{\beta_{(\rho\lambda)}} ,$$

$$m^{(\alpha\beta)} = B\,H^{\alpha\beta\rho\lambda}\,\beta_{(\rho\lambda)}$$

$$= B\,H^{\alpha\beta\rho\lambda}\,\underbrace{(\varkappa_{(\rho\lambda)} - \varphi_{\rho\varepsilon}\,b^{\varepsilon}_{\lambda})}_{\beta_{(\rho\lambda)}} = B\,H^{\alpha\beta\rho\lambda}\,\beta_{(\rho\lambda)} \qquad (6.14)$$

the equation for q^{α} remains unchanged.

Next we define the following matrix columns:

$$\underline{p} = \begin{pmatrix} p^1 \\ p^2 \\ p^3 \\ c^1 \\ c^2 \end{pmatrix}, \quad \underline{\sigma} = \begin{pmatrix} n^{11} \\ n^{12} \\ n^{21} \\ n^{22} \\ q^1 \\ q^2 \\ m^{11} \\ m^{12} \\ m^{21} \\ m^{22} \end{pmatrix}, \quad \underline{\varepsilon} = \begin{pmatrix} \varphi_{11} \\ \varphi_{12} \\ \varphi_{21} \\ \varphi_{22} \\ \gamma_1 \\ \gamma_2 \\ \varkappa_{11} \\ \varkappa_{12} \\ \varkappa_{21} \\ \varkappa_{22} \end{pmatrix}, \quad \underline{u} = \begin{pmatrix} v_1 \\ v_2 \\ v_3 \\ w_1 \\ w_2 \end{pmatrix} \qquad (6.15)$$

and write the kinematic equations (6.11) in terms of $\underline{\varepsilon}$ and $\underline{u}$. If we abbreviate the covariant derivatives by

$$\ldots/_{\alpha} : \quad \ldots/_{1}\,, \quad \ldots/_{2}\,,$$

we find the following matrix differential equation:

$$\begin{pmatrix}\varphi_{11}\\ \varphi_{12}\\ \varphi_{21}\\ \varphi_{22}\\ \hline \gamma_1\\ \gamma_2\\ \hline \kappa_{11}\\ \kappa_{12}\\ \kappa_{21}\\ \kappa_{22}\end{pmatrix} = \left(\begin{array}{cc|c|cc} \scriptstyle 1 & \scriptstyle 2 & \scriptstyle 3 & \scriptstyle 1 & \scriptstyle 2\\ |_1 & 0 & -b_{11} & 0 & 0\\ 0 & |_1 & -b_{12} & 0 & 0\\ |_2 & 0 & -b_{21} & 0 & 0\\ 0 & |_2 & -b_{22} & 0 & 0\\ \hline b_1^1 & b_1^2 & |_1 & 1 & 0\\ b_2^1 & b_2^2 & |_2 & 0 & 1\\ \hline 0 & 0 & 0 & |_1 & 0\\ 0 & 0 & 0 & \frac{1}{2}|_2 & \frac{1}{2}|_1\\ 0 & 0 & 0 & \frac{1}{2}|_2 & \frac{1}{2}|_1\\ 0 & 0 & 0 & 0 & |_2 \end{array}\right) \cdot \begin{pmatrix} v_1\\ v_2\\ \hline v_3\\ \hline w_1\\ w_2\end{pmatrix} \cdot \tag{6.16}$$

Using some easily understandable abbreviations for the sub-operators we end up with:

$$\underline{\varepsilon} = \begin{pmatrix} \underline{\varphi}_{\alpha\beta}\\ \underline{\gamma}_{\alpha}\\ \underline{\kappa}_{(\alpha\beta)}\end{pmatrix} = \begin{pmatrix} \underline{D}_{11} & \underline{B}_{12} & \underline{0}\\ \underline{B}_{21} & \underline{D}_{22} & \underline{I}\\ \underline{0} & \underline{0} & \underline{D}_{33}\end{pmatrix} \begin{pmatrix}\underline{v}_{\alpha}\\ \underline{v}_3\\ \underline{w}_{\alpha}\end{pmatrix} = \underline{D}_k\, \underline{u} \; . \tag{6.17}$$

Similar transformations for the equations of equilibrium will lead to:

$$-\underline{p} = -\begin{pmatrix} \underline{p}^{\alpha} \\ \underline{p}^{3} \\ \underline{c}^{\alpha} \end{pmatrix} = \begin{pmatrix} \underline{D}_{11}^{T} & -\underline{B}_{21}^{T} & \underline{0} \\ -\underline{B}_{12}^{T} & \underline{D}_{22}^{T} & \underline{0} \\ \underline{0} & -\underline{I} & \underline{D}_{33}^{T} \end{pmatrix} \begin{pmatrix} \underline{n}^{\alpha\beta} \\ \underline{q}^{\alpha} \\ \underline{m}^{(\alpha\beta)} \end{pmatrix} = \underline{D}_e \underline{\sigma} \,, \tag{6.18}$$

in which the transposed sub-operators of (6.17) appear. One of the many advantages of using consistent formulations lies in this symmetry on the sub-operator level of the dynamic and kinematic equations. Finally, all variables and equations of this consistent formulation are given in Table 1.

7. LINEAR KIRCHHOFF-LOVE-TYPE SHELL THEORY

7.1 Derivation of the Basic Equations

Any *Kirchhoff-Love*-Type shell theory is characterized by the assumption, that the shear angles γ^{α} in Fig. 8 are considered to be immaterial compared with the other strain measures:

$$\gamma^{\alpha} = 0 \,. \tag{6.19}$$

Introducing this constraint into (4.35):

$$\gamma_{\alpha} = w_{\alpha} + v_{3},_{\alpha} + b_{\alpha}^{\lambda} v_{\lambda} = 0 \longrightarrow w_{\alpha} = -v_{3},_{\alpha} - b_{\alpha}^{\lambda} v_{\lambda} \tag{6.20}$$

changes the components of this displacement vector $\underset{\sim}{w}$ into a set of

1. Principle of virtual displacements	
$\iint_F \mathbf{p}\cdot\delta\mathbf{v}\,dF + \oint_C (\tilde{\mathbf{n}}\,\delta\mathbf{v} + \mathbf{m}_t\cdot\delta\boldsymbol{\omega}_t)ds + [m_u\,\delta v_3]_C - \iint_F (\tilde{n}^{(\alpha\beta)}\delta\alpha_{(\alpha\beta)} + m^{(\alpha\beta)}\delta\omega_{(\alpha\beta)})dF = 0$	
2. External mechanical variables	
$\mathbf{p} = p^\alpha \mathbf{a}_\alpha + p^3 \mathbf{a}_3$	$\mathbf{v} = v_\alpha \mathbf{a}^\alpha + v_3 \mathbf{a}^3$
3. Internal mechanical variables	
$\tilde{n}^{(\alpha\beta)} = n^{\alpha\beta} + m^{(\alpha\varrho)} b_\varrho^\beta,\; m^{(\alpha\beta)} = m^{\alpha\beta}$	$\alpha_{(\alpha\beta)},\; \omega_{(\alpha\beta)}$
4. Boundary variables	
$\tilde{\mathbf{n}} = \tilde{n}_t \mathbf{t} + \tilde{n}_u \mathbf{u} + \tilde{n}_3 \mathbf{a}_3,$ $\mathbf{m}_t = m_t \mathbf{t}$	$\mathbf{v} = v_t \mathbf{t} + v_u \mathbf{u} + v_3 \mathbf{a}_3,$ $\boldsymbol{\omega}_t = \omega_t \mathbf{t}$
5. Field equation	
$-p^\beta = \tilde{n}^{\alpha\beta}\|_\alpha - 2m^{(\alpha\lambda)}\|_\alpha b_\lambda^\beta - m^{(\alpha\lambda)} b_\lambda^\beta\|_\alpha;$ $-p^3 = \tilde{n}^{\alpha\beta} b_{\alpha\beta} + m^{(\alpha\lambda)}\|_{\alpha\lambda} - m^{(\alpha\lambda)} b_\lambda^\beta b_{\alpha\beta}$	$\alpha_{(\alpha\beta)} = \frac{1}{2}(v_\alpha\|_\beta + v_\beta\|_\alpha - 2b_{\alpha\beta} v_3),$ $\omega_{(\alpha\beta)} = -(v_3\|_{\alpha\beta} + v_\lambda b_\alpha^\lambda\|_\beta + b_\alpha^\lambda v_\lambda\|_\beta + b_\beta^\lambda v_\lambda\|_\alpha - b_\alpha^\lambda b_{\lambda\beta} v_3)$
6. Strain energy density of isotropic shells	
$\pi_i = \frac{1}{2}(DH^{\alpha\beta\lambda\mu}\alpha_{(\alpha\beta)}\alpha_{(\lambda\mu)} + BH^{\alpha\beta\lambda\mu}\omega_{(\alpha\beta)}\omega_{(\lambda\mu)}) = \frac{1}{2}(\tilde{n}^{(\alpha\beta)}\alpha_{(\alpha\beta)} + m^{(\alpha\beta)}\omega_{(\alpha\beta)})$	
7. Constitutive relations of isotropic shells	
$\tilde{n}^{(\alpha\beta)} = \frac{\partial\pi_i}{\partial\alpha_{(\alpha\beta)}} = DH^{\alpha\beta\lambda\mu}\,\alpha_{(\lambda\mu)};$	$m^{(\alpha\beta)} = \frac{\partial\pi_i}{\partial\omega_{(\alpha\beta)}} = BH^{\alpha\beta\lambda\mu}\,\omega_{(\lambda\mu)}$
8. Boundary conditions	
$\tilde{\mathbf{n}}^\circ = (\tilde{n}^{(\alpha\beta)} - m^{(\alpha\lambda)} b_\lambda^\beta - k_1 m^{(\alpha\beta)}) u_\alpha t_\beta \mathbf{t}$ $+ ((\tilde{n}^{(\alpha\beta)} - m^{(\alpha\lambda)} b_\lambda^\beta) u_\alpha u_\beta - k_2 m^{(\alpha\beta)} u_\alpha t_\beta)\mathbf{u}$ $+ ((m^{\lambda\alpha}{}_{\|\lambda} + m^{(\alpha\beta)}{}_{\|\lambda} t^\lambda t_\beta) u_\alpha + m^{(\alpha\beta)}(u_\alpha t_\beta)\|_\lambda t^\lambda)\mathbf{a}_3;$ $\mathbf{m}_t^\circ = m^{\alpha\beta} u_\alpha u_\beta$ force singularity: $E_3 = -m^{\alpha\beta}[u_\alpha t_\beta]_{-0}^{+0}$	$\mathbf{v}^\circ = v_\alpha t^\alpha \mathbf{t} + v_\alpha u^\alpha \mathbf{u} + v_3 \mathbf{a}_3;$ $\boldsymbol{\omega}_t^\circ = -(v_{3,n} + b_\lambda^\beta u^\lambda v_\beta)\mathbf{t}$

Table 1: Mechanical variables and equations of a best first-order linear shell theory, including shear deformations; consistently formulated

dependent variables. The first strain tensor $\alpha_{(\alpha\beta)}$ remains unaltered by (6.20), while in the second one the component w_α can be exchanged by the components of the displacement vector $\underset{\sim}{v}$:

$$\alpha_{(\alpha\beta)} = \frac{1}{2}(v_{\alpha|\beta} + v_{\beta|\alpha} - 2b_{\alpha\beta}\, v_3) ,$$

$$\beta_{(\alpha\beta)} = \frac{1}{2}(-v_{3|\alpha\beta} - v_{3|\beta\alpha} - 2v_{\lambda|\alpha}\, b^{\lambda}_{\beta} - 2v_{\lambda|\beta}\, b^{\lambda}_{\alpha} - v_{\lambda}\, b^{\lambda}_{\alpha|\beta} - v_{\lambda}\, b^{\lambda}_{\beta|\alpha}$$

$$+ v_3\,(b_{\alpha\lambda}\, b^{\lambda}_{\beta} + b^{\lambda}_{\alpha}\, b_{\lambda\beta})) . \qquad (6.21)$$

From the principle of virtual work (6.12, 6.13) we find, that because of (6.19) the constitutive equation for q^{α} can be dropped, while those for $\tilde{n}^{(\alpha\beta)}$ and $m^{(\alpha\beta)}$ remain unchanged:

$$\tilde{n}^{(\alpha\beta)} = D\, H^{\alpha\beta\rho\lambda}\, \alpha_{(\rho\lambda)} , \quad m^{(\alpha\beta)} = B\, H^{\alpha\beta\rho\lambda}\, \beta_{(\rho\lambda)} . \qquad (6.22)$$

If we exclude the couple load $\underset{\sim}{c}$ on the middle surface from further considerations, we are able to express the presently dependent variable q^{β} by the covariant derivative of the moment tensor. From the last line of (6.3) we obtain:

$$c^{\beta} = 0 \longrightarrow q^{\beta} = m^{\alpha\beta}{}_{|\alpha} . \qquad (6.23)$$

If we also exchange the stress resultant tensor $n^{(\alpha\beta)}$ by the definition (6.2) of the pseudo stress resultant $\tilde{n}^{(\alpha\beta)}$

$$n^{\alpha\beta} = \tilde{n}^{(\alpha\beta)} - m^{(\alpha\lambda)}\, b^{\beta}_{\lambda} = \frac{1}{2}(\tilde{n}^{(\alpha\beta)} + \tilde{n}^{(\beta\alpha)} - m^{(\alpha\lambda)}\, b^{\beta}_{\lambda} - m^{(\lambda\alpha)}\, b^{\beta}_{\lambda}) , \qquad (6.24)$$

we find for the remaining equations of equilibrium a consistent formulation with respect to the choice of internal variables

$$\tilde{n}^{(\alpha\beta)}, m^{(\alpha\beta)} \longleftrightarrow \alpha_{(\alpha\beta)}, \beta_{(\alpha\beta)}$$

in (6.12) and with respect to the kinematic relations (6.21):

$$-p^{\beta} = \frac{1}{2}(\tilde{n}^{\alpha\beta}|_{\alpha} + \tilde{n}^{\beta\alpha}|_{\alpha} - 2m^{\lambda\alpha}|_{\alpha} b^{\beta}_{\lambda} - 2m^{\alpha\lambda}|_{\alpha} b^{\beta}_{\lambda} - m^{\lambda\alpha} b^{\beta}_{\lambda}|_{\alpha} - m^{\alpha\lambda} b^{\beta}_{\lambda}|_{\alpha}),$$

$$-p^{3} = \frac{1}{2}(m^{\alpha\lambda}|_{\alpha\lambda} + m^{\lambda\alpha}|_{\alpha\lambda} + b_{\alpha\beta}\tilde{n}^{\alpha\beta} + b_{\alpha\beta}\tilde{n}^{\beta\alpha} - b_{\alpha\beta}(m^{\alpha\lambda} b^{\beta}_{\lambda} + m^{\lambda\alpha} b^{\beta}_{\lambda})). \tag{6.25}$$

7.2 A Consistent Formulation

Also for this special formulation of the *Kirchhoff-Love*-Type shell theory we show the symmetry on the sub-operator level, which is a typical criterion of any consistent theory. For this purpose we define the following matrices of the independent external and internal variables:

$$\underline{p} = \begin{pmatrix} \underline{p}^{\alpha} \\ \underline{p}^{3} \end{pmatrix} = \begin{pmatrix} p^{1} \\ p^{2} \\ p^{3} \end{pmatrix}, \qquad \underline{u} = \begin{pmatrix} \underline{v}_{\alpha} \\ \underline{v}_{3} \end{pmatrix} = \begin{pmatrix} v_{1} \\ v_{2} \\ v_{3} \end{pmatrix},$$

$$\underline{\sigma} = \begin{pmatrix} \underline{n}^{(\alpha\beta)} \\ \\ \underline{m}^{(\alpha\beta)} \end{pmatrix} = \begin{pmatrix} n^{(11)} \\ n^{(12)} \\ n^{(21)} \\ n^{(22)} \\ \\ m^{(11)} \\ m^{(12)} \\ m^{(21)} \\ m^{(22)} \end{pmatrix} \qquad \underline{\varepsilon} = \begin{pmatrix} \underline{\alpha}_{(\alpha\beta)} \\ \\ \underline{\beta}_{(\alpha\beta)} \end{pmatrix} = \begin{pmatrix} \alpha_{(11)} \\ \alpha_{(12)} \\ \alpha_{(21)} \\ \alpha_{(22)} \\ \\ \beta_{(11)} \\ \beta_{(12)} \\ \beta_{(21)} \\ \beta_{(22)} \end{pmatrix} \tag{6.26}$$

If we now write the equations of equilibrium (6.25) and the correspon-ding kinematic relations (6.21) in terms of this matrix notation, we obtain:

$$-\underline{p} = \begin{pmatrix} \underline{D}_s^T & -\underline{D}^T \\ \\ \underline{B}^T & \underline{D}_p^T \end{pmatrix} \cdot \underline{\sigma} \qquad \underline{\varepsilon} = \begin{pmatrix} \underline{D}_s & -\underline{B} \\ \\ -\underline{D} & -\underline{D}_p \end{pmatrix} \cdot \underline{u}$$

$$= \underline{D}_e\, \underline{\sigma}\ , \qquad\qquad = \underline{D}_k\, \underline{u}\ , \tag{6.27}$$

in which we have abbreviated the operator matrices in the following manner:

$$\underline{D}_s = \frac{1}{2} \begin{pmatrix} 2|_1 & 0 \\ |_2 & |_1 \\ \\ |_2 & |_1 \\ 0 & 2|_2 \end{pmatrix} , \qquad \underline{B} = \begin{pmatrix} b_{11} \\ b_{12} \\ \\ b_{21} \\ b_{22} \end{pmatrix} ,$$

$$\underline{D} = \frac{1}{2}\begin{pmatrix} (4\big|_1 b^1_1 + 2b^1_1\big|_1) & (4\big|_1 b^2_1 + 2b^2_1\big|_1) \\ \begin{array}{l}(2\big|_1 b^1_2 + 2\big|_2 b^1_1 \\ \quad + b^1_1\big|_2 + b^1_2\big|_1)\end{array} & \begin{array}{l}(2\big|_1 b^2_2 + 2\big|_2 b^2_1 \\ \quad + b^2_1\big|_2 + b^2_2\big|_1)\end{array} \\ \begin{array}{l}(2\big|_1 b^1_2 + 2\big|_2 b^1_1 \\ \quad + b^1_1\big|_2 + b^1_2\big|_1)\end{array} & \begin{array}{l}(2\big|_1 b^2_2 + 2\big|_2 b^2_1 \\ \quad + b^2_1\big|_2 + b^2_2\big|_1)\end{array} \\ (4\big|_2 b^1_2 + b^1_2\big|_2) & (4\big|_2 b^2_2 + 2b^2_2\big|_2) \end{pmatrix},$$

$$\underline{D}_p = \frac{1}{2}\begin{pmatrix} (2\big|_{11} - 2b_{11}b^1_1 - 2b_{12}b^2_1) \\ \begin{array}{l}(\big|_{12} + \big|_{21} - b_{11}b^1_2 \\ \quad - b_{12}(b^1_1 + b^2_2) - b_{22}b^2_1)\end{array} \\ \begin{array}{l}(\big|_{12} + \big|_{21} - b_{11}b^1_2 \\ \quad - b_{12}(b^1_1 + b^2_2) - b_{22}b^2_1)\end{array} \\ (2\big|_{22} - 2b_{12}b^1_2 - 2b_{22}b^2_2) \end{pmatrix} \qquad (6.28)$$

1. Principle of virtual displacements	
$\iint_F (\boldsymbol{p}\cdot\delta\boldsymbol{v} + \boldsymbol{c}\cdot\delta\boldsymbol{\omega})dF + \oint_C (\boldsymbol{n}\,\delta\boldsymbol{v} + \boldsymbol{m}\cdot\delta\boldsymbol{\omega})ds - \iint_F (n^{\alpha\beta}\,\delta\varphi_{\alpha\beta} + m^{(\alpha\beta)}\,\delta\varkappa_{(\alpha\beta)} + q^{\alpha}\,\delta\gamma_{\alpha})dF = 0$	
2. External mechanical variables	
$\boldsymbol{p} = p^{\alpha}\boldsymbol{a}_{\alpha} + p^{3}\boldsymbol{a}_{3}$, $\boldsymbol{c} = c^{\rho}\,\varepsilon_{\rho\beta}\,\boldsymbol{a}^{\beta}$	$\boldsymbol{v} = v_{\alpha}\,\boldsymbol{a}^{\alpha} + v_{3}\,\boldsymbol{a}^{3}$; $\boldsymbol{\omega} = \omega_{\alpha}\boldsymbol{a}^{\alpha} = w_{\rho}\,\varepsilon^{\rho\beta}\,\boldsymbol{a}_{\beta}$
3. Internal mechanical variables	
$n^{\alpha\beta}$, q^{α}, $m^{(\alpha\beta)}$	$\varphi_{\alpha\beta}$, γ_{α}, $\varkappa_{(\alpha\beta)}$
4. Boundary variables	
$\boldsymbol{n} = n_t\,\boldsymbol{t} + n_u\,\boldsymbol{u} + n_3\,\boldsymbol{a}_3$, $\boldsymbol{m} = m_t\,\boldsymbol{t} + m_u\,\boldsymbol{u}$	$\boldsymbol{v} = v_t\,\boldsymbol{t} + v_u\,\boldsymbol{u} + v_3\,\boldsymbol{a}_3$, $\boldsymbol{\omega} = \omega_t\,\boldsymbol{t} + \omega_u\,\boldsymbol{u}$
5. Field equation	
$-p^{\beta} = n^{\alpha\beta}\|_{\alpha} - q^{\alpha}b_{\alpha}^{\beta}$, $-p^{3} = n^{\alpha\beta}b_{\alpha\beta} + q^{\alpha}\|_{\alpha}$, $-c^{\beta} = m^{\alpha\beta}\|_{\alpha} - q^{\beta}$	$\varphi_{\alpha\beta} = v_{\beta\|\alpha} - v_3\,b_{\alpha\beta}$, $\gamma_{\alpha} = w_{\alpha} + v_{3,\alpha} + v_{\lambda}\,b_{\alpha}^{\lambda}$, $\varkappa_{(\alpha\beta)} = \frac{1}{2}(w_{\alpha\|\beta} + w_{\beta\|\alpha})$
6. Strain energy density of isotropic shells	
$\pi_i = \frac{1}{2}(DH^{\alpha\beta\lambda\mu}\varphi_{\alpha\beta}\,\varphi_{\lambda\mu} - 2BH^{\alpha\beta\lambda\mu}b_{\lambda}^{\sigma}\varphi_{\mu\sigma}\,\varkappa_{(\alpha\beta)} + BH^{\alpha\beta\lambda\mu}\varkappa_{(\alpha\beta)}\varkappa_{(\lambda\mu)} + Gh\,a^{\alpha\lambda}\gamma_{\alpha}\gamma_{\lambda})$ $= \frac{1}{2}(n^{\alpha\beta}\,\varphi_{\alpha\beta} + m^{(\alpha\beta)}\,\varkappa_{(\alpha\beta)} + q^{\alpha}\,\gamma_{\alpha})$	
7. Constitutive relations of isotropic shells	
$n^{\alpha\beta} = \frac{\partial\pi_i}{\partial\varphi_{\alpha\beta}} = DH^{\alpha\beta\lambda\mu}\varphi_{\lambda\mu} - BH^{\alpha\rho\lambda\mu}b_{\rho}^{\beta}\varkappa_{(\lambda\mu)}$, $q^{\alpha} = \frac{\partial\pi_i}{\partial\gamma_{\alpha}} = Gh\,a^{\alpha\lambda}\gamma_{\lambda}$, $m^{(\alpha\beta)} = \frac{\partial\pi_i}{\partial\varkappa_{(\alpha\beta)}} = BH^{\alpha\beta\lambda\mu}(\varkappa_{(\lambda\mu)} - b_{\lambda}^{\sigma}\,\varphi_{\mu\sigma})$	
8. Boundary conditions	
$\boldsymbol{n}^{\circ} = n^{\alpha\beta}u_{\alpha}t_{\beta}\,\boldsymbol{t} + n^{\alpha\beta}\,u_{\alpha}u_{\beta}\,\boldsymbol{u} + q^{\alpha}u_{\alpha}\boldsymbol{a}^{3}$, $\boldsymbol{m}^{\circ} = m^{\alpha\beta}u_{\alpha}u_{\beta}\,\boldsymbol{t} - m^{\alpha\beta}\,u_{\alpha}t_{\beta}\,\boldsymbol{u}$	$\boldsymbol{v}^{\circ} = v_{\alpha}t^{\alpha}\,\boldsymbol{t} + v_{\alpha}\,u^{\alpha}\boldsymbol{u} + v_3\,\boldsymbol{a}_3$, $\boldsymbol{\omega}^{\circ} = \omega_{\alpha}t^{\alpha}\,\boldsymbol{t} + \omega_{\alpha}\,u^{\alpha}\boldsymbol{u} = w_{\beta}u^{\beta}\boldsymbol{t} - w_{\beta}t^{\beta}\boldsymbol{u}$

Table 2: Mechanical variables and equations of a best first-order linear Kirchhoff-Love-Type shell theory, consistently formulated

The constitutive equations (6.22) and the (consistently reformulated) boundary conditions (6.6) can also be transformed into operator equations comparable to (6.27) [11]. Table 2 finally reviews all variables and equations of this consistent formulation.

8. CONCLUDING REMARKS

In this paper we have tried to give a brief survey on general shell theory. Many details concerning the derivation and transformation of equations had to be left to the reader, many essential gaps were not filled. This is true for the large variety of different consistent formulations known today [1, 11] or for dual variables and dual equations, such as stress functions and compatibility equations, which can be used with great advantage [19, 20]. No attempt has been made, to treat physically or geometrically non-linear shell problems. Nevertheless we hope that the reader has gained some insight and may look forward to a deeper understanding of this field.

REFERENCES

[1] *Budiansky, B., Sanders, J.L.: On the Best First Order linear Shell Theory. Progress in Applied Mechanics, Macmillan, New York 1963, p. 129.*

[2] *Green, A.E., Naghdi, P.M., Wainright: A General Theory of a Cosserat Surface. Arch. Rat. Mech. Analysis 20 (1965), p. 287.*

[3] *Green, A.E., Zerna, W.: Theoretical Elasticity. 2. Edition, At the Clarendon Press, Oxford 1968.*

[4] *Koiter, W.T.: A consistent First Approximation in the General Theory of Thin Shells. Proc. IUTAM-Symp. on Theory of Thin Elastic Shells. North-Holland Publ. Comp., Amsterdam, 1959.*

[5] *Koiter, W.T.: On the foundations of the linear theory of thin elastic shells. Proc. Kon. Ned. Ak. Wet., Vol. 73 (1970), p. 169.*

[6] Koiter, W.T., Simmonds, J.G.: Foundations of Shell Theory. Rep. Nr. 473, Lab. Eng. Mech., Delft University of Technology, 1972.

[7] Zerna, W.: Herleitung der ersten Approximation der Theorie elastischer Schalen. Abhandl. Braunschw. Wissensch. Gesellschaft 19 (1967), p. 52.

[8] Green, A.E., Naghdi, P.M.: Non-isothermal Theory of Rods, Plates and Shells. Int. Journ. Solids Struct., 6 (1970), p. 209.

[9] John, F.: Estimates for the Derivatives of the Stresses in a Thin Shell and Interior Shell Equations. Comm. Pure Appl. Math., 18 (1965), p. 235.

[10] Sensenig, C.B.: A Shell Theory Compared with the Exact Three-dimensional Theory of Elasticity. Int. Journ. Eng. Science 6 (1968), p. 435.

[11] Basar, Y., Krätzig, W.B.: Mechanik der Flächentragwerke. Vieweg-Verlag (in Print).

[12] Cohen, H., De Silva, C.N.: Nonlinear Theory of Elastic Directed Surfaces. Journ. Math. Phys., 7 (1966), p. 960.

[13] Krätzig, W.B.: Optimale Schalengrundgleichungen und deren Leistungsfähigkeit. ZAMM 54 (1974), p. 265.

[14] Krätzig, W.B.: Die erste Approximation der Schalentheorie und deren Unschärfebereich. Konstruktiver Ingenieurbau in Forschung und Praxis, Werner-Verlag 1976, p. 33.

[15] Krätzig, W.B.: Thermodynamics of Deformations and Shell Theory. Techn. Report Nr. 71-3, Institut für Konstruktiven Ingenieurbau der RUB, Bochum 1971.

[16] Naghdi, P.M.: The Theory of Shells and Plates. Handbuch der Physik VI, A2, Springer-Verlag, Berlin-Heidelberg 1972.

[17] Naghdi, P.M.: Foundation of Elastic Shell Theory. Progress in Solid Mechanics, Vol. IV. North-Holland Publ. Comp., Amsterdam 1963.

[18] Reissner, E.: On the foundation of the theory of elastic shells. Proc. 11th Int. Congr. Appl. Mech., München 1964. Springer-Verlag 1966, S. 20.

[19] *Tonti, E.: On the Formal Structure of Physical Theories. Inst. Mat. Politecnico Milano 1975.*

[20] *Oden, J.T., Reddy, J.N.: On Dual-Complementary Variational Principles. Int. Journ. Eng. Sc., Vol. 12 (1974), p. 1.*

[21] *Naghdi, P.M.: Further results in the derivation of the general equations of elastic shells. Int. Journ. Engng. Sci. 2 (1964), p. 269.*

[22] *Pietraszkiewicz, W.: Introduction to the Non-Linear Theory of Shells. Ruhr-Universität Bochum, Mitteilungen aus dem Institut für Mechanik, Mai 1977.*

GENERAL THEORY OF SHELL STABILITY

by

W.T. Koiter

University of Technology, Delft

GENERAL THEORY OF SHELL STABILITY

W.T. Koiter

1. BASIC PRINCIPLES OF ELASTIC STABILITY

1.1 Energy Criterion for Conservative Systems

State I: *Fundamental* (pre-buckling) state of equilibrium to be investigated as to its stability; potential energy P_I.

State II: *Adjacent* configuration, obtained from fundamental state I by (additional) displacement field $\underset{\sim}{u}(\underset{\sim}{x})$; potential energy P_{II}.

Potential *energy functional:*

$$P[\underset{\sim}{u}(\underset{\sim}{x})] = P_{II} - P_I . \tag{1.1}$$

Basic energy criterion: Equilibrium in state I is *stable,* if $P[\underset{\sim}{u}]$ is *positive definite,* i.e. $P[\underset{\sim}{u}] > 0$ for *all* displacement fields $\underset{\sim}{u}(\underset{\sim}{x})$ with sufficiently small norm $0 < \|\underset{\sim}{u}\| < c$. Equilibrium in state I is *unstable,* if $P[\underset{\sim}{u}]$ is negative for *some* displacement field $\underset{\sim}{u}(\underset{\sim}{x})$ with arbitrarily small norm $\|\underset{\sim}{u}\|$.

Expansion of energy functional:

$$P[\underset{\sim}{u}] = P_2[\underset{\sim}{u}] + P_3[\underset{\sim}{u}] + P_4[\underset{\sim}{u}] + \dots \tag{1.2}$$

Criterion of second variation: Equilibrium in state I is *stable*, if $P_2[\underset{\sim}{u}]$ is *positive definite*. Equilibrium in state I is unstable, if $P_2[\underset{\sim}{u}]$ has negative values.

Critical case of neutral equilibrium (stability limit): $P_2[\underset{\sim}{u}]$ is *semi-definite positive*, i.e.

> Min $P_2[\underset{\sim}{u}] = 0$ is attained for non-vanishing displacement fields $\underset{\sim}{u}_h(\underset{\sim}{x})$, $h = 1,2,..n$, the *buckling modes*, and hence also for any linear combination $a_h \underset{\sim}{u}_h(\underset{\sim}{x})$.

Further *necessary* conditions for stability in *critical* case of neutral equilibrium:

$$(1.3) \qquad P_3[a_h \underset{\sim}{u}_h] \equiv 0, \quad P_4[a_h \underset{\sim}{u}_h] \geqq 0 .$$

Arbitrary displacement field $\underset{\sim}{u}(\underset{\sim}{x})$ may be written in form

$$(1.4) \qquad \underset{\sim}{u}(\underset{\sim}{x}) = a_h \underset{\sim}{u}_h(\underset{\sim}{x}) + \underset{\sim}{v}(\underset{\sim}{x}) , \qquad \underset{\sim}{v}(\underset{\sim}{x}) \perp \underset{\sim}{u}_j(\underset{\sim}{x}),\ j = 1,2,..n ,$$

where $P_2[\underset{\sim}{v}]$ is *positive definite*.

Hence we may *minimize* $P[a_h \underset{\sim}{u}_h + \underset{\sim}{v}]$, for sufficiently small amplitudes a_j, $j = 1,2,..n$, with respect to $\underset{\sim}{v}(\underset{\sim}{x})$. Result is

$$(1.5) \qquad [\underset{\sim}{v}(\underset{\sim}{x})]_{min} = a_h a_k \underset{\sim}{v}_{hk}(\underset{\sim}{x}) + 0[(a_h a_h)^{3/2}] ,$$

where $\underset{\sim}{v}_{hk}(\underset{\sim}{x}) = \underset{\sim}{v}_{kh}(\underset{\sim}{x})$ are *unique*. Result is

$$(1.6) \qquad \begin{aligned} \underset{(a_h = \text{const.})}{\text{Min}} P[a_h \underset{\sim}{u}_h + \underset{\sim}{v}] &= P_3[a_h \underset{\sim}{u}_h] + P_4[a_h \underset{\sim}{u}_h] - \\ &\quad - P_2[a_h a_k \underset{\sim}{v}_{hk}] + 0[(a_h a_h)^{5/2}] = \\ &= A_{ijk}\, a_i a_j a_k + A_{ijk\ell}\, a_i a_j a_k a_\ell + 0[(a_h a_h)^{5/2}] . \end{aligned}$$

$$(1.7) \qquad A_{ijk}\, a_i a_j a_k = P_3[a_h \underset{\sim}{u}_h] ,$$

$$(1.8) \qquad A_{ijk\ell}\, a_i a_j a_k a_\ell = P_4[a_h \underset{\sim}{u}_h] - P_2[a_h a_k \underset{\sim}{v}_{hk}] .$$

Further *necessary* conditions for stability in *critical* case of neutral equilibrium:

$$(1.9) \qquad A_{ijk}\, a_i a_j a_k \equiv 0, \quad A_{ijk\ell}\, a_i a_j a_k a_\ell \geqq 0 .$$

Sufficient conditions are:

$$A_{ijk}\, a_i a_j a_k \;\equiv\; 0\,; \quad A_{ijk\ell}\, a_i a_j a_k a_\ell \text{ is } \textit{positive definite.}$$

Physical limitations of energy criterion:

a) *elastic* material behaviour;

b) *conservative* external loads (e.g. dead loads, pressure loads).

Mathematical weakness: energy functional in *classical* threedimensional non-linear theory of elasticity is ***not*** twice continuously Fréchet differentiable. This weakness does ***not*** occur in one-dimensional theory of *rods* or two-dimensional theory of *quasi-shallow shells.*

1.2 The Energy Functional for Elastic Structures

Particle is identified by radius vector $\underset{\sim}{x}$, Cartesian coordinates x_i, $i = 1,2,3$, in *fundamental* state I.

Additional deformation due to displacement field $\underset{\sim}{u}(\underset{\sim}{x})$ from pre-buckling state I to adjacent state II is described by *symmetric strain tensor*

$$\gamma_{ij} = \frac{1}{2}\,[u_{i,j} + u_{j,i} + u_{h,i}\,u_{h,j}] = \theta_{ij} + \frac{1}{2}\,u_{h,i}\,u_{h,j}\,, \qquad (1.10)$$

where θ_{ij} is *linearized* strain tensor.

Elastic energy density W per unit volume of *undeformed* body:

$$W_{II} - W_I = \left(\frac{\partial W}{\partial \gamma_{ij}}\right)_I \gamma_{ij} + \frac{1}{2}\left(\frac{\partial^2 W}{\partial \gamma_{ij}\,\partial \gamma_{k\ell}}\right)_I \gamma_{ij}\,\gamma_{k\ell} + \cdots \qquad (1.11)$$

Mass density is ρ_0 in *undeformed* state and ρ in *fundamental* state I.

Symmetric *stress tensor* S_{ij} in *fundamental* state I:

(1.12) $$S_{ij} = \frac{\rho}{\rho_0}\left[\frac{\partial W}{\partial \gamma_{(ij)}}\right]_I = \frac{1}{2}\frac{\rho}{\rho_0}\left[\frac{\partial W}{\partial \gamma_{ij}} + \frac{\partial W}{\partial \gamma_{ji}}\right]_I .$$

Tensor of *elastic moduli* in *fundamental* state:

(1.13) $$E_{ijk\ell}{}^{I} = \frac{\rho}{\rho_0}\left[\frac{\partial^2 W}{\partial\gamma_{(ij)}\,\partial\gamma_{(k\ell)}}\right]_I .$$

Elastic energy functional:

(1.14) $$P^{e\ell}[\underset{\sim}{u}] = \int \left[S_{ij}u_{i,j} + \frac{1}{2} S_{ij}u_{h,i}u_{h,j} + \frac{1}{2} E_{ijk\ell}{}^{I}\gamma_{ij}\gamma_{k\ell} + \ldots\right] dv .$$

Linear term is *cancelled* by linear term in potential energy of external loads. *Total* potential energy functional for *dead loads:*

(1.15) $$P[\underset{\sim}{u}] = \int \left[\frac{1}{2} S_{ij}u_{h,i}u_{h,j} + \frac{1}{2} E_{ijk\ell}{}^{I}\gamma_{ij}\gamma_{k\ell} + \ldots\right] dv .$$

In case of other conservative loads add the *nonlinear* terms of potential energy of external loads.

Second variation of energy for *dead loads:*

(1.16) $$P_2[\underset{\sim}{u}] = \int \left[\frac{1}{2} S_{ij}u_{h,i}u_{h,j} + \frac{1}{2} E_{ijk\ell}{}^{I}\theta_{ij}\theta_{k\ell}\right] dv .$$

Approximation valid for *small strains.*

(1.17) $$E_{ijk\ell}{}^{I} \cong E_{ijk\ell} = \left[\frac{\partial^2 W}{\partial\gamma_{(ij)}\,\partial\gamma_{(k\ell)}}\right]_0 .$$

For *isotropic* materials

(1.18) $$E_{ijk\ell} = G\left[\delta_{ik}\delta_{j\ell} + \delta_{i\ell}\delta_{jk} + \frac{2\nu}{1-2\nu}\delta_{ij}\delta_{k\ell}\right] ,$$

where G is shear modulus, ν is Poisson's ratio, δ_{ij} is unit tensor.

Simplification in case of *linear pre-buckling state.* Loads described by *unit load system* multiplied by *load factor* λ. Stress tensor in fundamental state

(1.19) $$S_{ij} = \lambda S_{ij}^{(1)} .$$

Energy functional and ***second variation*** for ***small*** strains, ***linear*** pre-buckling state and ***dead loads:***

$$P[\underset{\sim}{u}] = \int \left[\frac{1}{2} \lambda S_{ij}^{(1)} u_{h,i} u_{h,j} + \underline{\frac{1}{2} E_{ijk\ell} \gamma_{ij} \gamma_{k\ell}}\right] dv \, , \tag{1.20}$$

$$P_2[\underset{\sim}{u}] = \int \left[\frac{1}{2} \lambda S_{ij}^{(1)} u_{h,i} u_{h,j} + \underline{\frac{1}{2} E_{ijk\ell} \theta_{ij} \theta_{k\ell}}\right] dv \, . \tag{1.21}$$

Cartesian coordinates in ***undeformed*** configuration may now be employed as independent variables.

Underlined terms are ***classical*** energy density in ***physically linear*** and ***geometrically nonlinear*** elasticity theory, and in ***fully linear*** elasticity theory respectively. ***Distinction*** between ***classical*** theory and ***buckling*** theory occurs only by presence of first term involving ***initial stresses*** in fundamental state. ***Classical approximations*** of underlined terms in theories of ***rods, plates and shells*** thus also applicable in buckling.

Further discussion restricted to ***small strains*** and ***linear pre-buckling state.*** This restriction is ***not*** quite necessary but it simplifies discussion considerably, and it applies to many fundamental problems.

1.3 Neutral Equilibrium and Buckling Modes

Introduce (fictitious) stresses of classical linear elasticity theory, defined by

$$\sigma_{ij} = E_{ijk\ell} \, \theta_{k\ell} \, . \tag{1.22}$$

Minimum condition for second variation of energy leads to ***equations of neutral equilibrium***

$$\sigma_{ij,i} + \lambda (S_{hi}^{(1)} u_{j,h})_{,i} = 0 \, , \tag{1.23}$$

geometric boundary conditions $u_i = 0$ on part of surface where displacements are specified, and ***dynamic boundary conditions***

(1.24) $\quad \sigma_{ij}\, n_i + \lambda\, S_{hi}{}^{(1)}\, u_{j,h}\, n_i = 0\,,$

on part of surface where tractions are specified (in case of dead loads). *Critical* load factor λ_1 is ***smallest*** (positive) ***eigenvalue*** of this linear eigenvalue problem. Associated linearly independent solutions for displacement vector $\underset{\sim}{u}$ are the ***buckling modes*** $\underset{\sim}{u}_h$, $h = 1,2..n$.

Orthonormalize buckling modes by

(1.25) $\quad \int \frac{1}{2} E_{ijk\ell}\, \theta_{ij}(\underset{\sim}{u}_m)\, \theta_{k\ell}(\underset{\sim}{u}_n) dv = \delta_{mn}$

Normalize unit load system such that $\lambda_1 = 1$.

Second variation for linear combination of buckling modes $\underset{\sim}{u} = a_h \underset{\sim}{u}_h$ at *arbitrary* load factor λ is now

(1.26) $\quad P_2[a_h \underset{\sim}{u}_h] = (1 - \lambda)\, a_h\, a_h\,.$

1.4 Stability at Critical Load

Substituting from $\underset{\sim}{u} = a_h \underset{\sim}{u}_h$ into

(1.27) $\quad P_3[\underset{\sim}{u}] = \int \frac{1}{2} E_{ijk\ell}\, \theta_{ij}\, u_{m,k}\, u_{m,\ell}\, dv\,,$

we may verify whether necessary condition for stability $P_3[a_h \underset{\sim}{u}_h] \equiv 0$ holds true. If this identity is *not* verified (the general case in a mathematical sense), equilibrium at the critical load is *unstable*.

Necessary condition $P_3[a_h \underset{\sim}{u}_h] \equiv 0$ is *often verified* as a consequence of *symmetries* in the structure. *In that case we have also*

(1.28) $\quad P_4[\underset{\sim}{u}] = \int \frac{1}{8} E_{ijk\ell}\, u_{m,i}\, u_{n,k}\, u_{n,\ell}\, dv \geqq 0\,,$

for any linear combination of buckling modes $\underset{\sim}{u} = a_h \underset{\sim}{u}_h$.

Now solve the ***minimum problem*** for $\underset{\sim}{v}(\underset{\sim}{x})$

$$P_2[\underset{\sim}{v}] + P_{21}[a_h \underset{\sim}{u}_h, \underset{\sim}{v}] \quad \text{is minimum,} \tag{1.29}$$

where $P_{21}[\underset{\sim}{y},\underset{\sim}{z}]$ denotes the functional of degree of 2 in $\underset{\sim}{y}$ and of degree 1 in $\underset{\sim}{z}$ which appears in the expansion of $P_3[\underset{\sim}{y} + \underset{\sim}{z}]$. The solution is $\underset{\sim}{v}(\underset{\sim}{x}) = a_h a_k \underset{\sim}{v}_{hk}(\underset{\sim}{x})$. The *further necessary condition* for stability at the critical load

$$P_4[a_h \underset{\sim}{u}_h] - P_2[a_h a_k \underset{\sim}{v}_{hk}] \geqq 0 \tag{1.30}$$

is now also ***sufficient,*** if the inequality sign holds for all $a_h \neq 0$.

1.5 Post-buckling Behaviour

For *small deflections* from the fundamental state and for *load factors* λ ***in the vicinity of the critical value*** unity, $|1-\lambda| \ll 1$, the post-buckling behaviour is described to a ***first approximation*** by the function

$$F(a_h ; \lambda) = (1-\lambda) a_i a_i + \begin{matrix} A_{ijk} a_i a_j a_k \\ A_{ijk\ell} a_i a_j a_k a_\ell \end{matrix} . \tag{1.31}$$

The *upper* expression with

$$A_{ijk} a_i a_j a_k = P_3[a_h \underset{\sim}{u}_h]$$

holds, if this cubic form ***does not vanish*** identically, and the *lower* expression is valid, if the cubic form is *identically zero.* In this case we have

$$A_{ijk\ell} a_i a_j a_k a_\ell = P_4[a_h \underset{\sim}{u}_h] - P_2[a_h a_k \underset{\sim}{v}_{hk}] . \tag{1.32}$$

Equilibrium configurations are characterized by ***stationary*** values for the function $F(a_h ; \lambda)$

with respect to the amplitudes a_h of the buckling modes. The equilibrium is *stable*, if and only if the stationary value is a *proper minimum.*

We regard the amplitudes a_h of the buckling modes as components of a vector $\tilde{a}$ in Euclidean n-space. We write $\tilde{a} = a\tilde{e}$, where $\tilde{e}$ is a unit vector in the *direction of the post-buckling path* and a is the *magnitude* of the deflection from the fundamental state. The *equations of equilibrium* are

$$(1.33) \qquad 2(1-\lambda)a e_i + \begin{matrix} 3a^2 A_{ijk} e_j e_k \\ 4a^3 A_{ijk\ell} e_j e_k e_\ell \end{matrix} = 0 \,.$$

The *stability condition* is expressed by

$$(1.34) \qquad 2(1-\lambda) + \begin{matrix} 6a\, A_{ijk}\, b_i b_j e_k \\ 12a^2 A_{ijk\ell} b_i b_j e_k e_\ell \end{matrix} > 0$$

for *all* unit vectors $\tilde{b}$. The solution $a = 0$ describes the fundamental state, stable for $\lambda < 1$ and unstable for $\lambda > 1$.

The *generalized additional deflection* $\Delta\varepsilon$ associated with the load factor λ, on an *appropriate scale,* is given by

$$(1.35) \qquad \Delta\varepsilon = -\frac{\partial F(a_h\,;\,\lambda)}{\partial \lambda} = a^2 \,.$$

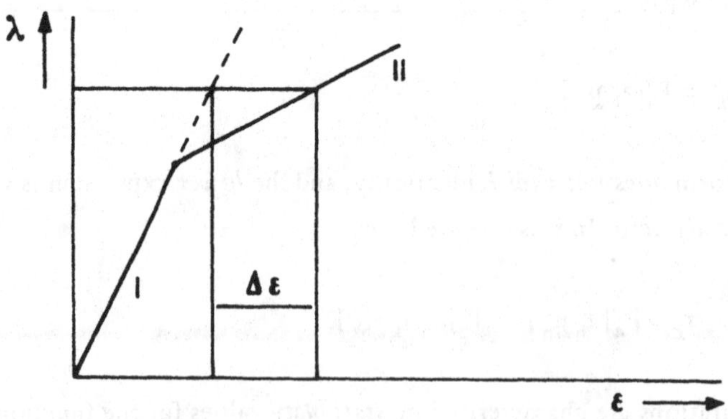

Generalized additional deflection in a rising post-buckling path

The *direction* $\underset{\sim}{e}$ of *post-buckling paths* coincide with unit vectors $\underset{\sim}{t}$ for which the cubic form $A_{ijk}t_i t_j t_k$, or the quartic form $A_{ijk\ell}t_i t_j t_k t_\ell$, takes a *stationary value* on the unit sphere $|\underset{\sim}{t}| = 1$.

Of *particular interest* is the post-buckling path for which the cubic or quartic form takes its *absolute minimum* on the sphere. Let $\underset{\sim}{t}^*$ denote this *minimizing direction*, and let A_3^* or A_4^* be the minimum value in question.

In the case of a *non-vanishing cubic form* A_3^* is necessarily *negative*. The magnitude of the deflection is

$$a = 2(\lambda - 1)/3A_3^* \ , \tag{1.36}$$

and we have $\lambda < 1$ in an *unstable post-buckling path of steepest descent.*

In the case of an identically vanishing cubic for, A_4^* may be either positive or negative. The magnitude of the deflection in the post-buckling path is

$$a = [(\lambda - 1)/2A_4^*]^{1/2} \ , \tag{1.37}$$

where $\lambda \gtrless 1$ accordingly as A_4^* is positive or negative. In the first case we have a *stable post post-buckling path of slowest rise,* in the second case again an *unstable post-buckling path of steepest descent.*

1.6 Influence of Geometric Imperfections

Buckling of a structure with a linear pre-buckling state is always of *bifurcation type.* Irregularities in the actual structure such as load excentricities and geometric imperfections may result both in a *nonlinear* behaviour before buckling, and possibly *premature* buckling as well. In shells the most damaging effect is usually due to *geometric* imperfections.

Geometric imperfections may be described in terms of a "displacement" field $\overset{\circ}{\underset{\sim}{u}}(\underset{\sim}{x})$. The most important consequence is that the additional strain tensor (1.10) is replaced by

(1.38) $$\gamma_{ij} = \theta_{ij} + \frac{1}{2} u_{h,i} u_{h,j} + \frac{1}{2}(\overset{\circ}{u}_{h,i} u_{h,j} + \overset{\circ}{u}_{h,j} u_{h,i}) .$$

The *dominant* additional term in the functional (1.20) is *bilinear* in the imperfections and the actual displacements from the fundamental state

(1.39) $$\int \lambda S_{ij}^{(1)} \overset{\circ}{u}_{h,i} u_{h,j} \, dv .$$

Decompose the imperfection field

(1.40) $$\overset{\circ}{\underset{\sim}{u}}(\underset{\sim}{x}) = \kappa \, \overset{\circ}{e}_h \underset{\sim}{u}_h(\underset{\sim}{x}) + \overset{\circ}{\underset{\sim}{v}}(\underset{\sim}{x}) , \quad \overset{\circ}{\underset{\sim}{v}} \perp \underset{\sim}{u}_j , \quad j = 1,2,..n ,$$

where $\overset{\circ}{e}_h$ are components of a *unit vector* $\overset{\circ}{\underset{\sim}{e}}$ in Euclidean n-space, and κ is the *magnitude* of the imperfection in the direction $\overset{\circ}{\underset{\sim}{e}}$.

The function $F(a_h ; \lambda)$ in (1.31) is replaced by

(1.41) $$F^*(a_h ; \lambda) = (1-\lambda) a_i a_i + \begin{matrix} A_{ijk} a_i a_j a_k \\ \\ A_{ijk\ell} a_i a_j a_k a_\ell \end{matrix} - 2\lambda\kappa \overset{\circ}{e}_i a_i .$$

Equation (1.33) is replaced by

(1.42) $$2(1-\lambda) a e_i + \begin{matrix} 3a^2 A_{ijk} e_j e_k \\ \\ 4a^3 A_{ijk\ell} e_j e_k e_\ell \end{matrix} - 2\lambda\kappa \overset{\circ}{e}_i = 0 .$$

The *stability condition* remains (1.34).

In case of *non-vanishing cubic form worst imperfection* is in direction of *post-buckling path of steepest descent* $\overset{\circ}{\underset{\sim}{e}} = \underset{\sim}{t}^*$. Buckling load factor $\lambda^* < 1$ of imperfect structure occurs at a *limit point* given by

(1.43) $$(1-\lambda^*)^2 = -6\lambda^* \kappa A_3^* .$$

In case of *vanishing* cubic form and *negative* A_4^* likewise limit point $\lambda^* < 1$, now given by

$$(1-\lambda^*)^{3/2} = (3/2)\,\lambda^*\,\kappa(-6A_4^*)^{1/2}\,. \tag{1.44}$$

In *both* cases of *unstable* bifurcation point of perfect structure *strong imperfection-sensitivity*.

2. *GENERAL THEORY OF SHELL BUCKLING*

Reference is made to the paper

W.T. Koiter. General Equations of Elastic Stability for Thin Shells. Proc. Symp. on the Theory of Shells to Honor Lloyd Hamilton Donnell (April 1966). Published by University of Houston (1967), pp. 187-227.

3. *BUCKLING OF CYLINDRICAL SHELLS UNDER AXIAL COMPRESSION AND EXTERNAL PRESSURE*

3.1. Neutral Equilibrium (linear buckling theory)

Starting-point is ***second variation*** of potential energy discussed in chapter 2.

$$
\begin{aligned}
P_2[\underset{\sim}{u}] = \int \{ & \frac{1}{2}\, hE^{\alpha\beta\lambda\mu}[\theta_{\alpha\beta}\,\theta_{\lambda\mu} + \frac{h^2}{12}\,\rho_{\alpha\beta}\,\rho_{\lambda\mu}] + \\
& + \frac{1}{2}\, N^{\alpha\beta}[(\theta_{\alpha}^{\ \kappa} - \omega^{\kappa}_{\cdot\alpha})(\theta_{\kappa\beta} - \omega_{\kappa\beta}) + \phi_\alpha\phi_\beta] + \qquad (3.1) \\
& + \frac{1}{2}\, P(w\theta_{\alpha}^{\ \alpha} - u^\alpha\phi_\alpha) \}\, dS\, .
\end{aligned}
$$

Introduce *Cartesian coordinates* x, y and put

$$\frac{\partial}{\partial x}() = \frac{1}{2}()', \ \frac{\partial}{\partial y} = \frac{1}{R}()^{\cdot}; \ u_x = u, \ u_y = v\ ;$$

$$\theta_{xx} = \frac{1}{R}\,u' \ \ \theta_{yy} = \frac{1}{R}(v^{\cdot} + w), \ \theta_{xy} = \frac{1}{2R}(u^{\cdot} + v')\ ;$$

$$\omega_{xy} = \frac{1}{2R}(v' - u^{\cdot}), \ \phi_x = \frac{1}{R}w', \ \phi_y = \frac{1}{R}(w^{\cdot} - v)\ ;$$

$$\rho_{xx} = \frac{1}{R^2}\,w'', \ \rho_{yy} = \frac{1}{R^2}(w^{\cdot} - v)^{\cdot}, \ \rho_{xy} = \frac{1}{R^2}(w'^{\cdot} - \frac{3}{4}v' + \frac{1}{4}u^{\cdot})\, .$$

$$P_2[\underset{\sim}{u}] = \int \frac{Eh}{2(1-\nu^2)}\{\theta_{xx}^{\ 2} + \theta_{yy}^{\ 2} + 2\nu\theta_{xx}\theta_{yy} + 2(1-\nu)\,\theta_{xy}^{\ 2} +$$

$$+ \frac{h^2}{12}[\rho_{xx}^{\ 2} + \rho_{yy}^{\ 2} + 2\nu\rho_{xx}\rho_{yy} + 2(1-\nu)\rho_{xy}^{\ 2}]\}\, dS\, +$$

$$+ \int \{ \frac{1}{2} N_{xx} [\underline{\theta_{xx}^2} + (\theta_{yx} - \omega_{yx})^2 + \phi_x^2] +$$

$$(3.2) \qquad + \frac{1}{2} N_{yy} [(\theta_{xy} - \omega_{xy})^2 + \underline{\theta_{yy}^2} + \phi_y^2] +$$

$$+ \frac{1}{2} P [w(\theta_{xx} + \theta_{yy}) - u\phi_x - v\phi_y] \} \, dS .$$

First line is *positive definite* in θ_{xx} and θ_{yy}. Hence we may *omit underlined terms.*

Equations of neutral equilibrium complicated by appearance of u and v in ρ_{yy} and ρ_{xy}.

Accuracy of second variation, *positive definite* in first two lines in θ and ρ, not affected by addition of terms in integrand

$$\frac{Eh^3}{24(1-\nu^2)} [- \frac{2(1-\nu)}{R} \theta_{xx} \rho_{yy} + \frac{2}{R} \theta_{yy} (\rho_{xx} + \rho_{yy}) +$$

$$(3.3) \qquad + \frac{1}{R^2} \theta_{yy}^2 + \frac{2(1-\nu)}{R} \theta_{xy} \rho_{xy} - \frac{3(1-\nu)}{2R^2} \theta_{xy}^2] .$$

Magnitude of this additional term is *bounded above* by 0.467 h/R times original first two lines. *Negligible error* in context of shell theory.

Second variation now reduced to

$$P_2 [\underset{\sim}{u}] = \int \frac{Eh}{2(1-\nu^2)R^2} [u'^2 + (v^{\cdot} + w)^2 + 2\nu u'(v^{\cdot} + w) + \frac{1}{2}(1-\nu)(u^{\cdot} + v')^2 +$$

$$+ \frac{h^2}{12R^2} \{ w''^2 + w^{\cdot\cdot 2} + 2\nu w'' w^{\cdot\cdot} + 2(1-\nu) w'^{\cdot 2} + 2w(w'' + w^{\cdot\cdot}) + w^2 \} +$$

$$(3.4) \qquad + 2(1-\nu) \frac{h^2}{12R^2} \{ u'v^{\cdot} - u^{\cdot}v' + u^{\cdot}w'^{\cdot} - u'w^{\cdot\cdot} + v^{\cdot}w'' - v'w'^{\cdot} \}] \, dS +$$

$$+ \int [\frac{N_{xx}}{2R^2} (v'^2 + w'^2) + \frac{N_{yy}}{2R^2} \{ u^{\cdot 2} + (w^{\cdot} - v)^2 \} +$$

$$+ \frac{P}{2R} (u'w - uw' + v^{\cdot}w - vw^{\cdot} + v^2 + w^2)] \, dS .$$

Introduce *load parameters* ($N_{yy} = -pR$!)

$$\lambda_x = (1-\nu^2)\,\frac{N_{xx}}{Eh} \qquad \lambda_y = (1-\nu^2)\,\frac{N_{yy}}{Eh} = -(1-\nu^2)\,\frac{pR}{Eh}\ . \tag{3.5}$$

Equations of *neutral equilibrium*

$$u'' + \frac{1}{2}(1-\nu)u^{\cdot\cdot} + \frac{1}{2}(1+\nu)v'^{\cdot} + \nu w' + \lambda_y(u^{\cdot\cdot} - w') = 0\ , \tag{3.6}$$

$$\frac{1}{2}(1+\nu)u'^{\cdot} + \frac{1}{2}(1-\nu)v'' + v^{\cdot\cdot} + w^{\cdot} + \lambda_x v'' = 0\ , \tag{3.7}$$

$$\begin{aligned} \nu u' + v^{\cdot} + w + \frac{h^2}{12R^2}[w'''' + 2w''^{\cdot\cdot} + w^{\cdot\cdot\cdot\cdot} + 2(w'' + w^{\cdot\cdot}) + w] - \\ - \lambda_x w'' - \lambda_y(u' + w^{\cdot\cdot} + w) = 0\ . \end{aligned} \tag{3.8}$$

3.2 Evaluation for Simply-supported Edges

Edge conditions for simply-supported edges at shell ends $x = 0$ and $x = \ell$ *satisfied* by:

$u = A\cos(px/R)\sin(qy/R)$, $v = B\sin(px/R)\cos(qy/R)$,
$w = C\sin(px/R)\sin(qy/R)$,

where $p = k\pi R/\ell$, k and q integers; A, B, C *constants*.

Do not confuse wave number p here with external pressure in previous section!

Buckling determinant

$$\begin{vmatrix} p^2 + \frac{1}{2}(1-\nu)q^2 + \lambda_y q & \frac{1}{2}(1+\nu)pq & -\nu p + \lambda_y p \\ \frac{1}{2}(1+\nu)pq & \frac{1}{2}(1-\nu)p^2 + q^2 + \lambda_x p^2 & -q \\ -\nu\phi + \lambda_y p & -q & \begin{array}{l} 1 + \frac{h^2}{12R^2}(p^2 + q^2 - 1)^2 + \\ + \lambda_x p^2 + \lambda_y(q^2-1) \end{array} \end{vmatrix} = 0$$

Careful expansion of determinant, neglecting *only* λ_x and λ_y *in comparison with unity,* yields equation

$$(1-\nu^2)p^4 + \frac{h^2}{12R^2}(p^2+q^2)^2(p^2+q^2-1)^2 + \tag{3.9}$$
$$+ \lambda_x p^2[(p^2+q^2)^2+q^2] + \lambda_y q^2[(p^2+q^2)^2-(3p^2+q^2)] = 0 .$$

Result is equivalent to earliest fully accurate equation obtained by Flügge 45 years ago. Present simpler form coincides with Danielson and Simmonds (1969). Present analysis most reliable one because *relative error* in critical stresses has been *proved* to be at most of order σ/E.

3.3 Special Cases

3.3.1 *Case* $q=0$

$$(1-\nu^2)p^4 + \frac{h^2}{12R^2}p^4(p^2-1)^2 + \lambda_x p^6 = 0$$

$$|\lambda_x| \ll 1 \quad \text{requires} \quad p^2 \gg 1, \; -\lambda_{x\,min} = (1-\nu^2)\frac{h}{cR}$$

$$\sigma_{x\,cr} = \frac{Eh}{cR} \quad \text{attained for} \quad p^2 = \frac{2cR}{h}, \qquad \boxed{c = \sqrt{3(1-\nu^2)}}$$

3.3.2 *Case* $q=1$

$$(1-\nu^2)p^4 + \frac{h^2}{12R^2}p^4(p^2+1)^2 + \lambda_x p^2[(p^2+1)^2+1] + \lambda_y[p^4-p^2] = 0$$

In *long* tubes we *may* have $p^2 \ll 1$. In that case

$$(1-\nu^2)p^2 + (2\lambda_x - \lambda_y) = 0$$

Euler buckling of long tube! See sketch!

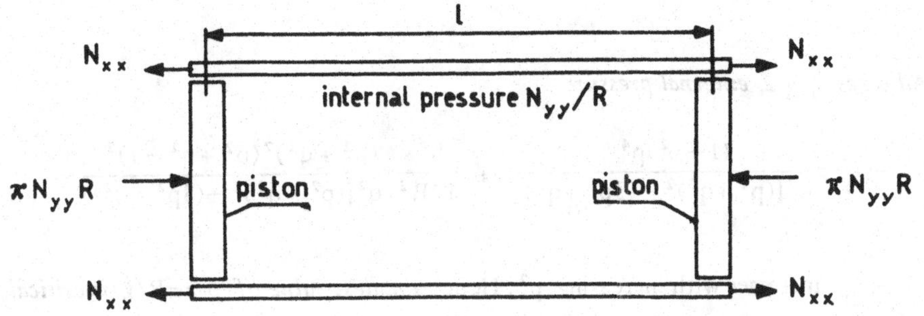

Column load $F = \pi N_{yy} R - 2\pi N_{xx} R$

Euler load $F = \dfrac{\pi^2 \cdot \pi R^3 hE}{\ell^2}$.

In case $q = 1$ we *cannot* have $p = O(1)$, but we *may* have $p^2 >> 1$. In that case *no* effect of internal or external pressure, and $-\sigma_{x\,cr} = Eh/cR$.

3.3.3 *Cases $q = 2, 3$ etc. but $q = O(1)$, axial compression*

$$-\lambda_x = \frac{(1-\nu^2)p^2}{(p^2+q^2)+q^2} + \frac{h^2}{12R^2}\,\frac{(p^2+q^2)^2(p^2+q^2-1)^2}{p^2[(p^2+q^2)^2+q^2]}$$

We may have either $p^2 \ll 1$ or $p^2 \gg 1$.
In case $p^2 \gg 1$ we have same result $-\lambda_x = (1-\nu^2)h/cR$ as for $q = 0$ and $q = 1$.
In case $p^2 \ll 1$ we obtain *(long shells!)*

$$-\lambda_x = (1-\nu^2)\frac{p^2}{q^2(q^2+1)} + \frac{h^2}{12R^2}\,\frac{q^2(q^2-1)^2}{p^2(q^2+1)} .$$

Minimizing with respect to p^2

$$p^2_{min} = \frac{h}{2cR} q^2(q^2-1), \qquad -\lambda_{x\ min} = (1-\nu^2) \frac{h}{cR} \frac{q^2-1}{q^2+1} .$$

3.3.4 *All cases $q \geqq 2$, external pressure*

$$-\lambda_y = \frac{(1-\nu^2)p^4}{q^2[(p^2+q^2)^2-(3p^2+q^2)]} + \frac{h^2}{12R^2} \frac{(p^2+q^2)^2(p^2+q^2-1)^2}{q^2[(p^2+q^2)^2-(3p^2+q^2)]}$$

$-\lambda_y$ increases with increasing p^2. Hence *smallest value of $p = \pi R/\ell$ is critical.* For *long shells* $p \to 0$

$$-\lambda_y = \frac{h^2}{12R^2} (q^2-1) . \qquad \textit{Ring formula}$$

3.3.5 *All cases p^2 and/or q^2 large, axial compression. Shallow buckling modes*

$$-\lambda_x = \frac{(1-\nu^2)p^2}{(p^2+q^2)^2} + \frac{h^2}{12R^2} \frac{(p^2+q^2)^2}{p^2} \qquad \textit{Donnell theory}$$

Minimum is $-\lambda_x = (1-\nu^2)h/cR$, attained for *all* combinations of wave numbers which satisfy

$$p^2+q^2 = p\sqrt{2cR/h} = pp_o . \qquad \textit{Semi-circle}$$

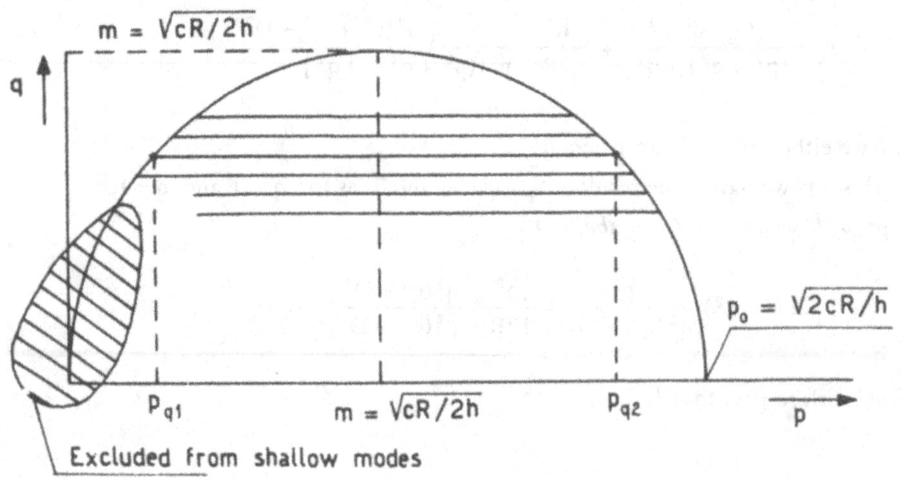

3.4 Nonlinear Theory for Shallow Buckling Modes

In view of large p edge conditions *unimportant* (provided they do not lead to premature buckling due to inadequate restraints). Buckling modes

$$\begin{matrix}\cos\\ \sin\end{matrix}\left[\begin{matrix}p_{q1}\\ p_{q2}\end{matrix}\; x/R\right]\quad\begin{matrix}\cos\\ \sin\end{matrix}\;[qy/R]$$

p_{q1}, p_{q2} from $p_q^2 + q^2 = p_q p_o$ (on circle), $p_{q1} p_{q2} = q^2$.

Evaluation of cubic terms leads to *non-zero* result for any *triple* of modes for which *seem or difference of wave numbers is zero,* both in axial and circumferential directions.

Because sine and cosine modes are the same except for *phase shift,* no need to consider fully general expression. *Restrict analysis to*

$$\begin{aligned} w = h[&b_o \cos(p_o x/R) + \\ &+ \Sigma \{c_{q1} \cos(p_{q1} x/R) + c_{q2} \cos(p_{q2} x/R)\} \cos(qy/R) + \\ &+ c_m \cos(mx/R) \cos(my/R)]. \end{aligned} \tag{3.10}$$

(Last term present, if $m = \sqrt{cR/2h}$ is integer).

Associated *in-plane displacements* u, v are

$$\begin{aligned} u = h[&-\frac{\nu}{p_o} b_o \sin(p_o x/R) - \sum_q \frac{p_{q1}(\nu p_{q1}^2 - q^2)}{(p_{q1}^2 + q^2)^2} c_{q1} \sin(p_{q1} x/R) \cos(qy/R) - \\ &- \sum_q \frac{p_{q2}(\nu p_{q2}^2 - q^2)}{(p_{q2}^2 + q^2)^2} c_{q2} \sin(p_{q2} x/R) \cos(qy/R) + \\ &+ \frac{1-\nu}{4m} c_m \sin(mx/R) \cos(my/R)], \end{aligned} \tag{3.11}$$

$$\begin{aligned} v = h[&-\sum_q \frac{q\{(2+\nu)p_{q1}^2 + q^2\}}{(p_{q1}^2 + q^2)^2} c_{q1} \cos(p_{q1} x/R) \sin(qy/R) - \\ &- \sum_q \frac{q\{(2+\nu)p_{q2}^2 + q^2\}}{(p_{q2}^2 + q^2)^2} c_{q2} \cos(p_{q2} x/R) \sin(qy/R) - \end{aligned} \tag{3.12}$$

$$- \frac{3+\nu}{4m} c_m \cos(mx/R)\sin(my/R)] .$$

Cubic terms in energy functional for *shallow* buckling modes are

$$P_3[\underset{\sim}{u}] = \int \frac{Eh}{2(1-\nu^2)R^3} [\{u' + \nu(v^{\cdot} + w)\}w'^2 +$$

$$(3.13) \qquad + \{\nu u' + v^{\cdot} + w\}w^{\cdot 2} +$$

$$+ (1-\nu)(u^{\cdot} + v')w'w^{\cdot}]dS .$$

Write N_{xx} in fundamental state $N_{xx} = -\lambda \frac{Eh^2}{cR}$.

Critical load factor is $\lambda = \lambda_1 = 1$.

Quadratic terms in energy functional

$$(3.14) \qquad P_2[\underset{\sim}{u}; \lambda] = \int (1-\lambda) \frac{Eh^2}{2cR^3} w'^2 dS .$$

Neglecting higher order terms, *total energy functional* is evaluated at

$$P_2[\underset{\sim}{u}; \lambda] + P_3[\underset{\sim}{u}] =$$

$$(3.15) \qquad = \frac{\pi}{4c} \frac{E\ell h^4}{R^2} [(1-\lambda)\{2p_o^2 b_o^2 + \sum_q (p_{q1}^2 c_{q1}^2 + p_{q2}^2 c_{q2}^2) + m^2 c_m^2\} +$$

$$+ 3c\{\sum_q q^2 b_o c_{q1} c_{q2} + \frac{1}{2} m^2 b_o c_m^2\}] .$$

Equations for *post-buckling behaviour:*

$$(3.16) \qquad 4(1-\lambda)p_o^2 b_o + 3c\{\sum_q q^2 c_{q1} c_{q2} + \frac{1}{2} m^2 c_m^2\} = 0 ,$$

$$(3.17) \qquad 2(1-\lambda)p_{q1}^2 c_{q1} + 3cq^2 b_o c_{q2} = 0 ,$$

$$2(1-\lambda)p_{q2}^{2}c_{q2} + 3c\,q^2 b_o c_{q1} = 0\,, \tag{3.18}$$

$$2(1-\lambda)m^2 c_m + 3c\,m^2 b_o c_m = 0\,. \tag{3.19}$$

Non-vanishing solution of (3.17) – (3.19), if $\boxed{b_o^2 = 4(1-\lambda)^2/9c.}$

We obtain $p_{q1}c_{q1} = \pm\, p_{q2}c_{q2}$ for $b_o = \mp\, 2(1-\lambda)/3c\,.$

$c_m = \begin{matrix}\text{arbitrary}\\ 0\end{matrix}$ for $b_o = \mp\, 2(1-\lambda)/3c.$

From (3.16) only one further condition.

Many post-buckling solutions. All unstable.

For all solutions *same overall shortening*

$$\boxed{\frac{\Delta\varepsilon}{\Delta\varepsilon_1} = \Delta\varepsilon\,\frac{cR}{h} = \frac{2}{3}(1-\lambda)^2} \tag{3.20}$$

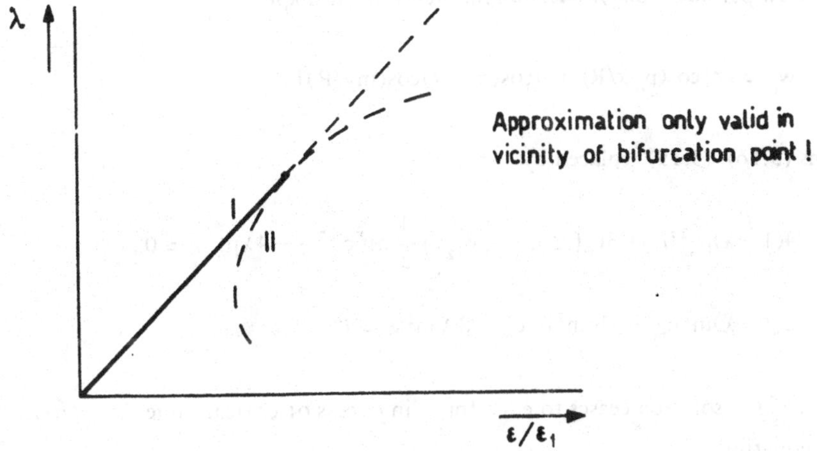

3.5 Influence of Imperfections

Examine first *axisymmetric* imperfections

$$w_o = \kappa h \cos(p_o x/R) .$$

Eq. (3.16) replaced by

$$(3.21) \qquad 4(1-\lambda)p_o^2 b_o + 3c \{ \sum_q q^2 c_{q1} c_{q2} + \frac{1}{2} m^2 c_m^2 \} - 4\lambda p_o^2 \kappa = 0 .$$

Eqs. (3.17) – (3.19) unchanged. Solution is

$$(3.22) \qquad b_o = \frac{\lambda}{1-\lambda} \kappa , \qquad c_{q1} = c_{q2} = c_m = 0 .$$

Bifurcation with respect to c_{q1}, c_{q2} (and also with respect to c_m for $\kappa < 0$) at $b_o^2 = 4(1-\lambda)^2/9c^2$.

Hence *critical load factor imperfect shell* $\lambda^* < 1$ given by

$$(3.23) \qquad (1-\lambda^*)^2 = \frac{3}{2} c \kappa \lambda^* .$$

Examine now *periodic nonsymmetric* imperfections in shape

$$w_o = \kappa h[\cos(p_o x/R) + 4\cos(mx/R)\cos(my/R)]$$

Equations (3.16), (3.19) replaced by

$$4(1-\lambda)p_o^2 b_o + 3c \{ \sum_q q^2 c_{q1} c_{q2} + \frac{1}{2} m^2 c_m^2 \} - 4\lambda p_o^2 \kappa = 0 ,$$

$$2(1-\lambda)m^2 c_m + 3c m^2 b_o c_m - 8\lambda m^2 \kappa = 0 .$$

For increasing λ solution ceases to exist for λ in excess of critical value λ^*, a *limit point*, given by equation

$$(1-\lambda^*)^2 = 6c\kappa\lambda^* .$$

Here only limit point for ***negative*** κ, where inward imperfection amplitude is 5/3 times outward imperfection amplitude.

More or less localized imperfection dimple is described by ($\kappa < 0$)

$$w_o = \kappa h[\cos(p_o x/R) + 4\cos(mx/R)\cos(my/R)] \exp[-\frac{\mu^2(x^2+y^2)}{2R^2}] .$$

Even for small μ^2/m^2 *rapid decay.* Neglecting μ^2/m^2 in comparison with unity, we have as approximate result that cubic terms in energy are reduced by an additional factor 2/3 in comparison with quadratic terms. Result is

$$(1-\lambda^*)^2 = -4c\,\kappa\lambda^* .$$

More or less localized imperfections are equally harmful as periodic imperfections with an amplitude reduced by a factor 2/3.

Example: $\kappa = 0.1 \rightarrow \lambda^* = 0.45$.

VIBRATIONS OF THIN ELASTIC SHELLS

by

M. Dikmen

Technical University, Istanbul

VIBRATIONS OF THIN ELASTIC SHELLS

M. Dikmen

INTRODUCTION

In this introductory Chapter, we first give some information about the history of shell dynamics, and then make remarks concerning the scope of these lectures. The description of the elastic shell as it shall be considered here is followed by the equations of linear shell dynamics in general, and by a particular differential operator.

1. Brief historical remarks:

The study of the vibrations of bells is in the origin of shell theory. The earliest known work is due to L. Euler[1] who imagined the bell subdivided into thin annuli behaving as curved bars, thus leaving out of consideration the effects in the planes of the meridians. More than twenty years later, Jacob Bernoulli[2] (the younger) proposed a model of shell essentially consisting of a double sheet of curved bars, placed at right angles, along parallels resp. meridians. All this was done before the discovery of the general equations of lasticity. Even after these equations were formulated, the rational theory of shells had to wait until the precursory work of H. Aron.[3] There followed E Mathieu,[4] who applied to shells a method similar to the one used by S. D. Poisson[5] for plates. By the same time, Lord Rayleigh[6,7] proposed a different theory. The decisive advance in shell theory was made by the fundamental works of A. E. H. Love[8,9,10] and the almost simultaneous contributions of A. B. Basset[11] and H. Lamb.[12]

By that time, interest in the statical engineering applications of a rational shell theory seems to be practically inexistant (*). This interest began to grow later, in the second decade of the twentieth century. For a time, less attention was then paid to dynamical problems. Then followed publications on theoretical as well as experimental research, still partly motivated by acoustical problems.[14-18] The important paper of W. Flügge[19] appeared in 1933 and his famous monograph[20] followed a year later. K. Federhofer published his results[21-25] between 1935 and 1939. Without a claim for completeness, we may cite, in chronological order, a number of papers[26-52] which were published in the period between 1946 and 1960, and dealt with more or less general problems of shell dynamics. From 1960 on, the literature in this field has reached such an extent and diversity that any attempt to establish even a summary bibliography would lead to an exceedingly long list.

2. Scope of the lectures:

The obvious time limitations of a short series of lectures make it imperative to severely restrict the subject, in order to keep it within somewhat manageable limits. So, we shall consider a very simple model of shell and restrict ourselves to fully linearized theory (see § § 3, 4, 5). We shall leave out of discussion, interesting though they are, topics such as anisotropy, inhomogeneity, variable thickness, stiffened shells, shells with holes, prestress, thermal effects, interactions with surrounding media and/or physical fields, as well as nonlinear behaviour of any kind. But even under these drastic restrictions, the subject

(*) For some details in this respect, see M. Dikmen.[13]

remains still too vast to permit an exhaustive treatment. We shall therefore concentrate on some *basic problems* and *methods*, rather than dwelling on a forcibly incomplete survey of particular problems. Besides, such an attempt would inevitably require some scrutiny of various linear theories, thus taking us far away from the core of the subject. It is believed however that indications given in appropriate places in the course of the text and references cited in various occasions provide an efficient link to further literature on general as well as particular problems of shell dynamics.

3. Description of the shell:

The shell which shall be considered throughout these lectures is of uniform thickness, made of homogeneous, isotropic material with elasticity modulus E and Poisson's ratio ν. It will be represented as usual in thin shell theory by its middle surface, which will consist in a finite and simply connected region G of a surface embedded in the three-dimensional euclidean space, and having a curve Γ as boundary. For simplicity, both the middle surface as well as its boundary will be assumed to be as smooth as required by the calculations. The thickness of the shell is 2h.

The points on the middle surface will be referred to a system of curvilinear coordinates, consisting of two families of parametric curves α and β. For simplicity, these families will be taken mutually orthogonal, forming an orthogonal system of curvilinear coordinates, but not coinciding necessarily with the lines of curvature of the surface, unless the contrary is specified. This chice does not restrict generality, within our purposes. While it is most convenient to use tensor formalism in general shell theory, it proves in many instances more efficient to work with explicitly written equations in physical components. This is the way we shall prefer in these lectures.

If the position vector (in the three-dimensional space) of a generic point of the middle surface is denoted by $\underset{\sim}{r}$, then the vectors $\underset{\sim}{a}_1 \equiv \frac{\partial \underset{\sim}{r}}{\partial \alpha}$, $\underset{\sim}{a}_2 \equiv \frac{\partial \underset{\sim}{r}}{\partial \beta}$ are tangent to the surface and can be used to express the metric on the surface by the first fundamental form

$$ds^2 = A^2\, d\alpha^2 + B^2 d\beta^2 \qquad (0.3.1)$$

where $A \equiv |\frac{\partial \underset{\sim}{r}}{\partial \alpha}|$ resp. $B \equiv |\frac{\partial \underset{\sim}{r}}{\partial \beta}|$ are the Lamé coefficients.

The curvature is described by the second fundamental form

$$-\frac{A^2}{R'_1}\, d\alpha^2 + 2\,\frac{AB}{R_{12}}\, d\alpha d\beta - \frac{B^2}{R'_2}\, d\beta^2 \qquad (0.3.2)$$

where R'_1 and R'_2 are the radii of curvature along the α resp. β lines. There exist the relations

$$\frac{1}{R'_1} + \frac{1}{R'_2} = \frac{1}{R_1} + \frac{1}{R_2}$$

$$\frac{1}{R_1 R_2} = \frac{1}{R'_1 R'_2} - \frac{1}{R^2_{12}} \qquad (0.3.3)$$

We use the particular notation R_1 and R_2 (instead of R'_1 and R'_2, respectively) for the principal radii of curvature.

The unit tangential vectors $\underset{\sim}{e}_1 \equiv \frac{1}{A}\frac{\partial \underset{\sim}{r}}{\partial \alpha}$, $\underset{\sim}{e}_2 \equiv \frac{1}{B}\frac{\partial \underset{\sim}{r}}{\partial \beta}$ and the unit normal vector $e_3 \equiv \underset{\sim}{n}$, define together, at each point of the middle surface a base (orthogonal trihedron). Any vector field on the middle surface can then be represented by its components along the axes thus determined. Specifically, the displacement vector $\underset{\sim}{U} = \underset{\sim}{U}(\alpha,\beta,t)$ can be resolved into components U_1, U_2 and U_3.

With the intention of using later (see Chapter I, Section 2) a dimensionless small parameter, we introduce here an auxiliary concept. As *characteristic length* relative to a given middle surface, as defined above, we take arbitrarily a quantity R which is of the same order as the lengths of the longest segments intercepted by Γ along the α *and* β lines. This supposes of course that such a quantity exists. Then, we may define

$$h^* \equiv \frac{h}{R} \qquad (0.3.4)$$

In *thin shells*, $h^* \ll 1$.

4. General form of the differential equations:

Let a shell, as defined in §3, be given in an unstressed rest (reference) configuration. Let ρ be the mass density of the material, $\underset{\sim}{f}$ the external forces per unit area of the middle surface (body forces and loads), and $\underset{\sim}{U}$ the displacement field (from the reference configuration).

Then, as long as rotatory inertia effects are not taken into account, any fully linear theory of the motion of thin shells exhibits the form

$$\sum_{j=1}^{3} L_{ij}\, U_j = K\left(f_i - 2h\rho \frac{\partial^2 U_i}{\partial t^2}\right) \qquad (0.4.1)$$

where the L_{ij} are linear partial differential operators and where K is a constant depending on the thickness of the shell, but also on its material constants. It is clear that this factor can be reduced to 1, by dividing both sides of (0.4.1) by K. But it proves more convenient to maintain it in the right side of the equation; this conforms also to the general usage.

For a complete specification of the motion, (0.4.1) must of course be supplemented by appropriate boundary conditions.

The effect of rotatory inertia could of course be taken into account, at least formally, by appending rotatory inertia terms to the eqs. (0.4.1) (see e.g. T. C. Lin and G. W. Morgan,[42] and P. M. Naghdi and R. M. Cooper.[43]). But then, for those high frequencies for which this effect becomes appreciable, the validity of the shell theory employed should also be verified.

5. Choice of a particular differential operator:

As it is well known, any "derived" shell theory contains by its very nature, discrepencies and hence inherent errors. It has often been argued that there is no universally suitable shell theory, but that there are more or less suitable theories for classes of problems. A shell theory can thus at best be a compromise between the accuracy it offers and the easiness in working with its equations. A systematic and fully satisfactory estimate of the errors remains still a difficult question, inspite of the important advances achieved in the last decade or so. In some instances, however, an asymptotic analysis may give some very valuable clues.

All what has been said above holds true also for the (relatively simple) linear theories. Choosing a particular differential operator remains therefore always a delicate question, the answer to which depends much of the nature of the individual problem under consideration.

In these lectures, we will employ a differential operator (operating on the displacement vector alone) given in the monograph by A. L. Gol'denveizer.[53] This operator presents the triple advantage of having been intensively discussed in the cited monograph, of having a rather wide domain of applicability and of having been very widely used (in the form given or in equivalent forms). On the other hand, it should be noted that this operator does not account for transverse shear effects and that some care might be needed when applied to shells with nonpositive total curvature. We will leave out of consideration the effect of rotatory inertia. We observe however, that the problems investigated in the following Chapters can be studied by means of some other linear theory by adapting the same methods as used here, although the results obtained may of course differ more or less.

We write now the operator, split into two parts:

$$\mathcal{L}_{ij} = L_{ij} + h^{*2} N_{ij} \tag{0.5.1}$$

Clearly, in this operator, the second part is the one which accounts for the strain distribution across thickness.

Here (*)

$$\begin{aligned}
L_{11} &= -\frac{1}{A}\frac{\partial}{\partial\alpha}\frac{1}{AB}\frac{\partial}{\partial\alpha}B - \frac{1-\nu}{2}\frac{1}{B}\frac{\partial}{\partial\beta}\frac{1}{AB}\frac{\partial}{\partial\beta}A - \frac{1-\nu}{R_1R_2}\\
L_{12} &= -\frac{1}{A}\frac{\partial}{\partial\alpha}\frac{1}{AB}\frac{\partial}{\partial\beta}A + \frac{1-\nu}{2}\frac{1}{B}\frac{\partial}{\partial\beta}\frac{1}{AB}\frac{\partial}{\partial\alpha}B\\
L_{13} &= \frac{1}{A}\frac{\partial}{\partial\alpha}\left(\frac{1}{R_1}+\frac{1}{R_2}\right) - \frac{1-\nu}{R_2'}\frac{1}{A}\frac{\partial}{\partial\alpha} - \frac{1-\nu}{R_{12}}\frac{1}{B}\frac{\partial}{\partial\beta}\\
L_{21} &= -\frac{1}{B}\frac{\partial}{\partial\beta}\frac{1}{AB}\frac{\partial}{\partial\alpha}B + \frac{1-\nu}{2}\frac{1}{A}\frac{\partial}{\partial\alpha}\frac{1}{AB}\frac{\partial}{\partial\beta}A
\end{aligned} \tag{0.5.2}$$

(*) The operator introduced here has opposite sign, as compared to the form given in by A. L. Gol'denveizer.[53]

$$L_{22} = -\frac{1}{B}\frac{\partial}{\partial\beta}\frac{1}{AB}\frac{\partial}{\partial\beta}A - \frac{1-\nu}{2}\frac{1}{A}\frac{\partial}{\partial\alpha}\frac{1}{AB}\frac{\partial}{\partial\alpha}B - \frac{1-\nu}{R_1R_2}$$

$$L_{23} = \frac{1}{B}\frac{\partial}{\partial\beta}\left(\frac{1}{R_1}+\frac{1}{R_2}\right) - \frac{1-\nu}{R_1'}\frac{1}{B}\frac{\partial}{\partial\beta} - \frac{1-\nu}{R_{12}}\frac{1}{A}\frac{\partial}{\partial\alpha}$$

$$L_{31} = -\left(\frac{1}{R_1}+\frac{1}{R_2}\right)\frac{1}{AB}\frac{\partial}{\partial\alpha}B + \frac{1-\nu}{AB}\frac{\partial}{\partial\alpha}\frac{B}{R_2'} + \frac{1-\nu}{AB}\frac{\partial}{\partial\beta}\frac{A}{R_{12}}$$

$$L_{32} = -\left(\frac{1}{R_1}+\frac{1}{R_2}\right)\frac{1}{AB}\frac{\partial}{\partial\beta}A + \frac{1-\nu}{AB}\frac{\partial}{\partial\beta}\frac{A}{R_1'} + \frac{1-\nu}{AB}\frac{\partial}{\partial\alpha}\frac{B}{R_{12}}$$

$$L_{33} = \frac{1}{R_1^2} + \frac{2\nu}{R_1R_2} + \frac{1}{R_2^2}$$

$$N_{11} = -\frac{R^2}{3}\Big\{\left(\frac{1}{R_1'}\frac{1}{A}\frac{\partial}{\partial\alpha} - \frac{1}{R_{12}}\frac{1}{B}\frac{\partial}{\partial\beta}\right)\frac{1}{AB}\left(\frac{\partial}{\partial\alpha}\frac{B}{R_1'} - \frac{\partial}{\partial\beta}\frac{A}{R_{12}}\right) +$$

$$+\frac{1-\nu}{2}\Big[\frac{1}{R_1'}\left(\frac{1}{R_1'}\frac{1}{B}\frac{\partial}{\partial\beta} + \frac{1}{R_{12}}\frac{1}{A}\frac{\partial}{\partial\alpha}\right) + \frac{1}{R_{12}}\times$$

$$\times\left(\frac{1}{R_2'}\frac{1}{A}\frac{\partial}{\partial\alpha} + \frac{1}{R_{12}}\frac{1}{B}\frac{\partial}{\partial\beta}\right)\Big]\frac{1}{AB}\frac{\partial}{\partial\beta}A + \frac{1-\nu}{R_1R_2}\left(\frac{1}{R_1'^2} + \frac{1}{R_2^2}\right)\Big\}$$

$$N_{12} = -\frac{R^2}{3}\Big\{\left(\frac{1}{R_1'}\frac{1}{A}\frac{\partial}{\partial\alpha} - \frac{1}{R_{12}}\frac{1}{B}\frac{\partial}{\partial\beta}\right)\frac{1}{AB}\left(\frac{\partial}{\partial\beta}\frac{A}{R_2'} - \frac{\partial}{\partial\alpha}\frac{B}{R_{12}}\right) -$$

$$-\frac{1-\nu}{2}\Big[\frac{1}{R_1'}\left(\frac{1}{R_1'}\frac{1}{B}\frac{\partial}{\partial\beta} + \frac{1}{R_{12}}\frac{1}{A}\frac{\partial}{\partial\alpha}\right) + \frac{1}{R_{12}}\times$$

$$\times\left(\frac{1}{R_2'}\frac{1}{A}\frac{\partial}{\partial\alpha} + \frac{1}{R_{12}}\frac{1}{B}\frac{\partial}{\partial\beta}\right)\Big]\frac{1}{AB}\frac{\partial}{\partial\alpha}B - \frac{1-\nu}{R_1R_2}\frac{1}{R_{12}}\left(\frac{1}{R_1} + \frac{1}{R_2}\right)\Big\}$$

$$N_{13} = -\frac{R^2}{3}\Big\{\left(\frac{1}{R_1'}\frac{1}{A}\frac{\partial}{\partial\alpha} - \frac{1}{R_{12}}\frac{1}{B}\frac{\partial}{\partial\beta}\right)\frac{1}{AB}\left(\frac{\partial}{\partial\alpha}\frac{B}{A}\frac{\partial}{\partial\alpha} + \frac{\partial}{\partial\beta}\frac{A}{B}\frac{\partial}{\partial\beta}\right) +$$

$$+\frac{1-\nu}{R_1R_2}\left(\frac{1}{R_1'}\frac{1}{A}\frac{\partial}{\partial\alpha} - \frac{1}{R_{12}}\frac{1}{B}\frac{\partial}{\partial\beta}\right)\Big\}$$

$$N_{21} = -\frac{R^2}{3}\Big\{\left(\frac{1}{R_2'}\frac{1}{B}\frac{\partial}{\partial\beta} - \frac{1}{R_{12}}\frac{1}{A}\frac{\partial}{\partial\alpha}\right)\frac{1}{AB}\left(\frac{\partial}{\partial\alpha}\frac{B}{R_1'} - \frac{\partial}{\partial\beta}\frac{A}{R_{12}}\right) -$$

$$-\frac{1-\nu}{2}\Big[\frac{1}{R_2'}\left(\frac{1}{R_2'}\frac{1}{A}\frac{\partial}{\partial\alpha} + \frac{1}{R_{12}}\frac{1}{B}\frac{\partial}{\partial\beta}\right) + \frac{1}{R_{12}}\times$$

$$\times \left(\frac{1}{R_1'}\frac{1}{B}\frac{\partial}{\partial\beta} + \frac{1}{R_{12}}\frac{1}{A}\frac{\partial}{\partial\alpha}\right)\Big]\frac{1}{AB}\frac{\partial}{\partial\beta}A - \frac{1-\nu}{R_1R_2}\frac{1}{R_{12}}\left(\frac{1}{R_1} + \frac{1}{R_2}\right)\Big\}$$

$$N_{22} = -\frac{R^2}{3}\Big\{\left(\frac{1}{R_2'}\frac{1}{B}\frac{\partial}{\partial\beta} - \frac{1}{R_{12}}\frac{1}{A}\frac{\partial}{\partial\alpha}\right)\frac{1}{AB}\left(\frac{\partial}{\partial\beta}\frac{A}{R_2'} - \frac{\partial}{\partial\alpha}\frac{B}{R_{12}}\right) +$$

$$+ \frac{1-\nu}{2}\Big[\frac{1}{R_2'}\left(\frac{1}{R_2'}\frac{1}{A}\frac{\partial}{\partial\alpha} + \frac{1}{R_{12}}\frac{1}{B}\frac{\partial}{\partial\beta}\right) + \frac{1}{R_{12}} \times$$

$$\times \left(\frac{1}{R_1'}\frac{1}{B}\frac{\partial}{\partial\beta} + \frac{1}{R_{12}}\frac{1}{A}\frac{\partial}{\partial\alpha}\right)\Big]\frac{1}{AB}\frac{\partial}{\partial\alpha}B + \frac{1-\nu}{R_1R_2}\left(\frac{1}{R_1^2} + \frac{1}{R_{12}^2}\right)\Big\}$$

$$N_{23} = -\frac{R^2}{3}\Big\{\left(\frac{1}{R_2'}\frac{1}{B}\frac{\partial}{\partial\beta} - \frac{1}{R_{12}}\frac{1}{A}\frac{\partial}{\partial\alpha}\right)\frac{1}{AB}\left(\frac{\partial}{\partial\alpha}\frac{B}{A}\frac{\partial}{\partial\alpha} + \frac{\partial}{\partial\beta}\frac{A}{B}\frac{\partial}{\partial\beta}\right) +$$

$$+ \frac{1-\nu}{R_1R_2}\left(\frac{1}{R_2'}\frac{1}{B}\frac{\partial}{\partial\beta} - \frac{1}{R_{12}}\frac{1}{A}\frac{\partial}{\partial\alpha}\right)\Big\}$$

$$N_{31} = \frac{R^2}{3}\Big\{\frac{1}{AB}\left(\frac{\partial}{\partial\alpha}\frac{B}{A}\frac{\partial}{\partial\alpha} + \frac{\partial}{\partial\beta}\frac{A}{B}\frac{\partial}{\partial\beta}\right)\frac{1}{AB}\left(\frac{\partial}{\partial\alpha}\frac{B}{R_1'} - \frac{\partial}{\partial\beta}\frac{A}{R_{12}}\right) +$$

$$+ \frac{1-\nu}{2}\frac{1}{AB}\Big[\frac{\partial}{\partial\alpha}B\left(\frac{1}{R_1'}\frac{1}{B}\frac{\partial}{\partial\beta} + \frac{1}{R_{12}}\frac{1}{A}\frac{\partial}{\partial\alpha}\right) -$$

$$- \frac{\partial}{\partial\beta}A\left(\frac{1}{R_2'}\frac{1}{A}\frac{\partial}{\partial\alpha} + \frac{1}{R_{12}}\frac{1}{B}\frac{\partial}{\partial\beta}\right)\Big]\frac{1}{AB}\frac{\partial}{\partial\beta}A + \frac{1-\nu}{AB}\left(\frac{\partial}{\partial\alpha}\frac{B}{R_1'} - \frac{\partial}{\partial\beta}\frac{A}{R_{12}}\right)\frac{1}{R_1R_2}\Big\}$$

$$N_{32} = \frac{R^2}{3}\Big\{\frac{1}{AB}\left(\frac{\partial}{\partial\alpha}\frac{B}{A}\frac{\partial}{\partial\alpha} + \frac{\partial}{\partial\beta}\frac{A}{B}\frac{\partial}{\partial\beta}\right)\frac{1}{AB}\left(\frac{\partial}{\partial\beta}\frac{A}{R_2'} - \frac{\partial}{\partial\alpha}\frac{B}{R_{12}}\right) +$$

$$+ \frac{1-\nu}{2}\frac{1}{AB}\Big[\frac{\partial}{\partial\beta}A\left(\frac{1}{R_2'}\frac{1}{A}\frac{\partial}{\partial\alpha} + \frac{1}{R_{12}}\frac{1}{B}\frac{\mu}{\partial\beta}\right) -$$

$$\frac{\partial}{\partial\alpha}B\left(\frac{1}{R_1'}\frac{1}{B}\frac{\partial}{\partial\beta} + \frac{1}{R_{12}}\frac{1}{A}\frac{\partial}{\partial\alpha}\right)\Big]\frac{1}{AB}\frac{\partial}{\partial\alpha}B - \frac{1-\nu}{AB}\left(\frac{\partial}{\partial\alpha}\frac{B}{R_{12}} - \frac{\partial}{\partial\beta}\frac{A}{R_2'}\right)\frac{1}{R_1R_2}\Big\}$$

$$N_{33} = \frac{R^2}{3}\Big\{\frac{1}{AB}\left(\frac{\partial}{\partial\alpha}\frac{B}{A}\frac{\partial}{\partial\alpha} + \frac{\partial}{\partial\beta}\frac{A}{B}\frac{\partial}{\partial\beta}\right)\frac{1}{AB}\left(\frac{\partial}{\partial\alpha}\frac{B}{A}\frac{\partial}{\partial\alpha} + \frac{\partial}{\partial\beta}\frac{A}{B}\frac{\partial}{\partial\beta}\right) +$$

$$+ \frac{1-\nu}{AB}\left(\frac{\partial}{\partial\alpha}\frac{B}{R_1R_2}\frac{1}{A}\frac{\partial}{\partial\alpha} + \frac{\partial}{\partial\beta}\frac{A}{R_1R_2}\frac{1}{B}\frac{\partial}{\partial\beta}\right)\Big\}$$

For completeness, we record the expressions which give, in terms of displacements, the tangential strains

$$\varepsilon_1 = \frac{1}{A}\frac{\partial U_1}{\partial\alpha} + \frac{1}{AB}\frac{\partial A}{\partial\beta}U_2 - \frac{U_3}{R_1'}$$

$$\bar{\omega} = \frac{A}{B}\frac{\partial}{\partial\beta}\left(\frac{U_1}{A}\right) + \frac{B}{A}\frac{\partial}{\partial\alpha}\left(\frac{U_2}{B}\right) + \frac{2U_3}{R_{12}} \qquad (0.5.3)$$

$$\varepsilon_2 = \frac{1}{B}\frac{\partial U_2}{\partial\beta} + \frac{1}{AB}\frac{\partial B}{\partial\alpha}U_1 - \frac{U_3}{R_2'}$$

and the bending strains

$$\varkappa_1 = \frac{1}{A}\frac{\partial}{\partial\alpha}\left(\frac{1}{A}\frac{\partial U_3}{\partial\alpha} + \frac{U_1}{R_1'} - \frac{U_2}{R_{12}}\right) + \frac{1}{AB}\frac{\partial A}{\partial\beta}\left(\frac{1}{B}\frac{\partial U_3}{\partial\beta} + \frac{U_2}{R_2'} - \frac{U_1}{R_{12}}\right) +$$

$$+ \frac{1}{2}\frac{1}{R_{12}}\frac{1}{AB}\left[\frac{\partial}{\partial\beta}(AU_1) - \frac{\partial}{\partial\alpha}(BU_2)\right]$$

$$\tau = \frac{1}{A}\frac{\partial}{\partial\alpha}\left(\frac{1}{B}\frac{\partial U_3}{\partial\beta} + \frac{U_2}{R_2'} - \frac{U_1}{R_{12}}\right) - \frac{1}{AB}\frac{\partial A}{\partial\beta}\left(\frac{1}{A}\frac{\partial U_3}{\partial\alpha} + \frac{U_1}{R_1'} - \frac{U_2}{R_{12}}\right) + \qquad (0.5.4)$$

$$+ \frac{1}{R_1'}\left(\frac{1}{B}\frac{\partial U_1}{\partial\beta} - \frac{1}{AB}\frac{\partial B}{\partial\alpha}U_2 + \frac{U_3}{R_{12}}\right) - \frac{1}{R_{12}}\left(\frac{1}{B}\frac{\partial U_2}{\partial\beta} + \frac{1}{AB}\frac{\partial B}{\partial\alpha}U_1 - \frac{U_3}{R_2'}\right)$$

$$\varkappa_2 = \frac{1}{B}\frac{\partial}{\partial\beta}\left(\frac{1}{B}\frac{\partial U_3}{\partial\beta} + \frac{U_2}{R_2'} - \frac{U_1}{R_{12}}\right) + \frac{1}{AB}\frac{\partial B}{\partial\alpha}\left(\frac{1}{A}\frac{\partial U_3}{\partial\alpha} + \frac{U_1}{R_1'} - \frac{U_2}{R_{12}}\right) +$$

$$+ \frac{1}{2}\frac{1}{R_{12}}\frac{1}{AB}\left[\frac{\partial}{\partial\alpha}(BU_2) - \frac{\partial}{\partial\beta}(AU_1)\right]$$

In the present case K is related to the thickness of the shell and to the material constants by

$$K \equiv \frac{1-\nu^2}{2Eh} \qquad (0.5.5)$$

CHAPTER I

FREE VIBRATIONS

In this Chapter, we consider those vibrations which may take place in a shell free of the effect of any external forces at any time, and subject to time-independent kinematic boundary conditions alone. Such vibrations are called *free vibrations*.

1. Standing waves:

If in the equations (0.4.1) we take $f_i = 0$, we are left with

$$\sum_{j=1}^{3} L_{ij} U_j = -2hK\rho \frac{\partial^2 U_i}{\partial t^2} \qquad (I.1.1)$$

Trying solutions of the form

$$U_i(\alpha,\beta,t) = u_i(\alpha,\beta)\ \psi(t) \qquad (I.1.2)$$

i.e. products of a function of time alone by a function of the variables α and β, we find

$$\sum_{j=1}^{3} L_{ij} u_j = -u_i\left[2h\rho K \frac{1}{\psi} \frac{d^2\psi}{dt^2} \right] \qquad (I.1.3)$$

Since the functions $u_i(\alpha,\beta)$ are independent of time, we conclude that the quantity in brackets must also be independent of time, and since ψ is a function of time only (*), this quantity must be a constant, which shall be denoted by $-\lambda$.

Hence

$$\frac{d^2\psi}{dt^2} + \frac{\lambda}{2h\rho K}\ \psi = 0 \qquad (I.1.4)$$

(*) ρ is considered to remain constant during deformation.

having as solution

$$\psi = e^{i\sqrt{\frac{\lambda}{2h\rho K}}\,t} \tag{I.1.5}$$

Or, if we put

$$\lambda \equiv 2h\rho K\ \omega^2 \tag{I.1.6}$$

then

$$\psi = e^{i\omega t} \tag{I.1.7}$$

and (I.1.3) reduces to

$$\sum_{j=1}^{3} L_{ij}\ u_j = \lambda u_i \qquad (i = 1,2,3) \tag{I.1.8}$$

For real ω, λ is positive, and ω is called the *frequency* of the vibration (measured as an angular velocity).

Waves of the form (I.1.2) are called *standing waves*. The time-independent kinematic boundary conditions are to be satisfied, accordingly, by the functions u_i.

The set of equations (I.1.8) and the boundary conditions together represent an *eigenvalue problem*. The values of λ for which there exist nontrivial solutions are the *eigenvalues*, the corresponding functions are the *eigenfunctions* (components of the *eigenvectors*), and the values of ω are the *eigenfrequencies*.

It is clear that due to the linearity of the system, any linear combination (superposition) of standing waves will again be solution of the system.

In particular, with the choice we made in Chapter I, § 5 for the differential operator,

$$
\begin{aligned}
&L_{11}u_1 + L_{12}u_2 + L_{13}u_3 + h^{*2}\left[N_{11}u_1 + N_{12}u_2 + N_{13}u_3\right] = \lambda u_1 \\
&L_{21}u_1 + L_{22}u_2 + L_{23}u_3 + h^{*2}\left[N_{21}u_1 + N_{22}u_2 + N_{23}u_3\right] = \lambda u_2 \\
&L_{31}u_1 + L_{32}u_2 + L_{33}u_3 + h^{*2}\left[N_{31}u_1 + N_{32}u_2 + N_{33}u_3\right] = \lambda u_3
\end{aligned}
\tag{I.1.9}
$$

with

$$\lambda = \frac{\rho(1-\nu^2)}{E}\,\omega^2 \tag{I.1.10}$$

2. Ideas underlying the asymptotic approach to shell problems:

The integration of the system of differential equations (I.1.9) proves to be rather difficult, except under particular circumstances of shell form and boundary. It can also hardly be expected to obtain insight into the character of the vibration, just by using these equations as such. On the other hand, the presence of the *small parameter* h^* in the differential equations suggests that asymptotic medhods can advantageously be applied.

Although not always substantiated rigorously, asymptotic methods may in many instances provide a satisfactory approximation. At the same time, they permit a *qualitative* study, by setting in relief the dominant character of the motion and thus giving rise to a classification.

The origins of the use of asymptotic methods in the statics of shells can be traced back to the works of H. Reissner, O. Blumenthal and E.Meissner. For the statics of a shell, an asymptotic method is expounded in great detail in A. L. Gol'denveizer's monograph,[53] to which the reader is referred for many useful details and stimulating ideas. The use of asymptotic methods in the dynamics of shells is more recent.[54-60] An asymptotic study of low frequency oscillations of convex shells of revolution by introducing two small parameters has been reported by P. E. Tovstik.[61] Later, we will have the occasion to cite still more references, as we are going to see further applications of the asymptotic approach.

The asymptotic approach we adopt here bases on the method given in the monograph[53] cited above. The results parallel those obtained by the same author in a number of papers.[54,55,59]

Asymptotic methods spring from the idea of formally expanding the solution in a series of powers of the small parameter, and cutting off this series at some term. The expression thus obtained is then expected to tend to the exact solution as the small parameter tends to zero.

3. Definitions and preliminaries:

We propose now using *formal solutions*

$$
\begin{aligned}
U_1 &= k^{\lambda}\, u_1(\alpha,\beta;k)\; e^{kf(\alpha,\beta)\,+\,k^b g(t;k)}\\
U_2 &= k^{\mu}\, u_2(\alpha,\beta;k)\; e^{kf(\alpha,\beta)\,+\,k^b g(t;k)}\\
U_3 &= k^{\nu}\, u_3(\alpha,\beta;k)\; e^{kf(\alpha,\beta)\,+\,k^b g(t;k)}
\end{aligned}
\qquad (I.3.1)
$$

with

$$k \equiv h^{*-a} \qquad (I.3.2)$$

where a is a nonnegative number. The expressions in (I.3.1) are in conformity with (I.1.2), but ψ is given now a particular structure somewhat more general than that in (I.1.7). The function g will be discussed later.

We assume that the quantities u_1, u_2, and u_3 are commensurate to (i.e. of the same order of magnitude as) a same power of h^*.

The asymptotic expansions of these functions are (*)

$$
\begin{aligned}
u_1 &\sim \overset{o}{u}_1(\alpha,\beta) + \frac{\overset{1}{u}_1(\alpha,\beta)}{k} + \frac{\overset{2}{u}_1(\alpha,\beta)}{k^2} + \dots\dots\\
u_2 &\sim \overset{o}{u}_2(\alpha,\beta) + \frac{\overset{1}{u}_2(\alpha,\beta)}{k} + \frac{\overset{2}{u}_2(\alpha,\beta)}{k^2} + \dots\dots\\
u_3 &\sim \overset{o}{u}_3(\alpha,\beta) + \frac{\overset{1}{u}_3(\alpha,\beta)}{k} + \frac{\overset{2}{u}_3(\alpha,\beta)}{k^2} + \dots\dots
\end{aligned}
\qquad (I.3.3)
$$

(*) Clearly, the indices at the top of the u_i do not indicate powers, but sequence of terms in the series.

where it is assumed that

$$\overset{o}{u}_1 \neq 0 \ , \ \overset{o}{u}_2 \neq 0 \ , \ \overset{o}{u}_3 \neq 0 \qquad (I.3.4)$$

It is obvious that this requirement does not restrict the generality of the discussion.

Now, consider, in general, an expression of the form

$$U = k^c\, u(\alpha,\beta;k)\, e^{kf(\alpha,\beta) + k^b g(t;k)} \qquad (I.3.5)$$

We have

$$\frac{\partial U}{\partial \alpha} = k^c\left(k\frac{\partial f}{\partial \alpha}u + \frac{\partial u}{\partial \alpha}\right)e^{kf + k^b g}$$
$$\frac{\partial U}{\partial \beta} = k^c\left(k\frac{\partial f}{\partial \beta}u + \frac{\partial u}{\partial \beta}\right)e^{kf + k^b g} \qquad (I.3.6)$$
$$\frac{\partial U}{\partial t} = k^c\left(k^b\frac{\partial g}{\partial t}u\right)e^{kf + k^b g}$$

where u,f and g are supposed to vary slowly.

f and g are called *functions of variability* (of space resp. time variables), u is called *function of intensity*, a the *index of variability* (with respect to space variables), c the *index of intensity*. The index of variability with respect to the time variable is then ab.

Since k is very large, it will be

$$\frac{\partial U}{\partial \alpha} \approx k\frac{\partial f}{\partial \alpha}U$$
$$\frac{\partial U}{\partial \beta} \approx k\frac{\partial f}{\partial \beta}U \qquad (I.3.7)$$

in general. In other words, differentiation with respect to α and β increases the index of intensity boy one. This has exceptions. Indeed, let us consider a parametric representation

$$\theta^\eta = \theta^\eta(q) \qquad (\eta = 1,2) \qquad (I.3.8)$$

of a curve fixed on a surface. The derivative of U with respect to the parameter q is

$$\frac{dU}{dq} = k^{c+1} \left[\left(\frac{\partial f}{\partial \theta^1} \frac{d\theta^1}{dq} + \frac{\partial f}{\partial \theta^2} \frac{d\theta^2}{dq}\right) u + \frac{1}{k} \frac{du}{dq}\right] e^{kf + k^b g} \qquad (I.3.9)$$

If the curve is a "contour line" of $f(\theta^\eta)$, then (I.3.7) may fail to hold. We shall not further pursue here this discussion and refer to A. L. Gol'denveizer's monograph.[53]

As to the consequence of differentiation with respect to time, we simply observe that

$$\frac{\partial U}{\partial t} = k^b \frac{\partial g}{\partial t} U \qquad (I.3.10)$$

and postpone further comments.

The procedure to be then followed is simple in essence, although it necessitates rather long calculations. The formal solutions (I.3.3) are substituted into equations (0.4.1) written by using the operator (0.5.1). There follows an expansion with respect to *decreasing powers* of k. A first approximation consists in retaining those terms only with the highest power of k, taking nevertheless care of having all three unknown functions $\overset{o}{u}_1$, $\overset{o}{u}_2$, $\overset{o}{u}_3$ appearing in the system thus left, and each pair of equations containing at least two of them. This condition is referred to as *condition of noncontradiction*. This procedure is explained and discussed in full detail in the monograph[53] cited above and to which the reader is referred. It is shown in the same reference (p. 329 and "Amendment" 12) that there is a consequence of the choice made for $f(\alpha,\beta)$ as a function independent of k, that may at first sight seem embarrassing: it excludes fractional values of $\frac{2}{a}$; since otherwise, the method becomes unsuitable. As we shall soon see, however, this fact is not of prime importance in our subsequent discussion. Thus, we retain this choice which has the advantage of simplicity. As to the function $g(t;k)$, on one hand, we are not bound with a similar difficulty, on the other hand, there is an essential advantage of keeping κ as argument, as we shall find out in short.

The system obtained by this procedure is

$$
\begin{aligned}
&k^{\lambda+2}\left(L^o_{11} + k^{-\frac{2}{a}} N^o_{11}\right)\overset{o}{u}_1 + k^{\mu+2}\left(L^o_{12} + k^{-\frac{2}{a}} N^o_{12}\right)\overset{o}{u}_2 + \\
&+ k^{\nu+1}\left(L^o_{13} + k^{-\frac{2}{a}} N^o_{13}\right)\overset{o}{u}_3 = -\frac{\rho(1-\nu^2)}{E}\, k^{\lambda+2b} \left(\frac{\partial g}{\partial t}\right)^2 \overset{o}{u}_1 \\
&k^{\lambda+2}\left(L^o_{21} + k^{-\frac{2}{a}} N^o_{21}\right)\overset{o}{u}_1 + k^{\mu+2}\left(L^o_{22} + k^{-\frac{2}{a}} N^o_{22}\right)\overset{o}{u}_2 + \\
&+ k^{\nu+1}\left(L^o_{23} + k^{-\frac{2}{a}} N^o_{23}\right)\overset{o}{u}_3 = -\frac{\rho(1-\nu^2)}{E}\, k^{\mu+2b} \left(\frac{\partial g}{\partial t}\right)^2 \overset{o}{u}_2 \qquad \text{(I.3.11)} \\
&k^{\lambda+2}\left(L^o_{31} + k^{2-\frac{2}{a}} N^o_{31}\right)\overset{o}{u}_1 + k^{\mu+2}\left(L^o_{32} + k^{2-\frac{2}{a}} N^o_{32}\right)\overset{o}{u}_2 + \\
&+ k^{\nu+1}\left(L^o_{33} + k^{4-\frac{2}{a}} N^o_{33}\right)\overset{o}{u}_3 = -\frac{\rho(1-\nu^2)}{E}\, k^{\nu+1+2b} \left(\frac{\partial g}{\partial t}\right)^2 \overset{o}{u}_3
\end{aligned}
$$

with (*)

$$L^o_{11} = -\frac{1}{A}\left(\frac{\partial f}{\partial \alpha}\right)^2 - \frac{1-\nu}{2}\frac{1}{B}\left(\frac{\partial f}{\partial \beta}\right)^2$$

$$L^o_{12} = -\frac{1+\nu}{2}\frac{1}{AB}\frac{\partial f}{\partial \alpha}\frac{\partial f}{\partial \beta}$$

$$L^o_{13} = \left(\frac{1}{R'_1} + \frac{\nu}{R'_2}\right)\frac{1}{A}\frac{\partial f}{\partial \alpha} - \frac{1-\nu}{R_{12}}\frac{1}{B}\frac{\partial f}{\partial \beta}$$

$$L^o_{21} = -\frac{1+\nu}{2}\frac{1}{AB}\frac{\partial f}{\partial \alpha}\frac{\partial f}{\partial \beta}$$

$$L^o_{22} = -\frac{1}{B^2}\left(\frac{\partial f}{\partial \beta}\right)^2 - \frac{1-\nu}{2}\frac{1}{A^2}\left(\frac{\partial f}{\partial \alpha}\right)^2$$

$$L^o_{23} = \left(\frac{1}{R'_2} + \frac{\nu}{R'_1}\right)\frac{1}{B}\frac{\partial f}{\partial \beta} - \frac{1-\nu}{R_{12}}\frac{1}{A}\frac{\partial f}{\partial \alpha}$$

$$L^o_{31} = -\left(\frac{1}{R'_1} + \frac{\nu}{R'_2}\right)\frac{1}{A}\frac{\partial f}{\partial \alpha} + \frac{1-\nu}{R_{12}}\frac{1}{B}\frac{\partial f}{\partial \beta}$$

$$L^o_{32} = -\left(\frac{1}{R'_2} + \frac{\nu}{R'_1}\right)\frac{1}{B}\frac{\partial f}{\partial \beta} + \frac{1-\nu}{R_{12}}\frac{1}{A}\frac{\partial f}{\partial \alpha}$$

$$L^o_{33} = \left(\frac{1}{R_1} + \frac{1}{R_2}\right)^2 - \frac{2(1-\nu)}{R_1 R_2}$$

(*)These expressions are given by A. L. Gol'denveizer,[53] with opposite sign (cf. footnote in Section 5 of the Introduction).

$$N^o_{11} = -\frac{R^2}{3}\Big\{\Big(\frac{1}{R_1'}\frac{1}{A}\frac{\partial f}{\partial\alpha} - \frac{1}{R_{12}}\frac{1}{B}\frac{\partial f}{\partial\beta}\Big)\Big(\frac{1}{R_1'}\frac{1}{A}\frac{\partial f}{\partial\alpha} - \frac{1}{R_{12}}\frac{1}{B}\frac{\partial f}{\partial\beta}\Big) + \frac{1-\nu}{2}\Big[\frac{1}{R_1'}\Big(\frac{1}{R_1'}\frac{1}{B}\frac{\partial f}{\partial\beta} + \frac{1}{R_{12}}\frac{1}{A}\frac{\partial f}{\partial\alpha}\Big) + \frac{1}{R_{12}}\Big(\frac{1}{R_2'}\frac{1}{A}\frac{\partial f}{\partial\alpha} + \frac{1}{R_{12}}\frac{1}{B}\frac{\partial f}{\partial\beta}\Big)\Big]\frac{1}{B}\frac{\partial f}{\partial\beta}\Big\}$$

$$N^o_{12} = -\frac{R^2}{3}\Big\{\Big(\frac{1}{R_1'}\frac{1}{A}\frac{\partial f}{\partial\alpha} - \frac{1}{R_{12}}\frac{1}{B}\frac{\partial f}{\partial\beta}\Big)\Big(\frac{1}{R_2'}\frac{1}{B}\frac{\partial f}{\partial\beta} - \frac{1}{R_{12}}\frac{1}{A}\frac{\partial f}{\partial\alpha}\Big) - \frac{1-\nu}{2}\Big[\frac{1}{R_1'}\Big(\frac{1}{R_1'}\frac{1}{B}\frac{\partial f}{\partial\beta} + \frac{1}{R_{12}}\frac{1}{A}\frac{\partial f}{\partial\alpha}\Big) + \frac{1}{R_{12}}\Big(\frac{1}{R_2'}\frac{1}{A}\frac{\partial f}{\partial\alpha} + \frac{1}{R_{12}}\frac{1}{B}\frac{\partial f}{\partial\beta}\Big)\Big]\frac{1}{A}\frac{\partial f}{\partial\alpha}\Big\}$$

$$N^o_{13} = -\frac{R^2}{3}\Big\{\Big(\frac{1}{R'}\frac{1}{A}\frac{\partial f}{\partial\alpha} - \frac{1}{R}\frac{1}{B}\frac{\partial f}{\partial\beta}\Big)\Big[\frac{1}{A^2}\Big(\frac{\partial f}{\partial\alpha}\Big)^2 + \frac{1}{B^2}\Big(\frac{\partial f}{\partial\beta}\Big)^2\Big]\Big\}$$

$$N^o_{21} = -\frac{R^2}{3}\Big\{\Big(\frac{1}{R'}\frac{1}{B}\frac{\partial f}{\partial\beta} - \frac{1}{R}\frac{1}{A}\frac{\partial f}{\partial\alpha}\Big)\Big(\frac{1}{R'}\frac{1}{A}\frac{\partial f}{\partial\alpha} - \frac{1}{R}\frac{1}{B}\frac{\partial f}{\partial\beta}\Big) - \frac{1-\nu}{2}\Big[\frac{1}{R_2'}\Big(\frac{1}{R_2'}\frac{1}{A}\frac{\partial f}{\partial\alpha} + \frac{1}{R_{12}}\frac{1}{B}\frac{\partial f}{\partial\beta}\Big) + \frac{1}{R_{12}}\Big(\frac{1}{R_1'}\frac{1}{B}\frac{\partial f}{\partial\beta} + \frac{1}{R_{12}}\frac{1}{A}\frac{\partial f}{\partial\alpha}\Big)\Big]\frac{1}{B}\frac{\partial f}{\partial\beta}\Big\} \tag{I.3.12}$$

$$N^o_{22} = -\frac{R^2}{3}\Big\{\Big(\frac{1}{R_2'}\frac{1}{B}\frac{\partial f}{\partial\beta} - \frac{1}{R_{12}}\frac{1}{A}\frac{\partial f}{\partial\alpha}\Big)\Big(\frac{1}{R_2'}\frac{1}{B}\frac{\partial f}{\partial\beta} - \frac{1}{R_{12}}\frac{1}{A}\frac{\partial f}{\partial\alpha}\Big) + \frac{1-\nu}{2}\Big[\frac{1}{R_2'}\Big(\frac{1}{R_2'}\frac{1}{A}\frac{\partial f}{\partial\alpha} + \frac{1}{R_{12}}\frac{1}{B}\frac{\partial f}{\partial\beta}\Big) + \frac{1}{R_{12}}\Big(\frac{1}{R_1'}\frac{1}{B}\frac{\partial f}{\partial\beta} + \frac{1}{R_{12}}\frac{1}{A}\frac{\partial f}{\partial\alpha}\Big)\Big]\frac{1}{A}\frac{\partial f}{\partial\alpha}\Big\}$$

$$N^o_{23} = -\frac{R^2}{3}\Big\{\Big(\frac{1}{R_2'}\frac{1}{B}\frac{\partial f}{\partial\beta} - \frac{1}{R_{12}}\frac{1}{A}\frac{\partial f}{\partial\alpha}\Big)\Big[\frac{1}{A^2}\Big(\frac{\partial f}{\partial\alpha}\Big)^2 + \frac{1}{B^2}\Big(\frac{\partial f}{\partial\beta}\Big)^2\Big]\Big\}$$

$$N^{o}_{31} = \frac{R^2}{3}\left\{\left(\frac{1}{R'_1}\frac{1}{A}\frac{\partial f}{\partial\alpha} - \frac{1}{R_{12}}\frac{1}{B}\frac{\partial f}{\partial\beta}\right)\left[\frac{1}{A^2}\left(\frac{\partial}{\partial\alpha}\right)^2 + \frac{1}{B^2}\left(\frac{\partial}{\partial\beta}\right)^2\right]\right\}$$

$$N^{o}_{32} = \frac{R^2}{3}\left\{\left(\frac{1}{R'_2}\frac{1}{B}\frac{\partial f}{\partial\beta} - \frac{1}{R_{12}}\frac{1}{A}\frac{\partial f}{\partial\alpha}\right)\left[\frac{1}{A^2}\left(\frac{\partial f}{\partial\alpha}\right)^2 + \frac{1}{B^2}\left(\frac{\partial f}{\partial\beta}\right)^2\right]\right\}$$

$$N^{o}_{33} = \frac{R^2}{3}\left\{\left[\frac{1}{A^2}\left(\frac{\partial f}{\partial\alpha}\right)^2 + \frac{1}{B^2}\left(\frac{\partial f}{\partial\beta}\right)^2\right]^2\right\}$$

The relations (I.3.11) still contain terms which are not multiplied by the highest power of k. To retain the highest power terms, we must distinguish between possible values of a.

It is expedient to impose lower and upper bounds to a:

$$0 < a < 1 \qquad \text{(I.3.13)}$$

The lower bound is obviously needed for the applicability of the method. The upper bound prevents excessive variability, in which case the shell theory used might be no more adequate.

We immediately see that $a = \frac{1}{2}$ occupies a prominent place. So we will have to distinguish between different cases, according to the values taken by a:

$$0 < a < \frac{1}{2}\ , \quad a = \frac{1}{2}\ , \quad \frac{1}{2} < a < 1 \qquad \text{(I.3.14)}$$

Here, a few words are in place, in view of the necessity of considering integral values only for $\frac{2}{a}$. It follows that the only possible values of a are $\frac{2}{n}$ $(n > 3)$ for $a < \frac{1}{2}$ and $\frac{2}{3}$ for $\frac{1}{2} < a < 1$, $(n = 4,5,6,\ldots)$.

We next realize that it must be $\lambda = \mu$. The relationship between λ, μ on one hand and ν on the other hand requires some more refinement in the classification.

Concerning b, we agree to prescribe an upper bound by way of the inequality

$$ab < 1 \qquad \text{(I.3.15)}$$

but leave b otherwise quasi-unspecified. The upper bound is imposed here again to avoid too rapid variations, while the absence of a lower bound allows including the static case, which occurs for $b = -\infty$.

Besides, it can easily be seen that there is some arbitrariness in the choice of λ, μ and ν, and that we could assign to ν, say, an arbitrary value. In fact, we are going to put conveniently $\nu = 0$.

4. The different cases of free vibrations:

The different cases which are obtained from the simple asymptotic analysis explained in the previous Section are shown in a Table. Following conclusions can be drawn from inspection of this Table.

The cases considered fall into two distinct categories, according to the values taken by b.

In the first category, i.e. for $b = 0$, we have $\lambda = \mu = -1$ for $\nu = 0$. In other words, the intensity of the tangential components of the displacement is of an order of magnitude smaller than that of the transverse motion. The vibrations, in this case, are said to be *quasi-transverse*. The time dependent factor in the expressions (I.3.1) is $e^{g(t;k)} = e^{i\omega t}$; i.e. $g = i\omega t$. As it can be seen at once from the Table, the integrals for $0 < a < \frac{1}{2}$ are *membrane integrals* , while for $\frac{1}{2} < a < 1$ (strictly speaking, for $a = \frac{2}{3}$) the equations uncouple for $\mathfrak{U}_3$, and we have *bending integrals*. The case $a = \frac{1}{2}$ is an intermediate one and the corresponding integrals may be called[59] *bending-planar integrals*. As it can be seen from the third column of the Table, the relations corresponding to this last case contain all the terms which appear in one or both of the other two cases of the same category. This suggests that although we have been considering integral values of $\frac{2}{a}$ only, the classification thus obtained might usefully be extended to the whole range of a, in (I.3.13).

In the second category, $b > 0$. We have then, for $b = 1$, $\lambda = \mu = 1$ (and $\nu = 0$). The intensities of all three components of the displacement are of the same order, if $b = \frac{1}{2}$. As to the function $g(t;k)$, we have

TABLE

a	λ,μ,ν	Equations for $\overset{o}{u}_1$, $\overset{o}{u}_2$, $\overset{o}{u}_3$	$\nu=0$	b
$0<a<\frac{1}{2}$	$\lambda=\mu=\nu-1$	$\overset{o}{L}_{11}\overset{o}{u}_1+\overset{o}{L}_{12}\overset{o}{u}_2+\overset{o}{L}_{13}\overset{o}{u}_3=0$ $\overset{o}{L}_{21}\overset{o}{u}_1+\overset{o}{L}_{22}\overset{o}{u}_2+\overset{o}{L}_{23}\overset{o}{u}_3=0$ $\overset{o}{L}_{31}\overset{o}{u}_1+\overset{o}{L}_{32}\overset{o}{u}_2+\overset{o}{L}_{33}\overset{o}{u}_3=-\frac{\rho(1-\nu^2)}{E}(\frac{\partial g}{\partial t})^2\overset{o}{u}_3$	$\lambda=\mu=-1$	0
$a=\frac{1}{2}$	$\lambda=\mu=\nu-1$	$\overset{o}{L}_{11}\overset{o}{u}_1+\overset{o}{L}_{12}\overset{o}{u}_2+\overset{o}{L}_{13}\overset{o}{u}_3=0$ $\overset{o}{L}_{21}\overset{o}{u}_1+\overset{o}{L}_{22}\overset{o}{u}_2+\overset{o}{L}_{23}\overset{o}{u}_3=0$ $\overset{o}{L}_{31}\overset{o}{u}_1+\overset{o}{L}_{32}\overset{o}{u}_2+(\overset{o}{L}_{33}+\overset{o}{N}_{33})\overset{o}{u}_3=-\frac{\rho(1-\nu^2)}{E}(\frac{\partial g}{\partial t})^2\overset{o}{u}_3$	$\lambda=\mu=-1$	0
$\frac{1}{2}<a<1$	$\lambda=\mu=\nu-1$	$\overset{o}{L}_{11}\overset{o}{u}_1+\overset{o}{L}_{12}\overset{o}{u}_2+\overset{o}{L}_{13}\overset{o}{u}_3=0$ $\overset{o}{L}_{21}\overset{o}{u}_1+\overset{o}{L}_{22}\overset{o}{u}_2+\overset{o}{L}_{23}\overset{o}{u}_3=0$ $\overset{o}{N}_{33}\overset{o}{u}_3=-\frac{\rho(1-\nu^2)}{E}(\frac{\partial g}{\partial t})^2\overset{o}{u}_3$	$\lambda=\mu=-1$	0
$0<a<\frac{1}{2}$	$\lambda=\mu=\nu+1$	$\overset{o}{L}_{11}\overset{o}{u}_1+\overset{o}{L}_{12}\overset{o}{u}_2=-\frac{\rho(1-\nu^2)}{E}(\frac{\partial g}{\partial t})^2\overset{o}{u}_1$ $\overset{o}{L}_{21}\overset{o}{u}_1+\overset{o}{L}_{22}\overset{o}{u}_2=-\frac{\rho(1-\nu^2)}{E}(\frac{\partial g}{\partial t})^2\overset{o}{u}_2$ $\overset{o}{L}_{31}\overset{o}{u}_1+\overset{o}{L}_{32}\overset{o}{u}_2=-\frac{\rho(1-\nu^2)}{E}(\frac{\partial g}{\partial t})^2\overset{o}{u}_3$	$\lambda=\mu=1$	1
$a=\frac{1}{2}$	$\lambda=\mu=\nu+1$	$\overset{o}{L}_{11}\overset{o}{u}_1+\overset{o}{L}_{12}\overset{o}{u}_2=-\frac{\rho(1-\nu^2)}{E}(\frac{\partial g}{\partial t})^2\overset{o}{u}_1$ $\overset{o}{L}_{21}\overset{o}{u}_1+\overset{o}{L}_{22}\overset{o}{u}_2=-\frac{\rho(1-\nu^2)}{E}(\frac{\partial g}{\partial t})^2\overset{o}{u}_2$ $\overset{o}{L}_{31}\overset{o}{u}_1+\overset{o}{L}_{32}\overset{o}{u}_2=-\frac{\rho(1-\nu^2)}{E}(\frac{\partial g}{\partial t})^2\overset{o}{u}_3$	$\lambda=\mu=1$	1
$\frac{1}{2}<a<1$	$\lambda=\mu=\nu+3-\frac{2}{a}$ $(\lambda=\mu=\nu)$	$\overset{o}{L}_{11}\overset{o}{u}_1+\overset{o}{L}_{12}\overset{o}{u}_2=0$ $\overset{o}{L}_{21}\overset{o}{u}_1+\overset{o}{L}_{22}\overset{o}{u}_2=0$ $\overset{o}{L}_{31}\overset{o}{u}_1+\overset{o}{L}_{32}\overset{o}{u}_2+\overset{o}{N}_{33}\overset{o}{u}_3=-\frac{\rho(1-\nu^2)}{E}(\frac{\partial g}{\partial t})^2\overset{o}{u}_3$	$\lambda=\mu=0$	$\frac{1}{2}$

$g(t;k) = (\frac{h}{R})^{ab}$ $i\omega t$, and hence $(\frac{\partial g}{\partial t})^2 = -\omega^2 (\frac{h}{R})^{ab}$. The first two equations given in the third column of the Table for each case, always uncouple and furnish the relation between $\overset{o}{u}_1$ and $\overset{o}{u}_2$ (the third equation can then be used for $\overset{o}{u}_3$). They correspond formally to the equations of plane elasticity. The integrals obtained this way may then be called[59] *planar integrals*. The corresponding vibrations may also be qualified[59] as *quasi-tangential*, in order to stress the fact that in such vibrations the tangential strains of the middle surface mainly occur because of the displacements U_1 and U_2. Since the function g is of slow variation, it is clear that, in the cases belonging to this category, ω takes rather high values.

The classification given here is akin to the one given by A. L. Gol'denveizer[59]. However, he is considering continuous ranges for b, while we limit here for simplicity the discussion to values 0, 1 and $\frac{1}{2}$ of b; besides, his b corresponds to our ab.

Before closing this Chapter, two remarks are in place:

Firstly, it is to be noted that in the preceding discussion it has been assumed that the coordinate curves on the surface do not coincide with a "contour line" of $f(\alpha,\beta)$. Otherwise, (I.3.7) would fail to hold. It is therefore expedient to have some information as to the "contour lines" of $f(\alpha,\beta)$. We briefly discuss this question for quasi-transverse vibrations. The function of variability $f(\alpha,\beta)$ must be such as to make equal to zero the determinant of the relations (homogeneous linear equations) appearing in the third column of the Table, for the components of $\overset{o}{\underset{\sim}{u}}$. We shall deal later specifically with the determinant for quasi-transverse vibrations with $b = 0$, to obtain the *frequency equation* which we shall use on several occasions. But, here, we anticipate by giving the results obtained by calculating the determinant when a coordinate line, say a β line, is a contour line, for quasi-transverse vibrations, in the case of membrane integrals,

$$\frac{1}{R_2'^2} = \frac{\rho}{E}\,\omega^2 \tag{I.4.1}$$

in the case of bending-planar integrals,

$$\frac{1}{R_2'^2}+\frac{R^2}{3}\,\frac{1}{(1-\nu^2)}\,\frac{1}{A^4}\left(\frac{\partial f}{\partial\alpha}\right)^4=\frac{\rho}{E}\,\omega^2 \tag{I.4.2}$$

and in the case of bending integrals,

$$\left(\frac{1}{A}\frac{\partial f}{\partial\alpha}\right)^4=\frac{3\rho(1-\nu^2)}{R^2}\;\omega^2 \tag{I.4.3}$$

For low variability (i.e. $a<\frac{1}{2}$), (I.4.1) shows that $\frac{\partial f}{\partial\alpha}$, i.e. the dependence of f on α remains indeterminate. In turn, we can conclude that the β lines are, in this case, lines along all of which the curvature has constant value $\pm\sqrt{\frac{\rho}{E}}\,\omega$ (for a shell with given ρ and E), since ω is the same for the whole shell.

For high variability (i.e. $a>\frac{1}{2}$), (I.4.3) constitutes a differential equation which we can solve for $f(\alpha,\beta)$; e.g.,

$$f=(1)^{\frac{1}{4}}\sqrt[4]{\frac{3\rho(1-\nu^2)}{R^2}\,\omega^2}\int_{\alpha_0}^{\alpha}A(\alpha')d\alpha' \tag{I.4.4}$$

a relation which can be valid only if A also does not depend on β.

The second remark relates to the variability index. We have adopted a single index of variability a in both variables α and β. But there is no compelling reason (other than simplicity) for doing so, and we could take instead of the factor $e^{kf(\alpha,\beta)}$ in the solution (I.3.1), a factor of the form $e^{k\phi(\alpha)+k^d\psi(\beta)}$, with $d\neq 1$, say $0<d<1$, to fix ideas. Then, we would have

$$\frac{\partial U}{\partial\alpha}\approx k\,\frac{\partial\phi}{\partial\alpha}\,U \qquad \frac{\partial U}{\partial\beta}\approx k^d\,\frac{\partial\psi}{\partial\beta}\,U \tag{I.4.5}$$

Then, there may appear integrals other than those we have found (and for which $\lambda=\mu$)[55].

Finally, we note that as a approaches 0, the tangential inertial forces may have appreciable effect, even for $b=0$.

CHAPTER II
THE CENTRAL SET OF EQUATIONS

The results of Chapter I, permit to write a reduced set of differential equations for each case given in the Table, separately. Here, we shall do this in the case $a = \frac{1}{2}$, $b = 0$, and we will indicate that the set thus obtained has actually a wider domain of applicability and occupies a *central* place in the study of transverse vibrations of thin elastic shells. This set shall therefore be called the *central set of equations*. We also derive the *frequency equation* corresponding to this set.

1. The central set of equations:

We write at once the set of equations

$$-\frac{1}{A^2}\frac{\partial^2 U_1}{\partial\alpha^2} - \frac{1-\nu}{2}\frac{1}{B^2}\frac{\partial^2 U_1}{\partial\beta^2} - \frac{1+\nu}{2}\frac{1}{AB}\frac{\partial^2 U_2}{\partial\alpha\partial\beta} + \left(\frac{1}{R_1'} + \frac{\nu}{R_2'}\right)\frac{1}{A}\frac{\partial U_3}{\partial\alpha} - \frac{1-\nu}{R_{12}}\frac{1}{B}\frac{\partial U_3}{\partial\beta} = 0$$

$$-\frac{1+\nu}{2}\frac{1}{AB}\frac{\partial^2 U_1}{\partial\alpha\partial\beta} - \frac{1}{B^2}\frac{\partial^2 U_2}{\partial\beta^2} - \frac{1-\nu}{2}\frac{1}{A^2}\frac{\partial^2 U_2}{\partial\alpha^2} + \left(\frac{\nu}{R_1'} + \frac{1}{R_2'}\right)\frac{1}{B}\frac{\partial U_3}{\partial\beta} - \frac{1-\nu}{R_{12}}\frac{1}{A}\frac{\partial U_3}{\partial\alpha} = 0 \qquad \text{(II.1.1)}$$

$$-\left(\frac{1}{R_1'} + \frac{\nu}{R_2'}\right)\frac{1}{A}\frac{\partial U_1}{\partial\alpha} - \left(\frac{\nu}{R_1'} + \frac{1}{R_2'}\right)\frac{1}{B}\frac{\partial U_2}{\partial\beta} + \frac{1-\nu}{R_{12}}\frac{1}{B}\frac{\partial U_1}{\partial\beta} + \frac{1-\nu}{R_{12}}\frac{1}{A}\frac{\partial U_2}{\partial\alpha} + \left[\left(\frac{1}{R_1} + \frac{1}{R_2}\right)^2 - \frac{2(1-\nu)}{R_1R_2}\right]U_3 + h^{*2}\frac{R^2}{3}\left[\frac{1}{A^2}\frac{\partial^2}{\partial\alpha^2} + \frac{1}{B^2}\frac{\partial^2}{\partial\beta^2}\right]^2 U_3 = -\frac{\rho(1-\nu^2)}{E}\frac{\partial^2 U_3}{\partial t^2}$$

simply by retaining in the operator (0.5.1) only those terms which have their counterpart in the equations given in the third column of the Table of Chapter I, for the case $a = \frac{1}{2}$, $b = 0$.

The differential operator thus obtained is suitable for a rather wide range of problems in the case of statics, as it has been discussed at some length by A. L. Gol'denveizer.[53] The system (II.1.1) contains actually all terms which appear in one of the three cases $0 < a < \frac{1}{2}$, $a = \frac{1}{2}$, $\frac{1}{2} < a < 1$ (for $b = 0$), but terms which are irrelevant for the cases $0 < a < \frac{1}{2}$ or $\frac{1}{2} < a < 1$ will not contribute significantly in these latter cases. Thus it is to be expected that this system will prove efficient in all cases of quasi-transverse vibrations.

We record here, also the approximate expressions for the kinematic quantities in quasi-transverse vibrations:

$$\varepsilon_1 = \frac{1}{A}\frac{\partial U_1}{\partial \alpha} + \frac{1}{AB}\frac{\partial A}{\partial \beta}U_2 - \frac{U_3}{R_1'}$$

$$\bar{\omega} = \frac{A}{B}\frac{\partial}{\partial \beta}\left(\frac{U_1}{A}\right) + \frac{B}{A}\frac{\partial}{\partial \alpha}\left(\frac{U_2}{B}\right) + \frac{2U_3}{R_{12}} \qquad \text{(II.1.2)}$$

$$\varepsilon_2 = \frac{1}{B}\frac{\partial U_2}{\partial \beta} + \frac{1}{AB}\frac{\partial B}{\partial \alpha}U_1 - \frac{U_3}{R_2'}$$

$$\varkappa_1 = \frac{1}{A}\frac{\partial}{\partial \alpha}\frac{1}{A}\frac{\partial U_3}{\partial \alpha} + \frac{1}{AB}\frac{\partial A}{\partial \beta}\frac{1}{B}\frac{\partial U_3}{\partial \beta}$$

$$\tau = \frac{1}{A}\frac{\partial}{\partial \alpha}\frac{1}{B}\frac{\partial U_3}{\partial \beta} - \frac{1}{AB}\frac{A}{\partial \beta}\frac{1}{A}\frac{\partial U_3}{\partial \alpha} + \left(\frac{1}{R_1} + \frac{1}{R_2}\right)\frac{1}{R_{12}}U_3 \qquad \text{(II.1.3)}$$

$$\varkappa_2 = \frac{1}{B}\frac{\partial}{\partial \beta}\frac{1}{B}\frac{\partial U_3}{\partial \beta} + \frac{1}{AB}\frac{\partial B}{\partial \alpha}\frac{\partial U_3}{\partial \alpha}$$

2. The frequency equation:

An estimate for ω^2, for quasi-transverse vibrations can be obtained by taking out of (I.3.11) the terms which do not have counterparts in the central set of equations (II.1.1), and setting equal to zero the determinant of the homogeneous system of linear equations (for $\overset{\circ}{u}_1$, $\overset{\circ}{u}_2$, $\overset{\circ}{u}_3$)

thus obtained, without assigning a specific value to a:

$$\begin{vmatrix} L^o_{11} & L^o_{12} & L^o_{13} \\ L^o_{21} & L^o_{13} & L^o_{23} \\ L^o_{31} & L^o_{32} & L^o_{33}+N_{33}\, k^{4-\frac{2}{a}} - \rho\,\frac{1-\nu^2}{E}\,\omega^2 \end{vmatrix} = 0 \qquad \text{(II.2.1)}$$

This is the *frequency equation*. Expanding the determinant, we obtain

$$\begin{aligned} &\frac{1}{R'^2_2}\left[\frac{1}{A^4}\left(\frac{\partial f}{\partial\alpha}\right)^4\right] + \frac{4}{R_{12}}\frac{1}{R'_2}\left[\frac{1}{A^3}\left(\frac{\partial f}{\partial\alpha}\right)^3\right]\left[\frac{1}{B}\left(\frac{\partial f}{\partial\beta}\right)\right] + \\ &+ 2\left(\frac{1}{R'_1R'_2}+\frac{1}{R^2_{12}}\right)\left[\frac{1}{A^2}\left(\frac{\partial f}{\partial\alpha}\right)^2\right]\left[\frac{1}{B^2}\left(\frac{\partial f}{\partial\beta}\right)^2\right] + \\ &+ \frac{4}{R_{12}}\frac{1}{R'_1}\left[\frac{1}{A}\left(\frac{\partial f}{\partial\alpha}\right)\right]\left[\frac{1}{B^3}\left(\frac{\partial f}{\partial\beta}\right)^3\right] + \frac{1}{R'^2_1}\left[\frac{1}{B^4}\left(\frac{\partial f}{\partial\beta}\right)^4\right] + \\ &+ \frac{R^2}{3}\frac{1}{1-\nu^2}k^{4-\frac{2}{a}}\left\{\left[\frac{1}{A^2}\left(\frac{\partial f}{\partial\alpha}\right)^2\right]+\left[\frac{1}{B^2}\left(\frac{\partial f}{\partial\beta}\right)^2\right]\right\}^4 = \frac{\rho}{E}\omega^2\left\{\left[\frac{1}{A^2}\left(\frac{\partial f}{\partial\alpha}\right)^2\right]+\left[\frac{1}{B^2}\left(\frac{\partial f}{\partial\beta}\right)^2\right]\right\}^2 \end{aligned} \qquad \text{(II.2.2)}$$

This equation covers all three cases given in the Table of Chapter I, for $b = 0$.

If the coordinate lines on the middle surface are lines of curvature, the frequency equation takes the simpler form

$$\begin{aligned} &\left[\frac{1}{R_2}\left(\frac{1}{A}\frac{\partial f}{\partial\alpha}\right)^2 + \frac{1}{R_1}\left(\frac{1}{B}\frac{\partial f}{\partial\beta}\right)^2\right]^2 + \\ &+ \frac{R^2}{3(1-\nu^2)}k^{4-\frac{2}{a}}\left[\left(\frac{1}{A}\frac{\partial f}{\partial\alpha}\right)^2 + \left(\frac{1}{B}\frac{\partial f}{\partial\beta}\right)^2\right]^4 = \frac{\rho}{E}\omega^2\left[\left(\frac{1}{A}\frac{\partial f}{\partial\alpha}\right)^2 + \left(\frac{1}{B}\frac{\partial f}{\partial\beta}\right)^2\right]^2 \end{aligned} \qquad \text{(II.2.3)}$$

We shall use this equation later (see Chapter IV), in order to obtain estimates for ω, when we are to discuss the density of nodal lines, in the cases $0 < a < \frac{1}{2}$, $a = \frac{1}{2}$, and $\frac{1}{2} < a < 1$.

If $a = \frac{1}{2}$, then (II.2.2) reduces to

$$
\begin{aligned}
&\frac{1}{R_2'^2}\left(\frac{1}{A}\frac{\partial f}{\partial \alpha}\right)^4 + \frac{4}{R_{12}}\frac{1}{R_2'}\left(\frac{1}{A}\frac{\partial f}{\partial \alpha}\right)^3\left(\frac{1}{B}\frac{\partial f}{\partial \beta}\right) + \\
&+ 2\left(\frac{1}{R_1'R_2'} + \frac{1}{R_{12}^2}\right)\left(\frac{1}{A}\frac{\partial f}{\partial \alpha}\right)^2\left(\frac{1}{B}\frac{\partial f}{\partial \beta}\right)^2 + \\
&+ \frac{4}{R_{12}}\frac{1}{R_1'}\left(\frac{1}{A}\frac{\partial f}{\partial \alpha}\right)\left(\frac{1}{B}\frac{\partial f}{\partial \beta}\right)^3 + \frac{1}{R_1'^2}\left(\frac{1}{B}\frac{\partial f}{\partial \beta}\right)^4 + \\
&+ \frac{R^2}{3}\frac{1}{1-\nu^2}\left[\left(\frac{1}{A}\frac{\partial f}{\partial \alpha}\right)^2 + \left(\frac{1}{B}\frac{\partial f}{\partial \beta}\right)^2\right]^4 = \frac{\rho}{E}\omega^2\left[\left(\frac{1}{A}\frac{\partial f}{\partial \alpha}\right)^2 + \left(\frac{1}{B}\frac{\partial f}{\partial \beta}\right)^2\right]^2
\end{aligned}
\tag{II.2.4}
$$

and (II.2.3) becomes

$$
\begin{aligned}
&\left[\frac{1}{R_2}\left(\frac{1}{A}\frac{\partial f}{\partial \alpha}\right)^2 + \frac{1}{R_1}\left(\frac{1}{B}\frac{\partial f}{\partial \beta}\right)^2\right]^2 + \\
&+ \frac{R^2}{3(1-\nu^2)}\left[\left(\frac{1}{A}\frac{\partial f}{\partial \alpha}\right)^2 + \left(\frac{1}{B}\frac{\partial f}{\partial \beta}\right)^2\right]^4 = \frac{\rho}{E}\omega^2\left[\left(\frac{1}{A}\frac{\partial f}{\partial \alpha}\right)^2 + \left(\frac{1}{B}\frac{\partial f}{\partial \beta}\right)^2\right]^2
\end{aligned}
\tag{II.2.5}
$$

When ω is given, the frequency equation provides a relation (a partial differential equation) between $\frac{\partial f}{\partial \alpha}$ and $\frac{\partial f}{\partial \beta}$, which can be used in constructing the function of variability. The asymptotic integration can then be carried through by the known method. We shall however not dwell on this asymptotic integration.

CHAPTER III
EDGE EFFECT

The solutions introduced in Chapter I do not, in general, satisfy the boundary conditions. It turns out however, that at least under the assumptions of validity of the central set of equations, in a sufficiently narrow band along the boundary, expressions can be constructed which serve to match the boundary conditions and reduce to the asymptotic solutions previously found, when receding into the inner region. These expressions together represent the *edge effect*. The edge effect in the dynamics of shells has been investigated by V. V. Bolotin.[62] *Vibration zones* also can be studied by systematically using the frequency equation. Such zones have been investigated by V. N. Moskalenko,[63] in the particular case of shells of revolution.

1. Assumptions concerning geometric quantities:

As noticed above, the edge effect shall be considered here, in those cases in which it remains confined to a sufficiently narrow band along the boundary. Nonetheless, the band shall be sufficiently wide to locate the edge effect. Thus, it will be reasonable to think that the geometric quantities such as A, B, R_1' and R_2' (hence R_{12}) remain approximately constant in that region.

2. The frequency equation with constant coefficients:

For the purposes of the present Chapter, we try a function of variability of the type

$$f(\alpha,\beta) = \chi\alpha + \eta\beta \tag{III.2.1}$$

where χ and η are *constants* to be determined.

Thus

$$\frac{\partial f}{\partial \alpha} = \chi \qquad \frac{\partial f}{\partial \beta} = \eta \tag{III.2.2}$$

We also introduce tentative solutions

$$U_1 = k^{-1} u_1 e^{kf}, \quad U_2 = k^{-1} u_2 e^{kf}, \qquad U_3 = u_3 e^{kf} \tag{III.2.3}$$

where the u_i are considered to remain constant.

Then, the frequency equation (II.2.4) takes the form

$$\left(\frac{1}{R'_2}\xi^2 + \frac{1}{R'_1}\zeta^2\right)^2 + \frac{4}{R_{12}}\left(\frac{1}{R'_2}\xi^2 + \frac{1}{R'_1}\zeta^2\right)\xi\zeta + \frac{2}{R_{12}^2}\xi^2\zeta^2 + \frac{R^2}{3(1-\nu^2)}(\xi^2+\zeta^2)^4 = \frac{\rho}{E}\omega^2(\xi^2+\zeta^2)^2 \tag{III.2.4}$$

with

$$\xi \equiv \frac{\chi}{A} \qquad \zeta \equiv \frac{\eta}{B} \tag{III.2.5}$$

If in particular, the coordinate curves coincide with the lines of curvature, eq. (III.2.4) reduces to

$$\left(\frac{1}{R}\xi^2 + \frac{1}{R}\zeta^2\right)^2 + \frac{R^2}{3(1-\nu^2)}(\xi^2+\zeta^2)^4 = \frac{\rho}{E}\omega^2(\xi^2+\zeta^2)^2 \tag{III.2.6}$$

and the frequency equation may be rewritten in the simple form

$$\omega^2 = \frac{E}{\rho}\left|\frac{R^2}{3(1-\nu^2)}(\xi^2+\zeta^2)^2 + \frac{\left(\frac{\xi^2}{R_2}+\frac{\zeta^2}{R_1}\right)^2}{(\xi^2+\zeta^2)^2}\right| \tag{III.2.7}$$

3. Governing equation for the edge effect:

The frequency equation (III.2.4), or its simpler version (III.2.6), provides an algebraic relation between ξ and ζ , when ω is given, and is symmetric in both these quantities.

Now, in a given vibration motion, ω is known, and is the same over the whole shell; i.e. in the inner region as well as on the boundary strip. It corresponds to a pattern of standing waves which, within the geometric approximated by sine waves, corresponding to particular values $\xi^2 = -(\frac{k_1}{k})^2$ and $\zeta^2 = -(\frac{k_2}{k})^2$, satisfying (III.2.4), or (III.2.6).

We take the boundary to be clamped or hinged. The boundary curve is then a *nodal line*, i.e. a line with all its points at rest during the motion. We assume, that this part of the boundary coincides with one of the coordinate curves, say α = const. Then, k_2 is determined completely.

We shall moreover assume, for the sake of simplicity, that the boundary in question coincides with a line of curvature, so that we can use the simpler eq. (III.2.6). But, otherwise, the procedure of calculating the edge effect remains the same, for the more general case of (III.2.4).

Thus, we confine our attention to eq. (III.2.6), which yields

$$\omega^2 = \frac{E}{\rho}\left|\frac{h^2}{3(1-\nu^2)}(k_1^2+k_2^2)^2 + \frac{\left(\frac{k_1^2}{R_2}+\frac{k_2^2}{R_1}\right)^2}{(k_1^2+k_2^2)^2}\right| \tag{III.3.1}$$

The frequency equation takes now the form

$$\frac{h^2}{3(1-\nu^2)}(\hat{\xi}^2 - k_2^2)^4 + \left(\frac{\hat{\xi}^2}{R_2} - \frac{k_2^2}{R_1}\right)^2 - K(\hat{\xi}^2 - k_2^2)^2 = 0 \tag{III.3.2}$$

where we have introduced

$$\hat{\xi} \equiv k\xi \qquad \hat{\zeta} \equiv k\zeta \tag{III.3.3}$$

and

$$K \equiv \frac{h^2}{3(1-\nu^2)}(k_1^2 + k_2^2)^2 + \frac{(\frac{k_1^2}{R_2} + \frac{k_2^2}{R_1})^2}{(k_1^2 + k_2^2)^2} \tag{III.3.4}$$

Alternatively, the left side of eq. (III.3.2) can be rearranged in the form of a polynomial of eigth degree in $\hat{\xi}$:

$$\hat{\xi}^8 - 4k_2^2\,\hat{\xi}^6 + \left[6k_2^4 + \frac{3(1-\nu^2)}{h^2}\frac{1}{R_2^2} - \frac{3(1-\nu^2)}{h^2}K\right]\hat{\xi}^4 -$$
$$- \left[4k_2^6 + \frac{6(1-\nu^2)}{h^2}\frac{k_2^2}{R_1R_2} - \frac{6(1-\nu^2)}{h^2}K k_2^2\right]\hat{\xi}^2 + \tag{III.3.5}$$
$$+ \left[k_2^8 + \frac{3(1-\nu^2)}{h^2}\frac{1}{R_1^2}k_2^4 - \frac{3(1-\nu^2)}{h^2}K k_2^4\right] = 0$$

We know that k_1^2 corresponds to two pure imaginary roots $\xi = \pm ik_1$. The polynomial on the left side of eq. (III.3.5) is divisible therefore by $(\hat{\xi}^2 + k_1^2)$, and we are left with the sixth degree algebraic equation

$$\hat{\xi}^6 - (k_1^2 + 4k_2^2)\hat{\xi}^4 +$$
$$+ \left[k_1^4 + 4k_1^2k_2^2 + 6k_2^4 + \frac{3(1-\nu^2)}{h^2}\frac{1}{R_2^2} - \frac{3(1-\nu^2)}{h^2}K\right]\hat{\xi}^2 +$$
$$+ \left[-k_1^6 - 4k_1^4k_2^2 - 6k_1^2k_2^4 - 4k_2^6 - \frac{3(1-\nu^2)}{h^2}\frac{k_1^2}{R_2^2} + \right. \tag{III.3.6}$$
$$\left. - \frac{6(1-\nu^2)}{h^2}\frac{k_2^2}{R_1R_2} + \frac{3(1-\nu^2)}{h^2}K k_1^2 + \frac{6(1-\nu^2)}{h^2}K k_2^2\right] = 0$$

This equation is completely equivalent to the sixth degree equation given by V. V. Bolotin,[62] and can be written also in the form

$$\hat{\xi}^6 - (k_1^2 + 4k_2^2)\,\hat{\xi}^4 +$$
$$+ k_2^2\,\{2k_1^2 + 5k_2^2 - \frac{3(1-\nu^2)}{h^2}\frac{1}{R_1^2}\,\frac{2k\,(\frac{R_1}{R_2})\left[1-(\frac{R_1}{R_2})\right] + k_2^2\left[1-(\frac{R_1}{R_2})^2\right]}{(k_1^2 + k_2^2)^2}\}\,\hat{\xi}^2 -$$

(III.3.7)

$$- k_2^4 \{k_1^2 + 2k_2^2 - \frac{3(1-\nu^2)}{h^2} \frac{1}{R_1^2} \frac{k_1^2[1-(\frac{R_1}{R_2})^2] + 2k_2^2[1-(\frac{R_1}{R_2})]}{(k_1^2 + k_2^2)^2}\} = 0$$

4. Classification of the edge effects:

We adopt the classification given by V. V. Bolotin,[62] which can be based on the discussion of the nature of the roots of eq. (III.3.6) viewed as an equation of third degree for $\hat{\xi}^2$.

1°) All three roots are distinct, real and positive. There will correspond three distinct, real, negative roots $(\hat{\xi}_3, \hat{\xi}_4, \hat{\xi}_5)$, and three distinct, real, positive roots $(\hat{\xi}_6, \hat{\xi}_7, \hat{\xi}_8)$. The latter, do not give a decaying solution and shell be discarded. The general solution of the approximate system of differential equations, with constant coefficients is thus, e.g. for the transversal component,

$$U_3 = u_3 \Big[C_1 \sin (\frac{k_1}{k} \frac{\alpha}{A^{-1}}) + C_2 \cos (\frac{k_1}{k} \frac{\alpha}{A^{-1}}) + \qquad \text{(III.4.1)}$$

$$+ C_3 e^{-\hat{\xi}_3 \frac{\alpha}{A^{-1}} \frac{1}{k}} + C_4 e^{-\hat{\xi}_4 \frac{\alpha}{A^{-1}} \frac{1}{k}} + C_5 e^{-\hat{\xi}_5 \frac{\alpha}{A^{-1}} \frac{1}{k}} \Big]$$

All boundary conditions can be satisfied by adjusting four integration constants, while the remaining one serves as a normalizing factor. This type of edge effect is said to be *nonoscillatory*.

2°) The roots $(\hat{\xi}^2)$ are again all real and positive, but they are not all distinct.

The edge effect remains nonoscillatory in this case too. If, for instance, there happens to be one double root $\hat{\xi}_4 = \hat{\xi}_5$, then the solution becomes, e.g. for the transversal component,

$$U_3 = u_3 \Big[C_1 \sin (\frac{k_1}{k} \frac{\alpha}{A^{-1}}) + C_2 \cos (\frac{k_1}{k} \frac{\alpha}{A^{-1}}) + C_3 e^{-\frac{\hat{\xi}_3}{k} \frac{\alpha}{A^{-1}}} + \qquad \text{(III.4.2)}$$

$$+ C_4 e^{-\frac{\hat{\xi}_4}{k} \frac{\alpha}{A^{-1}}} + C_5 e^{-\frac{\hat{\xi}_4}{k} \frac{\alpha}{A^{-1}}} (\frac{\hat{\xi}_4}{k} \frac{\alpha}{A^{-1}})\Big]$$

3°) There is one real positive root and two complex conjugate roots ($\hat{\xi}^2$). Then, $\hat{\xi}_3, \hat{\xi}_4, \hat{\xi}_5$ will be obtained, say, in the form $\hat{\xi}_3 = -\delta_3$, $\hat{\xi}_4 = -\delta_4 + i\gamma_4$, $\hat{\xi}_5 = -\delta_4 - i\gamma_4$ and the solution, e.g. for the transversal components, becomes

$$U_3 = u_3 \Big[C_1 \sin\Big(\frac{k_1}{k}\frac{\alpha}{A^{-1}}\Big) + C_2 \cos\Big(\frac{k_1}{k}\frac{\alpha}{A^{-1}}\Big) + C_3\, e^{-\frac{\delta_3}{k}\frac{\alpha}{A^{-1}}} + C_4\, e^{-\frac{\delta_4}{k}\frac{\alpha}{A^{-1}}} \sin\Big(\frac{\gamma_4}{k}\frac{\alpha}{A^{-1}}\Big) + C_5\, e^{-\frac{\delta_4}{k}\frac{\alpha}{A^{-1}}} \cos\Big(\frac{\gamma_4}{k}\frac{\alpha}{A^{-1}}\Big)\Big] \qquad \text{(III.4.3)}$$

In such a case, the edge effect is called *oscillatory*.

4°) In all other cases, there is no adequate solution to be found, which satisfies the boundary conditions and tends to the solution in the inner region. The edge effect is then said to be *degenerate*.

The discussion above shows that in order to have a non-degenerate edge effect, the algebraic equation (III 3.6), or its equivalent version (III.3.7), must have positive or complex conjugate roots. According to a theorem of Descartes, this occurs whenever the coefficients of the consecutive terms in this equation alternate in sign. Now, it can easily be shown that the expression in braces in the coefficient of $\hat{\xi}^2$ remains always superior in value to the expression in braces in the term of degree zero. Thus, the requirement is

$$k_1^2 + 2k_2^2 - \frac{3(1-\nu^2)}{h^2}\,\frac{1}{R_1^2}\,\frac{k_1^2\Big[1-\Big(\frac{R_1}{R_2}\Big)^2\Big] + 2k_2^2\Big[1-\Big(\frac{R_1}{R_2}\Big)\Big]}{(k_1^2+k_2^2)^2} > 0 \qquad \text{(III.4.3)}$$

This requirement is certainly fulfilled if $R_1 \geq R_2$. In particular, this holds in shells of zero Gaussian curvature, near a non-asymptotic edge. If $R_1 < R_2$, the problem requires a closer scrutiny. For further details, we refer to the paper[62] by V. V. Bolotin.

A more recent investigation of the edge effect, going beyond the simple form of solution (III.2.1)-(III.2.3), has been given by C. N. Chernyshev.[64]

5. Vibration zones:

We consider here again the frequency equation (II.2.4) or, in the case of coordinate curves coinciding with the lines of curvature, (II.2.5), and assume $\frac{\partial f}{\partial \beta}$, say, to be known over the middle surface. Then, for given ω, eq. (II.2.4), or (II.2.5), can be used as a fourth degree algebraic equation to determine $\frac{\partial f(\alpha,\beta)}{\partial \alpha}$ at each point of the middle surface and the solution can be used as a differential equation to determine $f(\alpha,\beta)$. According to the nature of this solution, it will be possible to distinguish between *vibration zones* of different types. We will discuss here only one type of such a zone.

We put

$$\left(\frac{1}{A}\frac{\partial f}{\partial \alpha}\right)^2 \equiv x \qquad \left(\frac{1}{B}\frac{\partial f}{\partial \beta}\right)^2 \equiv c \tag{III.5.1}$$

According to our assumption above, $c(\alpha,\beta)$ is known everywhere, while x is unknown. Further, we assume $\frac{\partial f}{\partial \beta}$ to be pure imaginary; i.e., $c < 0$.

If the coordinate curves coincide with the lines of curvature, we obtain from (II.2.5),

$$\left(\frac{x}{R_2} + \frac{c}{R_1}\right)^2 + \frac{R^2}{3(1-\nu^2)}(x+c)^4 = (\rho/E)\omega^2(x+c)^2 \tag{III.5.2}$$

or, rearranging to have a polynomial of fourth degree as left side member,

$$\frac{R^2}{3(1-\nu^2)}x^4 + 4\frac{R^2}{3(1-\nu^2)}cx^3 +$$

$$+\left(\frac{1}{R_2^2} + 6\frac{R^2}{3(1-\nu^2)}c^2 - \frac{\rho}{E}\omega^2\right)x^2 +$$

$$+\left(\frac{2}{R_1R_2}+4\frac{R^2}{3(1-\nu^2)}c^2-\frac{2\rho}{E}\omega^2\right)cx+$$

(III.5.3)

$$+\left(\frac{1}{R_1^2}+\frac{R^2}{3(1-\nu^2)}c^2-\frac{\rho}{E}\omega^2\right)c^2=0$$

If all four roots of this equation are real and positive, then there is no oscillating solution along α lines. The corresponding solution is of the edge effect type and the part of the shell consisting of points corresponding to this solution behaves like an elastic frame for the rest of the shell. Such a subregion is called *zone of zero type* (*).[63]

To obtain criteria for the existence of a zone of this type, we invoke here again a corollary of Descartes' theorem, and notice that the equation (III.5.3) is complete in the sense that it contains terms of all degrees up to the highest.Then, for such an equation to have all its roots real and positive, it is necessary (but not sufficient) that the coefficients of consecutive terms be alternating in sign. Applied to eq. (III.5.3), this theorem yields the necessary conditions

$$\frac{1}{R_2^2}+6\frac{R^2}{3(1-\nu^2)}c^2>\frac{\rho}{}\omega^2$$

$$\frac{1}{R_1R_2}+2\frac{R^2}{3(1-\nu^2)}c^2>\frac{\rho}{E}\omega^2 \qquad \text{(III.5.4)}$$

$$\frac{1}{R_1^2}\quad\frac{R^2}{3(1-\nu^2)}c^2>\frac{\rho}{E}\omega^2$$

noting that c is a negative quantity, by hypothesis.

These inequalities impose upper bounds for the frequency, respectively, lower bounds for the curvatures.

The second of these inequalities can be rewritten as

$$K>\frac{\rho}{E}\omega^2-2\frac{R^2}{3(1-\nu^2)}c^2 \qquad \text{(III.5.5)}$$

(*)The function f in [63] differs by a factor i.

where $\mathbb{K}$ stands for the total (Gaussian) curvature $\frac{1}{R_1 R_2}$.

On the other hand, addition of all three inequalities (III.5.4), after multiplication of the second inequality by 2, yields

$$\mathbb{B}^2 > 3\,\frac{\rho}{E}\,\omega^2 - 11\,\frac{R^2}{3(1-\nu^2)}\,c^2 \qquad \text{(III.5.6)}$$

where $\mathbb{B}$ stands for the mean curvature $\frac{1}{R_1} + \frac{1}{R_2}$.

CHAPTER IV

NODAL LINES

The purpose of this short Chapter is twofold: to study the trend of the density of nodal lines and to show an application of group theory when symmetries exist.

1. Density of nodal lines:

The key equation is here too the frequency equation, (II.2.2) or its simpler version (II.2.3). We immediately find the following estimates for ω:

$$\frac{\rho}{E}\omega^2 = k^o \frac{\left[\frac{1}{R_2}\left(\frac{1}{A}\frac{\partial f}{\partial \alpha}\right)^2 + \frac{1}{R_1}\left(\frac{1}{B}\frac{\partial f}{\partial \beta}\right)^2\right]^2}{\left[\left(\frac{1}{A}\frac{\partial f}{\partial \alpha}\right)^2 + \left(\frac{1}{B}\frac{\partial f}{\partial \beta}\right)^2\right]^2} \qquad (0 < a < \tfrac{1}{2})$$

$$\frac{\rho}{E}\omega^2 = k^o \frac{\left[\frac{1}{R_2}\left(\frac{1}{A}\frac{\partial f}{\partial \alpha}\right)^2 + \frac{1}{R_1}\left(\frac{1}{B}\frac{\partial f}{\partial \beta}\right)^2\right]^2}{\left[\left(\frac{1}{A}\frac{\partial f}{\partial \alpha}\right)^2 + \left(\frac{1}{B}\frac{\partial f}{\partial \beta}\right)^2\right]^2} + \frac{R^2}{3(1-\nu^2)}\left[\left(\frac{1}{A}\frac{\partial f}{\partial \alpha}\right)^2 + \left(\frac{1}{B}\frac{\partial f}{\partial \beta}\right)^2\right]^2 \qquad (a = \tfrac{1}{2})$$

$$\frac{\rho}{E}\omega^2 = k^{4-\frac{2}{a}} \frac{R^2}{3(1-\nu^2)}\left[\left(\frac{1}{A}\frac{\partial f}{\partial \alpha}\right)^2 + \left(\frac{1}{B}\frac{\partial f}{\partial \beta}\right)^2\right]^2 \qquad (\tfrac{1}{2} < a < 1)$$

It can be seen that, while in the interval $(0,\frac{1}{2}]$ the increase of α is not accompanied by an essential increase of ω^2, this latter increases as $k^{4-\frac{2}{a}}$ in the interval $(\frac{1}{2},1)$ of α.

The increase of the index of variability a indicates an increase in the number of nodal lines. We see that in shells, the increase of the number of nodal lines is not always accompanied by an increase of eigenvalues. In certain cases, there may even occur an *inversion*. In fact.

it has been shown[54] that, e.g. in shells of zero curvature when there is only one system of nodal lines coinciding with the asymptotic lines of the middle surface, the estimate for ω^2 can reach a minimum for $\alpha = \frac{1}{4}$, commensurate with k^{-1}.

2. Symmetry properties of the nodal pattern and group theory:

We consider, in the present Section, shells with middle surfaces which exhibit symmetries in the undeformed configuration. Our purpose is to describe a way for studying the symmetries of the displacement field $\underset{\sim}{U}$, in the case of standing waves. In this type of study, the group theory imposes itself in a quite natural way.

We recall now some definitions and basic facts of the group theory:[65]

A *representation* of a group is a matrix group (square matrices) onto which the group to be represented is homomorphic. The number of rows (or columns) in a representation matrix is called its *dimension*. If a matrix representation cannot be cast into block form by a similarity transformation (i.e. transformation under a change of basis), then it is said to be *irreducible*.

On the other hand, consider an eigenvalue problem. If more than one linearly independent eigenvector may belong to one eigenvalue, this eigenvalue is said to be *degenerate*. Clearly, every linear combination of degenerate eigenvectors is again an eigenvector with the same eigenvalue. All these combinations can be expressed as linear combinations of a linearly independent set, i.e. a basis. The vectors of this basis span under the operations of a group an invariant subspace of the space of all eigenvectors, and form thus a basis for an irreducible representation of the group.

Eigenvectors belonging to the same irreducible space belong to the same eigenvalue. The reciprocal is not always true. Two sets of eigenvectors not mixed under the symmetry operations of the mechanical system considered, may happen to have the same eigenvalue, due to dynamical peculiarities. Such cases are called *accidentally degenerate*.

We summarize now the main result, in the following statement:[66]

Provided there are no accidental degeneracies, each set of degenerate eigenvectors of the eigenvalue problem spans a vector space which transforms according to an irreducible representation of the group of symmetry operations of the system considered.

The eigenvalue problem of thin shells has been described in Chapter I, Section 1. The statement given above can efficiently be used in investigating the symmetries of nodal patterns and classifying modes of vibration. R. Perrin and T. Charnley[66] have applied group theory to classification of bell modes. They discard some possibility of accidental degeneracy, for which there is no experimental evidence in bells. The practical implication of the theorem stated above is that the modes of vibration of a shell can be classified according to the behaviour of symmetry operations of the undeformed configuration. In the particular case of a shell of revolution, a rotation through any angle about the axis of revolution is a symmetry operation, as well as its inverse. The rotation through the angle zero is the identity operation. Furthermore, taking the mirror image of the shell (more exactly, of its middle surface) in any plane containing the axis of revolution is also a symmetry operation. All these symmetry operations together form an infinite group. Specifically, in a bell, which is in general a shell of revolution possessing no other symmetries, the possible multiplicities of the vibration spectrum are equal to the dimensions of the irreducible representations of this group. Hence the classification of the modes of vibration of a bell will depend upon the pattern of nodal meridians. Different numbers of nodal circles may of course be possessed by different members of the same symmetry family. The modes with no nodal meridians divide into two families: the symmetric "breathing" modes where every parallel circle remains circular and with center on the axis of revolution but with periodically varying radius, and the antisymmetric modes which would correspond to torsional vibrations of the bell about its axis of symmetry. All other families turn out to have an even number of nodal meridians.

CHAPTER V

SPECTRAL THEOREMS

We have found in Chapter IV, Section 1, means of estimating the density of nodal lines, when ω is given. In the present Chapter we consider the *distribution* of the frequencies and their *density*.

1. The eigenvalue problem and further properties of the differential operator:

We pose the eigenvalue problem, by using the set of equations (I.1.8)

$$\sum_{j=1}^{3} (L_{ij} + h^{*2} N_{ij}) U_j = \lambda U_i \tag{V.1.1}$$

where we now omit all terms N_{ij} except N_{33}, in the light of the discussion in Chapter I, and imposing the boundary conditions

$$u_1\big|_\Gamma = u_2\big|_\Gamma = u_3\big|_\Gamma = \frac{\partial u_3}{\partial n}\bigg|_\Gamma = 0 \tag{V.1.2}$$

where the derivative in $(V.1.2)_4$ is with respect to the normal to Γ *on* G. The set (V.1.2) relates to a clamped boundary (*).

For simplicity, the coordinate curves are supposed to coincide with the lines of curvature.

For the purposes of this Chapter and of the next one, we introduce the inner product defined as

$$(v,w) = \iint_G v\,\bar{w}\,AB\,d\alpha d\beta \tag{V.1.3}$$

for functions, and

(*) If the boundary were hinged, then the derivative in $(V.1.2)_4$ should be replaced by the second order derivative.

$$(\underset{\sim}{v},\underset{\sim}{w}) = \sum_{i=1}^{3} (v_i,w_i) \tag{V.1.4}$$

for function vectors $\underset{\sim}{u} = \{v_i\}$, $\underset{\sim}{w} = \{w_i\}$, $(i = 1,2,3)$.

The product

$$J \equiv ((L \quad h^{*2}N)\, \underset{\sim}{u},\underset{\sim}{u}) \tag{V.1.5}$$

with an obvious notation for the two parts of the operator L_{ij} in (V.1.1), is called its *quadratic functional*. This product is equal to the strain energy of the shell, within a positive factor $\frac{Eh}{1-\nu^2}$, and is therefore positive for any $\underset{\sim}{u}$.

Using the theory of linear operators,[68,69] it can be shown that the operator L_{ij} (and the problem (V.1.1)-(V.1.2)) is *self-adjoint* and that it has a *positive point spectrum*. We shall not enter here into operator theoretical refinements and proofs; but, we give a few definitions which we shall need in what follows.

A self-adjoint operator (operating on a Hilbert space) A, is said to have no *point spectrum* if for each scalar λ, the equation $(A - \lambda I)f = 0$ has only the trivial solution $f = 0$. It is said to have a *pure point spectrum* if the totality of vectors which are a solution of $(A - \lambda I)f = 0$ for some λ (depending on f) span the space. If A has no point spectrum, it is said to have a *purely continuous spectrum*.[68]

Coming back to the operator L_{ij}, we have already noticed that it has a positive point spectrum; i.e. its eigenvalues lie all on the positive part of the real axis. In the next Section, we shall see how these eigenvalues are distributed.

2. Distribution of eigenvalues:

The distribution of the eigenvalues of the problem (V.1.1)-(V.1.2) has been intensively investigated in several papers,[70-76] by introducing an asymptotic distribution formula.

Let $N(\lambda)$ denote the number of all eigenvalues not exceeding a given λ. Then the fundamental result is expressed in the following

Theorem 1: For given λ and $h \to 0$, the asymptotic formula

$$N(\lambda) = \frac{\sqrt{3}}{8\pi^2 h}\left[\iint_G \int_0^{2\pi} Re\sqrt{\lambda - \Omega(\theta,\alpha,\beta)}\, d\theta\, AB\, d\alpha d\beta + O(h^\gamma)\right] \qquad (V.2.1)$$

is valid.

Here

$$\Omega(\theta,\alpha,\beta) \equiv (1-\nu^2)\left(\frac{\sin^2\theta}{R_1} + \frac{\cos^2\theta}{R_2}\right)^2 \qquad (V.2.2)$$

and

$$0 \le \theta \le 2\pi \qquad (V.2.3)$$

$\gamma > 0$, and the constant in the O term is independent of λ for $0 \le \lambda \le \lambda_o$.

Further, there is

Theorem 2: The asymptotic formula (V.2.1) is valid for any boundary conditions for which the system (V.1.1) is self-adjoint, provided that the quadratic functional agrees with the quadratic functional of the problem (V.1.1)-(V.1.2).

We omit the rather lengthy proofs of these theorems, and refer the reader to the original papers.[75,76]

3. Relation to the spectrum of the membrane operator:

If we put $h^* = 0$ in the differential equations (V.1.1) and discard the last two boundary conditions in (V.1.2), then we are left with the *membrane problem*:

$$\sum_{j=1}^{3} L_{ij} u_j = \lambda u_i \qquad (V.3.1)$$

$$u_1\big|_\Gamma = u_2\big|_\Gamma = 0 \qquad (V.3.2)$$

The operator L also is self-adjoint.

Following theorem holds

Theorem: The set Ω of values taken by the function $\Omega(\theta,\alpha,\beta)$ belongs to the continuous spectrum of the operator L.

4. Ellipticity of the operator L - λI:

There are various criteria[77,78] for the ellipticity of a system of N partial differential equations for N unknown functions

$$\sum_{j=1}^{N} P_{ij}\, u_j = p_j \qquad (i = 1,\dots,N) \tag{V.4.1}$$

The most general definition is due to A. Douglis and L. Nirenberg:[79]

The system (V.4.1) is called *elliptic in the sense of Douglis and Nirenberg,* if there exist non-negative numbers t_j and s_i $(i,j = 1,\dots,N)$, such that P_{ij} is of order $\leq t_j + s_i$ and

$$\det \mid P^{o}_{ij}\ (\xi) \mid \neq 0 \qquad \text{if} \quad 0 \neq \xi = R_n \tag{V.4.2}$$

where P^{o}_{ij} denotes the part of order $t_j + s_i$ exactly in P_{ij}.

In our case, the calculated determinant equals

$$\frac{1-\nu}{2}\left[(1-\nu^2)\ \left(\frac{\eta_1^2}{R_1} + \frac{\eta_2^2}{R_2}\right)^2 - \lambda(\eta_1^2 + \eta_2^2)^2\right] \tag{V.4.3}$$

with

$$\eta_1 \equiv \frac{1}{A}\,\xi_1$$
$$\eta_2 \equiv \frac{1}{A}\,\xi_2 \tag{V.4.4}$$

Taking $s_1 = s_2 = 1$, $s_3 = 0$ and $t_1 = t_2 = 1$, $t_3 = 0$, one can see that L - λI is elliptic in the sense of Douglis and Nirenberg, provided that $\lambda \notin \Omega$.

5. Density of distribution of the eigenfrequencies:

We have

$$\lambda \equiv \frac{\rho(1-\nu^2)}{E}\,\omega^2 \tag{V.5.1}$$

as introduced in Chapter I, Section 1.

To obtain the derivative of the distribution function N with respect to ω, at some value $\omega = \omega_0$, we first transform the variable λ according to (V.5.1), and write

$$\frac{\Delta N}{\Delta\omega} = \frac{N(\omega_0 + \Delta\omega) - N(\omega_0)}{\Delta\omega} \tag{V.5.2}$$

where N is now understood to depend on ω through $N[\lambda(\omega)]$.

Thus

$$\frac{\Delta N}{\Delta\omega} = \frac{\sqrt{3}}{8\pi^2 h}\frac{(1-\nu^2)\rho}{E}\{\iint_G[\int_0^{2\pi} \omega\, Re(\frac{(1-\nu^2)\rho}{E}\omega_*^2 - \Omega(\theta,\alpha,\beta))^{-\frac{1}{2}}\, d\theta] \times$$

$$\times AB\, d\alpha d\beta + \frac{O(h^\gamma)}{\Delta\omega}; \tag{V.5.3}$$

$$\omega_0 < \omega_* < \omega_0 + \Delta\omega$$

Whenever $\frac{O(h^\gamma)}{\Delta\omega} \to 0$, then

$$\frac{8\pi^2 Eh}{\sqrt{3}\rho(1-\nu^2)}\frac{dN}{d\omega} \equiv I(\omega) = \iint_G\{\int_0^{2\pi} \omega\, Re(\frac{\rho(1-\nu^2)}{E}\omega^2 - \Omega(\theta,\alpha,\beta))^{-\frac{1}{2}} d\theta\}\, AB\, d\alpha d\beta \tag{V.5.4}$$

For a discussion of the density of frequencies, we refer to a paper by A. G. Aslanian, Z. N. Kuzina, V. B. Lidskii and V. N. Tulovskii.[75] The density of frequencies in shells of revolution has been investigated by P. E. Tovstik.[72]

CHAPTER VI
THE INVERSE PROBLEM

In Chapters I to V, we have been studying the free vibrations of a *given* shell. The concept of *inverse problem* requires some comments. Essentially, the eigenvalues are considered as given, and it is then required to deduce from this data conclusions as to the shape of the shell (*). It turns out that knowing the sequence of eigenvalues does in general provide unsufficient information only, as the much simpler case of a plane membrane already suggests.[80,81,82] So it is necessary to impose some restriction to the problem mentioned. In the present Chapter, we shall specifically be concerned with the problem of finding the shape of a shell (of given constitution and thickness) corresponding to a given sequence of eigenvalues, when the boundary curve Γ is also given and the shape (hence the eigenvalues) of a "neighboring" shell (of the same thickness and constitution) is known.

1. General formulation:

The starting point is the eigenvalue problem of thin elastic shells as described in Chapter V,

$$\begin{aligned} L\,\underset{\sim}{u} - \lambda \underset{\sim}{u} &= 0 \qquad \text{(in G)} \\ B\,\underset{\sim}{u} &= 0 \qquad \text{(on } \Gamma\text{)} \end{aligned} \tag{VI.1.1}$$

where L stands for the operator (V.1.1).

The second row of (VI.1.1) stands for the adequate, linear and homogeneous, kinematic boundary conditions.

We have seen, in Chapter V, that the operator L is self-adjoint. It

(*) We continue to consider shells of uniform thickness and made of homogeneous, isotropic, elastic metarial.

can therefore be written in the form

$$L\,\underset{\sim}{u} = \sum_{|s|=0}^{2} (-1)^{|s|} D^s \left[a_s(\alpha,\beta;\theta)\, D^s \underset{\sim}{u}\right] \qquad \text{(VI.1.2)}$$

where D^s denotes[78] differentiation of order s, i.e.

$$D^s \equiv \frac{\partial^s}{\partial^p\alpha\ \partial^q\beta} \qquad \text{(VI.1.3)}$$

and

$$|s| \equiv p + q \qquad \text{(VI.1.4)}$$

The differentiation is understood to act on each component of the function vector $\underset{\sim}{u}$. The a_s are 3x3 square matrices.

For convenience, we shall assume that the boundary Γ of the shell coincides piecewise with coordinate lines, so that we can adopt instead of $(VI.1.1)_2$, the more particular form

$$D^k\,\underset{\sim}{u} = 0 \qquad (\text{on } \Gamma) \qquad \text{(VI.1.5)}$$

with k = 0,1.

We imagine now, that the given and desired shapes differ slightly, that the systems of coordinate curves on these shells can be regarded as convected, and that both these systems can be brought in correspondence by making the geometric quantities (such as Lamé coefficients and curvatures) depend on a parameter θ, which itself will in general depend on α and β ; i.e. $\theta(\alpha,\beta)$. We agree to call this parameter θ_o in the case of the given shape.

Now, for θ_o, let the sequence of eigenvalues be given:

$$0 < \lambda_{o1} < \lambda_{o2} < \ldots \qquad \text{(VI.1.6)}$$

The task is to determine the function $\theta(\alpha,\beta)$ in such a way that the ordered sequence (VI.1.6) be modified appropriately, to yield the given sequence

$$0 < \lambda_1 < \lambda_2 < \ldots \tag{VI.1.7}$$

2. Perturbation and moment problems:

It is convenient to assume that the modification in the eigenvalues will be of the type

$$\lambda_i = \lambda_{oi} + \varepsilon\lambda_{1i} \tag{VI.2.1}$$

where ε is a small number (small as compared to one).

We assume that θ as well as the operator L and the eigenfunctions $\underset{\sim}{u}_i$ can be expanded in series of powers of ε:

$$\theta = \theta_o + \varepsilon\theta_1 + \ldots.. \tag{VI.2.2}$$

$$L = L_o + \varepsilon L_1 + \ldots.. \tag{VI.2.3}$$

$$\underset{\sim}{u}_i = \underset{\sim}{u}_{oi} + \varepsilon \underset{\sim}{u}_{1i} + \ldots.. \tag{VI.2.4}$$

with

$$L_1 \underset{\sim}{u} = \sum_{|s|=0}^{2} (-1)^{|s|} D^s\left[\theta_1 \frac{\partial a_s}{\partial\theta} D^s \underset{\sim}{u}\right] \tag{VI.2.5}$$

Substituting (VI.2.1) and (VI.2.2)-(VI.2.4) into (VI.1.1), we find

$$\begin{aligned} L_o \underset{\sim}{u}_{1i} - \lambda_{oi} \underset{\sim}{u}_{1i} &= \lambda_{1i} \underset{\sim}{u}_{oi} - L_1 \underset{\sim}{u}_{oi} \\ B \underset{\sim}{u}_i &= 0 \end{aligned} \tag{VI.2.6}$$

since

$$\begin{aligned} L_o \underset{\sim}{u}_{oi} - \lambda_{oi} \underset{\sim}{u}_{oi} &= 0 \\ B \underset{\sim}{u}_{oi} &= 0 \end{aligned} \tag{VI.2.7}$$

by virtue of (VI.1.1) and the requirement that the (normalized) $\underset{\sim}{u}_{oi}$ are eigenfunctions of the problem with $\varepsilon = 0$.

Eqs. (VI.2.6) provide with relationship between the λ_{1i} and the θ_1

(through (VI.2.5)). The sets of differential equations $(VI.2.6)_1$ and $(VI.2.7)_1$ have the same structure (on their left side), except that the first set is not homogeneous. This set has a nontrivial solution if its right side is orthogonal to the solutions of the set $(VI.2.7)_1$. Hence

$$\lambda_{1i} = \frac{(L_1 \underset{\sim}{u}_{oi}, \underset{\sim}{u}_{oi})}{(\underset{\sim}{u}_{oi}, \underset{\sim}{u}_{oi})} \tag{VI.2.8}$$

We may now use the expression (VI.2.5) for L_1, in the scalar product appearing in the numerator of (VI.2.8), together with an integration by parts, to obtain

$$\lambda_{1i} = \iint_G \theta_1(\alpha,\beta)\, g_i(\alpha,\beta) AB\, d\alpha d\beta \tag{VI.2.9}$$

where

$$g_i \equiv \sum_{|s|=0}^{2} (-1)^{|s|} \frac{\partial a_s}{\partial \theta} (D^s \underset{\sim}{u}_{oi})^2 \tag{VI.2.10}$$

is a function which can be determined from the data of the problem, and thus can be considered here as *given*.

The inverse problem is thus reduced to a *moment problem*, i.e. determining the function θ from the relations (VI.2.9), where the g_i and λ_{1i} are given, by keeping the boundary conditions.

The moment problem over a finite domain, and when moments of *all* orders are considered, is called *Hausdorff problem*, and has been investigated in two dimensions thoroughly by T. H. Hildebrandt and I. J. Schoenberg.[83] When the sequence is cut off at some order, then we have to do with the *truncated moment problem*. Existence theorems for this problem in two dimensions have been given by M. Dikmen.[84]

The inverse problem for shells of revolution has been investigated by L. I. Ainola.[85]

CHAPTER VII

WAVE PROPAGATION

In Chapters I to VI, we studied the *standing waves*. In the present Chapter, we briefly discuss the *propagation of waves*. We begin with some general remarks, and conduct then a rapid investigation of *progressing waves* in thin shells, leading to a geometric optics analogy. The question of the *propagation of discontinuities* is then briefly mentioned.

1. Some definitions:

Families of *undistorded progressing waves* are defined as families of solutions

$$\mathcal{U}_i = \mathcal{U}_i[\phi(\alpha,\beta,t)] \qquad \text{(VII.1.1)}$$

Here, the vector function $\mathcal{U}_i$ determines the *wave from,* and ϕ is called the *phase function*. A more general class is that of *relatively undistorded progressing waves*, defined by

$$\mathcal{U}_i = G(\alpha,\beta,t)\ \hat{\mathcal{U}}_i(\phi) \qquad \text{(VII.1.2)}$$

It is possible to generalize this formulation, by introducing series and considering the wave form as a *distribution,* to allow for discontinuities also. For more information and through discussion, we refer the reader to R. Courant and D. Hilbert.[86]

2. Monochromatic wave solution:

We simply try a solution of the form

$$\begin{aligned} U_1 &= k^{-1}\ u_1(\alpha,\beta)e^{i[k_o\psi(\alpha,\beta)-\omega t]} \\ U_2 &= k^{-1}\ u_2(\alpha,\beta)e^{i[k_o\psi(\alpha,\beta)-\omega t]} \\ U_3 &= \phantom{k^{-1}}\ u_3(\alpha,\beta)e^{i[k_o\psi(\alpha,\beta)-\omega t]} \end{aligned} \qquad \text{(VII.2.1)}$$

where

$$k_o \equiv \sqrt[4]{\frac{3(1-\nu^2)\rho}{Eh^2}\,\omega^2} \qquad \text{(VII.2.2)}$$

and can also be expressed by

$$k_o \equiv \sqrt[4]{\frac{2\rho h}{D}\,\omega^2} \qquad \text{(VII.2.3)}$$

with

$$D \equiv \frac{2Eh^3}{3(1-\nu^2)} \qquad \text{(VII.2.4)}$$

as the flexural stiffness of the shell. The presence of k_o provides the necessary factor in the limiting case of a plate.

$\psi(\alpha,\beta)$ is the phase function.

According to the definition given in the previous Section, (VII.2.1) represents a relatively undistorded progressing wave , the phase being chosen as

$$\phi(\alpha,\beta,t) = \psi(\alpha,\beta) - \frac{\omega}{k_o}\,t \qquad \text{(VII.2.5)}$$

On the other hand, it can easily be seen that this solution is *formally* of the same character as (I.3.1), provided one takes $b = 0$ and establishes the correspondence

$$f \to i\,\frac{k_o}{k}\,\psi = i\left(\frac{R}{h}\right)^{\frac{1}{2}} \sqrt[4]{\frac{3(1-\nu^2)\rho}{Eh^2}\,\omega^2}\;\psi(\alpha,\beta) \qquad \text{(VII.2.6)}$$

or

$$\psi \to -i\,\frac{k}{k_o}\,f = -\left(\frac{h}{R}\right)^{\frac{1}{2}} \sqrt[4]{\frac{Eh^2\omega^2}{3(1-\nu^2)\rho}}\;f(\alpha,\beta) \qquad \text{(VII.2.7)}$$

Hence, the frequency equation of Chapter II can immediately be used here too, noting that

$$\frac{\partial f}{\partial\alpha} \to i\left(\frac{h}{R}\right)^{\frac{1}{2}} \sqrt[4]{\frac{3(1-\nu^2)\rho}{Eh^2}\,\omega^2}\;\frac{\partial\psi}{\partial\alpha}$$

$$\frac{\partial f}{\partial\beta} \to i\left(\frac{h}{R}\right)^{\frac{1}{2}} \sqrt[4]{\frac{3(1-\nu^2)\rho}{Eh^2}\,\omega^2}\;\frac{\partial\psi}{\partial\beta} \qquad \text{(VII.2.8)}$$

to obtain

$$n^4 + \frac{E}{\rho\omega^2} \frac{\left[\frac{1}{R_2}\left(\frac{n_1}{A}\right)^2 + \frac{1}{R_1}\left(\frac{n_2}{B}\right)^2\right]^2}{n^4} = 1 \qquad \text{(VII.2.9)}$$

where

$$n_1 \equiv \frac{\partial\psi}{\partial\alpha}$$
$$n_2 \equiv \frac{\partial\psi}{\partial\beta} \qquad \text{(VII.2.10)}$$

are the components of a surface vector, the *phase gradient*, with square length given by

$$n \equiv \frac{n_1 n_1}{A^2} + \frac{n_2 n_2}{B^2} \qquad \text{(VII.2.11)}$$

In obtaining eq. (VII.2.9), we have taken the lines of curvature as coordinate curves without impairing generality, since we are not interested here on the investigation of some edge effect.

Eq. (VII.2.9) is reminiscent of *Fresnel's equation* in crystal optics,[87] and indeed n may be regarded here too as an *index of refraction*, thus establishing an analogy between wave propagation in shells and *geometrical optics*. The second term in eq. (VII.2.9) corresponds to an anisotropy, due to the curvature of the shell. This anisotropy vanishes in the case of a plate, yielding $n = 1$. The same term is also responsible for *frequency dispersion*, which also disappears in the case of a plate. For further implications of the geometric optics analogy, we refer to the paper by O. A. Germogenova.[88]

As already mentioned, the solution discussed in this Section does not take care of edge effects. These effects can be investigated separately.

3. Propagation of discontinuities:

The problem of wave propagation in thin shells can be approached also by adopting a concept originally due to J. Hadamard.[89] This leads to consider a wave on the middle surface as a propagating curve across which certain field variables and/or their derivatives exhibit some discontinuities. The field in question may be e.g. the field of displacements with discontinuities in first or second order derivatives. This approach has been efficiently used by H. Cohen and S. L. Suh,[90] H. Cohen and A. B. Berkal,[91,92] to study the dynamic behaviour of thin elastic shells. We refer the reader to the papers cited, not having place here to develop the necessary preliminaries from the theory of discontinuity propagation.

COMPLEMENTS

There are a few closing remarks which are intended to complement the necessarily succint account of some basic problems and methods of shell dynamics given in the previous Chapters.

a) Loaded shells: The theory of free vibrations is mathematically formulated as a system of homogeneous differential equations accompanied by a system of homogeneous boundary conditions. Now, loads (and a.o.pulse loads) may act on a shell. Depending on whether these loads act on the face(s) or on the edge of the shell, we face a problem of inhomogeneous differential equations with homogeneous boundary conditions, resp. a problem of homogeneous differential equations with inhomogeneous boundary conditions. As is known, any one of these problems can be transformed into the other one by simple processes, and are so far mathematically equivalent.[86] In solving, say the problem of surface loads, then advantage is taken of the fact that the eigenfunctions form a complete system, and can thus be used to expand given functions in series. For more details in this kind of problems, quite in general, we refer the reader to R. Courant and D. Hilbert,[86] and for applications in thin shell theory to papers on the subject in periodical literature.

b) Thick Shells: We have discussed the dynamics of thin shells. There is a theory of thick elastic shells, using the three-dimensional elasticity theory. Solutions and comparisons to thin shell theory can be found in a number of papers.[93-98,48,51,52]

c) Additional references: A number of titles [99-105] have been added to the list of references. There the reader can find access to further references.

REFERENCES

1. Euler, L., Tentamen de sono campanarum, *Nov. Comm. Acad. Sci. Petropolitanae*, 10, 261, 1766.
2. Bernoulli, J., Essai théorique sur la vibration des plaques élastiques, *Nov. Acta Petropolitanae*, 5, 1789.
3. Aron, H., Das Gleichgewicht und die Bewegung éiner unendlich dünnen, beliebig gekrümmten, elastischen Schale, *J. für reine und ang. Math.*, 78, 136, 1874.
4. Mathieu, E., Mémoire sur le mouvement vibratoire des cloches, *J. de l'Ecole Polytechnique*, 51, 177, 1882.
5. Poisson, S. D., Mémoire sur l'équilibre et le mouvement des corps élastiques, *Mem. de l'Acad. Paris*, 8, 545, 1829.
6. Lord Rayleigh, On the bending and vibration of thin elastic shells, especially of cylindrical form, *Proc. Roy. Soc. London*, Ser. A 45, 105, 1888.
7. Lord Rayleigh, *Theory of Sound*, Vols. I, II, second ed., Macmillan, London, 1894 and 1896.
8. Love, A. E. H., The small free vibrations and deformation of a thin elastic shell, *Phil . Trans. Roy. Soc. London*, Ser. A 179, 491, 1888.
9. Love, A. E. H., Note on the present state of the theory of thin elastic shells, *Proc. Roy. Soc. London*, Ser. A 149, 100, 1891.
10. Love, A. E. H., *A Treatise on the Mathematical Theory of Elasticity*, fourth ed., Cambridge University Press, Cambridge, 1927.
11. Basset, A. B., On the extension and flexion of cylindrical and spherical thin elastic shells, *Phil. Trans. Roy. Soc. London*, Ser. A 181, 433, 1890.
12. Lamb, H., On the deformation of an elastic shell, *Proc. London Math. Soc.*, 21, 119, 1890.

13. Dikmen, M., *Direkte Schalentheorie*, Teil I, Lecture Notes, Braunschweig, 1970.

14. Yamashita, K., and Aoki, I., On the frequencies of the sound emitted by japanese hanging bells, *Mem. Coll. Sci. Kyoto* (A), 15, 323, 1932.

15. Aoki, I., Experimental studies on the sound of a japanese temple-bell. *Mem. Coll. Sci. Kyoto* (A), 15, 311, 1932.

16. Aoki, I., Study of the sound emitted by the japanese Dohachi. *Mem. Coll. Sci. Kyoto* (A), 16, 377, 1933.

17. Strutt, M. J. O., Eigenschwingungen einer Kegelschale. *Ann. Phys.*, 17, 729, 1933.

18. van Urk, A. Th. und Hut, G. B., Messung der Radialschwingungen von Aluminiumkegelschalen, *Ann. Phys.*, 17, 915, 1933.

19. Flügge, W., Schwingungen zylindrischer Schalen, *Z. angew. Math. Mech.*, 13, 425, 1933.

20. Flügge, W., *Statik und Dynamik der Schalen*, third ed., Springer, Berlin/Göttingen/Heidelberg, 1964 (first ed. 1934).

21. Federhofer, K., Über die Eigenschwingungen der geschlossenen Kugelschale bei gleichförmigem Oberflächendrucke, *Z. angew. Math. Mech.*, 15, 26, 1935.

22. Federhofer, K., Über die Eigenschwingungen der axial gedrückten Kreiszylinderschale, *Sitzungsber. Akad. Wiss. Wien*, Math.-Nat. Kl. IIa, 145, 681, 1936.

23. Federhofer, K., Zur Berechnung der Eigenschwingungen der Kugelschale, *Sitzungsber. Akad. Wiss. Wien*, Math.-Nat. Kl., 146, 57, 1937.

24. Federhofer, K., Eigenschwingungen der Kegelschale, *Ing.-Arch.*, 9, 288, 1938.

25. Federhofer, K., Zur Schwingzahlberechnung des dünnwandigen Hohlreifens, *Ing.-Arch.*, 10, 125, 1939.

26. Reissner, E., On vibrations of shallow spherical shells, *J. Appl. Phys.*, 17, 1038, 1946.

27. Reissner, E., Stresses and small displacements of shallow spherical shells, *J. Math. and Phys.*, 25, 279, 1947.

28. Arnold, R. N. and Warburton, G. B., Flexural vibrations of the walls of thin cylindrical shells having freely supported ends, *Proc. Roy.*

Soc., A 197, 238, 1949.

29. Krylov, A. N., Radial vibrations of a hollow cylinder, in *Collected Works*, Vol. 3, Pt. II, Akad. Nauk SSR Press, Moscow/Leningrad, 1949.

30. Reissner, E., On axisymmetrical deformation of thin shells of revolution, in *Proc. Symp. on Appl. Math. Am. Math. Soc.*, 3, 27, 1950.

31. Niordson, F. I. N., Transmission of shock waves in thinwalled cylindrical tubes, *K. Tekn. Hogsk. Handl.*, 57, 1952.

32. Niordson, F. I. N., Vibrations of a cylindrical tube containing flowing fluid, *K. Tekn. Hogsk. Handl.*, 73, 1953.

33. Mindlin, R. D. and Bleich, H. H., Response of an elastic cylindrical shell to a transverse step shock wave, *J. Appl. Mech.*, 20, 189, 1953.

34. Arnold, R. N. and Warburton, G. B., The flexural vibrations of thin cylinders, *Proc. Inst. Mech. Eng. London*, A 167, 62, 1953.

35. Baron, M. L. and Bleich, H. H., Tables of frequencise and modes of free vibrations of infinitely long thin cylindrical shells, *J. Appl. Mech.*, 21, 178, 1954.

36. Reissner, E., On transverse vibrations of thin, shallow elastic shells, *Quart. Appl. Math.*, 13, 169, 1955.

37. Reissner, E., On axi-symmetrical vibrations of shallow spherical shells, *Quart. Appl. Math.*, 13, 279, 1955.

38. Yu, Y. Y., Free vibrations of thin cylindrical shells having finite lengths with freely supported and clamped edges, *J. Appl. Mech.*, 22, 547, 1955.

39. Cremer, L., Theorie der Luftschalldämmung zylindrischer Schalen, *Acustica*, 5, 245, 1955.

40. Huth, J. H. and Cole, J. D., Elastic stress waves produced by pressure loads on a spherical shell, *J. Appl. Mech.*, 22, 473, 1955.

41. Weibel, E., *The Strains and the Energy in Thin Elastic Shells of Arbitrary Shape for Arbitrary Deformation*, Diss. Zürich, 1955.

42. Lin, T. C. and Morgan, G. W., A study of axisymmetric vibrations of cylindrical shells as affected by rotatory inertia and transverse shear, *J. Appl. Mech.*, 23, 255, 1956.

43. Naghdi, P. M. and Cooper, R. M., Propagation of elastic waves in cylindrical shells, including the effects of transverse shear and rotatory inertia, *J. Acoust. Soc. Am.*, 28, 56, 1956.

44. Cooper, R. M. and Naghdi, P. M., Propagation of nonaxially symmetric waves in elastic cylindrical shells, *J. Acoust. Soc. Am.*, 29, 1365, 1957.

45. Johnson, M. W. and Reissner, E., On transverse vibrations of shallow spherical shells, *Quart. Appl. Math.*, 15, 367, 1957.

46. Mann-Nachbar, P., The interaction of an acoustic wave and an elastic spherical shell, *Quart. Appl. Math.*, 15, 83, 1957.

47. Mirsky, I. and Herrmann, G., Nonaxially symmetric motions of cylindrical shells, *J. Acoust. Soc. Am.*, 29, 1116, 1957.

48. Herrmann, G. and Mirsky, I., Three-dimensional and shell-theory analysis of axially symmetric motions in cylinders, *J. Appl. Mech.*, 23, 563, 1956.

49. Mirsky, I. and Herrmann, G., Axially symmetric motions of thick cylindrical shells, *J. Appl. Mech.*, 25, 97, 1958.

50. Yu, Y. Y., Vibrations of thin cylindrical shells analysed by means of Donnell-type equations, *J. Aerospace Sci.*, 25, 699, 1958.

51. Gazis, D. C., Exact analysis of the plane-strain vibrations of thick-walled hollow cylinders, *J. Acoust. Soc. Am.*, 30, 786, 1958.

52. Gazis, D. C., Three-dimensional investigation of the propagation of waves in hollow circular cylinders, *J. Acoust. Soc. Am.*, 31, 568, 1959.

53. Gol'denveizer, A. L., *Theory of Elastic Thin Shells*, Pergamon Press, New York, 1961.

54. Gol'denveizer, A. L., Asymptotic properties of eigenvalues in problems of the theory of thin elastic shells, *Appl. Math. Mech. (PMM)*, 25, 1077, 1961.

55. Gol'denveizer, A. L., Qualitative analysis of free vibrations of an elastic thin shell, *Appl. Math. Mech. (PMM)*, 30, 110, 1965.

56. Ross, Jr., E. W., Asymptotic analysis of the axisymmetric vibrations of shells, *J. Appl. Mech.*, 33, 85, 1966.

57. Ross, Jr., E. W., Transition solutions for axisymmetric shell vibration, *J. Math. Phys.*, 45, 335, 1966.

58. Ross, Jr., E. W., On inextensional vibrations of thin shells, *J. Appl. Mech.*, 35, 516, 1968.

59. Gol'denveizer, A. L., Classification of integrals of the dynamic equations of the linear two-dimensional theory of shells, *Appl. Math. Mech. (PMM)*, 37, 559, 1973.

60. Chernyshev, C. N., On some properties of integrals of the dynamic equations of shell theory, *Mechanics of Solids (Mekh. Tverd. Tela)*, 8, 55, 1973.

61. Tovstik, P. E., Low frequency oscillations of convex shells of revolution, *Mechanics of Solids (Mekh. Tverd. Tela)*, 10, 95, 1975.

62. Bolotin, V. V., The edge effect in the oscillations of elastic shells, *Appl. Math. Mech. (PMM)*, 24, 1257, 1960.

63. Moskalenko, V. N., On the frequency spectra of natural vibrations of shells of revolution, *Appl. Math. Mech. (PMM)*, 36, 279, 1972.

64. Chernyshev, G. N., Asymptotic method of investigation of short-wave oscillations of shells, *Appl. Math. Mech. (PMM)*, 39, 319, 1975.

65. Boardman, A. D., O'Connor, D. E. and Young, P. A., *Symmetry and its Applications in Science*. Mc Graw-Hill, London/New York, 1973.

66. Perrin, R. and Charnley, T., Group theory and the bell, *J. Sound and Vibr.*, 31, 411, 1973.

67. Wigner, E. P., *Group Theory*, Academic Press, New York/London, 1959.

68. Lorch, E. R., *Spectral Theory*, Oxford University Press, New York, 1962.

69. Dunford, N. and Schwartz, J. T., *Linear Operators*, Vols. I, II, III, Interscience, New York, 1957, 1963, 1971.

70. Bolotin, V. V., On the density of distribution of natural frequencies of thin elastic shells. *Appl. Math. Mech. (PMM)*,538, 27, 1963.

71. Aslanyan, A. G. and Lidskii, V. B., Spectrum of a system describing the vibrations of a shell of revolution, *Appl. Math. Mech. (PMM)*, 35, 701, 1971.

72. Tovstik, P. E., On the density of the vibration frequencies of thin shells of revolution, *Appl. Math. Mech. (PMM)*, 36, 270, 1972.

73. Gulgazaryan, G. R., The spectrum of a moment-free operator in the theory of thin shells of revolution, *Differential Equations*, 10,28, 1975.

74. Gulgazaryan, G. R., Lidskii, V. B. and Eskin, G. I., Spectrum of a membrane system in the case of a thin shell of arbitrary outline,

Siber. Math. J., 14, 681, 1975.

75. Aslanian, A. G., Kuzina, Z. N., Lidskii, V. B. and Tulovskii, V. N., Distribution of the natural frequencies of a thin elastic shell of arbitrary outline, *Appl. Math. Mech. (PMM)*, 37, 571, 1973.

76. Aslanian, A. G. and Tulovskii, V. N., Asymptotic distribution of the eigenfrequencies of elastic shells, *Mathematical Physics*, 18, 120, 1973.

77. Hörmander, L., *Linear Partial Differential Operators*, Springer, Berlin/Heidelberg/New York, 1969.

78. Miranda, C., *Partial Differential Equations of Elliptic Type*, Springer, Berlin/Heidelberg/New York, 1970.

79. Douglis, A. and Nirenberg, L., Interior estimates for elliptic systems of partial differential equations, *Comm. Pure Appl. Math.*, 8, 503, 1955.

80. Kac, M., Can one hear the shape of a drum? *Amer. Math. Montly.*, 73, 1, 1966.

81. Stewartson, K. and Waechter, R. T., On hearing the shape of a drum: further results, *Proc. Camb. Phil. Soc.*, 69, 353, 1971.

82. Waechter, R. T., On hearing the shape of a drum: an extension to higher dimensions, *Proc. Camb. Phil. Soc.*, 72, 439, 1972.

83. Hildebrandt, T. H. and Schoenberg, I. J., On linear functional operations and the moment problem for a finite interval in one or several dimensions, *Ann. of Math.*, 34, 317, 1933.

84. Dikmen, M., Bazı moment problemleri (english summary), in *Mustafa Inan Anısına*, Istanbul Teknik Üniversitesi, Istanbul, 1971.

85. Ainola, L. I., On the inverse problem of natural vibrations of elastic shells, *Appl. Math. Mech. (PMM)*, 35, 317, 1971.

86. Courant, R. and Hilbert, D., *Methods of Mathematical Physics*, Vol. II, Interscience, New York, 1966.

87. Landau, L. D. and Lifshitz, E. M., *Electrodynamics of Continuous Media*, Pergamon Press, Oxford, 1963.

88. Germogenova, O. A., Geometrical theory for flexure waves in shells, *J. Acoust. Soc. Am.*, 53, 535, 1973.

89. Hadamard, J., *Leçons sur la Propagation des Ondes et les Equations*

de l'Hydrodynamique, Hermann, Paris, 1903.

90. Cohen, H. and Suh, S. L., Wave propagation in elastic surfaces, *J. Math. and Mech.*, 19, 1117, 1970.

91. Cohen, H. and Berkal, A. B., Wave propagation in elastic membranes, *Journal of Elasticity*, 2, 45, 1972.

92. Cohen, H. and Berkal, A. B., Wave propagation in elastic shells, *Journal of Elasticity*, 2, 35, 1972.

93. Armenakas, A. E., On the accuracy of some dynamic shell theories, *J. Eng. Mech. Div. ASCE.*, 93, 95, 1967.

94. Armenakas, A. E., Gazis, D. C. and Herrmann, G., *Free Vibrations of Circular Cylindrical Shells*, Pergamon Press, Oxford, 1969.

95. Greenspon, J. E., Vibrations of a thick-walled cylindrical shell. Comparison of the exact theory with approximate theories, *J. Acoust. Soc. Am.*, 32, 571, 1960.

96. Nigul, U., Three-dimensional shell theory of axially symmetric transient waves in a semi-infinite cylindrical shell, *Arch. Mech. Stosowanej*, 19, 1967.

97. Graff, K. F., *Wave Motion in Elastic Solids*, Clarendon Press, Oxford, 1975.

98. Murty, A. V. K. A., Frequency spectrum of shells, *AIAA Journal*, 13, 1109, 1975.

99. Steele, C. R., Bending waves in shells, *Quart. Appl. Math.*, 34, 385, 1977.

100. Lizarev, A. D., Natural vibration modes of spherical shells, *Mechanics of Solids (Mekh. Tverd. Tela)*, 10, 157, 1975.

101. Forsberg, K., Axisymmetric and beam-type vibrations of thin cylindrical shells, *AIAA Journal*, 7, 221, 1969.

102. Forsberg, K., Influence of boundary conditions on the modal characteristics of cylindrical shells, *AIAA Journal*, 2, 2150, 1964.

103. Yamaki, N., Flexural vibrations of thin cylindrical shells, *Rep. Inst. High Speed Mech.*, Japan, 1970.

104. Bergman, R. M., Investigation of the free vibrations of noncircular cylindrical shells, *Appl. Math. Mech. (PMM)*, 37, 1068, 1973.

105. Charnley, T. and Perrin, R., Torsional vibrations of bells, *J. Sound and Vibr.*, 40, 227, 1975.

FINITE ROTATIONS IN THE NONLINEAR THEORY OF THIN SHELLS

by

W. Pietraszkiewicz

Institute of Fluid-Flow Machinery, PAN, Gdánsk

FINITE ROTATIONS IN THE NONLINEAR THEORY OF THIN SHELLS

W. Pietraszkiewicz

Abstract

The theory of finite rotations in thin shells is developed and many shell relations in terms of finite rotations are presented. Three forms of geometric boundary conditions and energetically compatible static boundary conditions are constructed. Various sets of Eulerian and Lagrangean shell equations are discussed and their consistent simplification within the first-approximation geometrically non-linear theory of isotropic elastic shells is given. A classification of shell problems with small, moderate, large and finite rotations is proposed and appropriate sets of simplified shell equations are presented.

FINITE ROTATIONS IN THE NON-LINEAR THEORY OF THIN SHELLS

Wojciech Pietraszkiewicz
Institute of Fluid - Flow Machinery
of the Polish Academy of Sciences
ul. Fiszera 14, 80-952 Gdańsk, Poland

1. INTRODUCTION

The appearance of finite rotations of material elements is the fundamental kinematic property of any non-linear theory of shells. Finite rotations may appear even when strains are small everywhere in the shell, which is obvious from a trivial example of rolling of a sheet of paper into a cylinder.

According to the Cauchy theorem, the deformation of a neighbourhood of any material particle of a continua can be decomposed into a rigid-body translation, a pure stretch along principal directions of strain and a rigid-body rotation of the principal directions. Within the continuum mechanics the rotational part of deformation is conventionally described by a rotation tensor $\underset{\sim}{R}$ as defined, for example, by Truesdell and Noll [1]. An equivalent description of rotations, by means of a finite rotation vector $\underset{\sim}{\Omega}$, was used by Shamina [2] in discussion of continuum compatibility conditions. Rotation parameters appear explicitly in many three - dimensional relations, being particularly important in the theory of constitutive equations. They play also a fundamental role in analytic mechanics of a rigid - body motion [3].

The rotational part of deformation should play even more important role in any non-linear theory of thin bodies, in particular in the non-linear theory of thin shells, what was recognized long ago. Unfortunately, the shell literature is not free from confusions about analytic representation of the finite rotations. Usually the linearized rotations or displacement gradients are used, apparently on the intuitive grounds, to describe rotations of material elements also within the non-linear range of the shell deformation.

Within the Kirchhoff-Love type non-linear theory of shells Simmonds and Danielson [4,5] employed the finite rotation vector $\underset{\sim}{\Omega}$ of the principal directions of strain as the basic kinematic variable of the shell theory. Novozhilov and Shamina [6] used the notion of the total finite rotation vector $\underset{\sim}{\Omega}_t$ of the boundary to derive various forms of geometric boundary conditions. However, in these works the finite rotation vectors were introduced in a descriptive manner, without giving the analytic relations for them in terms of basic kinematical parameters of the shell deformation, such as displacements of the shell middle surface or components of the deformation gradient tensor.

A general theory of finite rotations in shells was developed in the author's thesis [7] (see also [8]). Many simplified relations , valid under the Kirchhoff-Love constraints, were derived in [7] as particular cases of the exact relations. Their independent derivation was also presented in [9]. The notion of a finite rotation prooved to be very helpful in deriving alternative forms of geometric and static boundary conditions, in obtaining various modified forms of equilibrium equations and in providing a new consistent classification of various approximate variants of geometrically non-linear theory of elastic shells and plates with restricted rotations.

In this report we shall develope the theory of finite rotations in thin shells and discuss some associated problems of shell theory , in which the rotations play an important role. It is assumed that the behaviour of a thin shell can be described with a sufficient accuracy by the behaviour of the shell middle surface. This is accomplished by imposing the Kirchhoff-Love constraints on the shell deformation. The change of the shell thickness during deformation is taken into account only in constitutive equations of an isotropic elastic shell under small strains. Such simplified approach is justified within the first - approximation geometrically non-linear theory of thin elastic shells [10], which is the ultimate goal of this work.

2. STRAINS AND ROTATIONS IN THIN SHELLS

2.1. Notation and preliminary relations

We shall adopt here, as far as possible, the system of notations used by Koiter [11] and by the author [7,9,12].

Let $\underset{\sim}{r}(\theta^\alpha) = x^k(\theta^\alpha)\underset{\sim}{i}_k$ and $\bar{\underset{\sim}{r}}(\theta^\alpha) = \bar{x}^k(\theta^\alpha)\underset{\sim}{i}_k$, $k = 1,2,3$, be position vectors of the shell middle surface in the reference and deformed configurations, respectively. Here θ^α , $\alpha = 1,2$, is a pair of surface convected coordinates and x^k and $\bar{x}^k$ are components of $\underset{\sim}{r}$ and $\bar{\underset{\sim}{r}}$ with respect to a Cartesian frame.

With the reference surface M we associate a standard surface covariant base vectors $\underset{\sim}{a}_\alpha = \underset{\sim}{r}_{,\alpha}$, covariant components of the metric tensor $a_{\alpha\beta} = \underset{\sim}{a}_\alpha \cdot \underset{\sim}{a}_\beta$ with determinant $a = |a_{\alpha\beta}|$, a unit vector normal to M , $\underset{\sim}{n} = \frac{1}{2}\epsilon^{\alpha\beta}\underset{\sim}{a}_\alpha \times \underset{\sim}{a}_\beta$, and covariant components of the curvature tensor $b_{\alpha\beta} = \underset{\sim}{a}_{\alpha,\beta} \cdot \underset{\sim}{n}$. Here comma $(\ )_{,\alpha}$ denotes partial differentiation with respect to surface coordinates θ^α , while $\epsilon^{\alpha\beta}$ are contravariant components of the permutation tensor. Contravariant components $a^{\alpha\beta}$ of the metric tensor satisfying the relations $a^{\alpha\gamma}a_{\beta\gamma} = \delta^\alpha_\beta$ are used to raise the indices at M . Similar geometric quantities associated with deformed surface $\bar{M}$ are marked by a dash: $\bar{\underset{\sim}{a}}_\alpha$, $\bar{a}_{\alpha\beta}$, $\bar{a}$, $\bar{\underset{\sim}{n}}$, $\bar{b}_{\alpha\beta}$, $\bar{\epsilon}^{\alpha\beta}$, $\bar{a}^{\alpha\beta}$ etc. The surface covariant differentiation at M or $\bar{M}$ is denoted by $(\ )_{|\alpha}$ or $(\ )_{||\alpha}$, respectively.

After the surface deformation the basic vectors $\bar{\mathbf{a}}_\alpha$, $\bar{\mathbf{n}}$ are expressed in terms of the geometry of M and displacement vector $\mathbf{u} = u_\alpha \mathbf{a}^\alpha + w\mathbf{n}$ by the relations [9,11,12]

$$\bar{\mathbf{a}}_\alpha = l_{\lambda\alpha}\mathbf{a}^\lambda + \phi_\alpha \mathbf{n} = \mathbf{a}_\alpha + \mathbf{u}_{,\alpha} \quad , \quad \bar{\mathbf{n}} = n_\alpha \mathbf{a}^\alpha + n\mathbf{n} \tag{2.1.1}$$

where

$$l_{\alpha\beta} = a_{\alpha\beta} + \theta_{\alpha\beta} - \omega_{\alpha\beta}$$

$$\theta_{\alpha\beta} = \frac{1}{2}(u_{\alpha|\beta} + u_{\beta|\alpha}) - b_{\alpha\beta}w \quad , \quad \phi_\alpha = w_{,\alpha} + b^\lambda_\alpha u_\lambda \tag{2.1.2}$$

$$\omega_{\alpha\beta} = \frac{1}{2}(u_{\beta|\alpha} - u_{\alpha|\beta}) = \epsilon_{\alpha\beta}\phi \quad , \quad \phi = \frac{1}{2}\epsilon^{\alpha\beta}u_{\beta|\alpha}$$

$$n_\mu = \sqrt{\frac{a}{\bar{a}}}\,\epsilon^{\alpha\beta}\epsilon_{\lambda\mu}\phi_\alpha l^\lambda_{\cdot\beta} \quad , \quad n = \frac{1}{2}\sqrt{\frac{a}{\bar{a}}}\,\epsilon^{\alpha\beta}\epsilon_{\lambda\mu}l^\lambda_{\cdot\alpha}l^\mu_{\cdot\beta} \tag{2.1.3}$$

The components of the Lagrangean surface strain tensor and the tensor of change of surface curvature are defined by

$$\gamma_{\alpha\beta} = \frac{1}{2}(\bar{a}_{\alpha\beta} - a_{\alpha\beta}) = \theta_{\alpha\beta} + \frac{1}{2}(\theta^\lambda_\alpha - \omega^\lambda_{\cdot\alpha})(\theta_{\lambda\beta} - \omega_{\lambda\beta}) + \frac{1}{2}\phi_\alpha\phi_\beta \tag{2.1.4}$$

$$\begin{aligned}\kappa_{\alpha\beta} &= -(\bar{b}_{\alpha\beta} - b_{\alpha\beta}) = \\ &= -\{n(\phi_{\alpha|\beta} + b^\lambda_\beta l_{\lambda\alpha}) + n_\lambda(l^\lambda_{\cdot\alpha|\beta} - b^\lambda_\beta\phi_\alpha) - b_{\alpha\beta}\}\end{aligned} \tag{2.1.5}$$

These strain measures satisfy the following compatibility conditions

$$\begin{aligned}&\epsilon^{\alpha\beta}\epsilon^{\lambda\mu}[\kappa_{\beta\lambda|\mu} + \bar{a}^{\kappa\nu}(b_{\kappa\lambda} - \kappa_{\kappa\lambda})\gamma_{\nu\beta\mu}] = 0 \\ &K\gamma^\kappa_\kappa + \epsilon^{\alpha\beta}\epsilon^{\lambda\mu}[\gamma_{\alpha\mu|\beta\lambda} - b_{\alpha\mu}\kappa_{\beta\lambda} + \frac{1}{2}(\kappa_{\alpha\mu}\kappa_{\beta\lambda} + \bar{a}^{\kappa\nu}\gamma_{\kappa\alpha\mu}\gamma_{\nu\beta\lambda})] = 0\end{aligned} \tag{2.1.6}$$

Here K is the Gaussian curvature of M and

$$\gamma_{\nu\beta\mu} = \gamma_{\nu\beta|\mu} + \gamma_{\nu\mu|\beta} - \gamma_{\beta\mu|\nu} \tag{2.1.7}$$

The base vectors $\mathbf{a}_\alpha$ and $\mathbf{a}^\alpha$ span at $M \in M$ a two-dimensional vector space V. The tensor product of two surface vectors $\mathbf{u}, \mathbf{v} \in V$ defines the second-order surface tensor $\mathbf{u} \otimes \mathbf{v} = \mathbf{T} \in T_2 = V \otimes V$. Tensor products of the base vectors $\mathbf{a}_\alpha \otimes \mathbf{a}_\beta$, $\mathbf{a}^\alpha \otimes \mathbf{a}_\beta$ etc. are bases of the tensor space T_2. For $\mathbf{T} = T^\alpha_{\cdot\beta}\mathbf{a}_\alpha \otimes \mathbf{a}^\beta$ the transposition is defined by $\mathbf{T}^T = T^\alpha_{\cdot\beta}\mathbf{a}^\beta \otimes \mathbf{a}_\alpha$, the contraction by $\mathrm{tr}\mathbf{T} = T^\alpha_{\cdot\alpha}$, the multiplication by $\mathbf{TS} = T^\alpha_{\cdot\lambda}S^\lambda_{\cdot\beta}\mathbf{a}_\alpha \otimes \mathbf{a}^\beta$ and the scalar product by $\mathbf{T} \cdot \mathbf{S} = T^\alpha_{\cdot\lambda}S^\lambda_{\cdot\alpha}$. The detailed presentation of tensor algebra and analysis in absolute notation is given by Lichnerowicz [13], Bowen and Wang [14] and Truesdell [15]. In the notation used here the metric and curvature tensors of M are given by

$$\mathbf{a} = a_{\alpha\beta}\mathbf{a}^\alpha \otimes \mathbf{a}^\beta \quad , \quad \mathbf{b} = b_{\alpha\beta}\mathbf{a}^\alpha \otimes \mathbf{a}^\beta \tag{2.1.8}$$

In the neighbourhood of M let us introduce a spatial normal coordinate system θ^i, $i = 1,2,3$, such that at M there is $\theta^3 = 0$ and $\theta^3 \equiv \zeta$ is the distance to a point $P \in P \subset E$ of the three-dimensional Euclidean point space E and $-h/2 \le \zeta \le +h/2$, where h is the constant shell thickness in the reference configuration. Then for the position vector of P and for the spatial base vectors at P we have

$$\begin{aligned} &\mathbf{p} = \mathbf{r} + \zeta\mathbf{n} \quad , \quad \mathbf{g}_\phi = \mathbf{p}_{,\phi} = \mu^\alpha_\phi \mathbf{a}_\alpha \quad , \quad \mathbf{g}_3 = \mathbf{p}_{,3} = \mathbf{n} \\ &\mu^\alpha_\phi = \mathbf{a}^\alpha \cdot \mathbf{g}_\phi = \delta^\alpha_\phi - \zeta\delta^\beta_\phi b^\alpha_\beta \end{aligned} \tag{2.1.9}$$

The base vectors $\mathbf{g}_i$ span at $P \in P$ a three-dimensional vector space W. The tensor product of two space vectors defines a second-order space tensor $\mathbf{L} \in L_2 = W \otimes W$. The space metric and translation tensors are given by

$$\begin{aligned} &\mathbf{1} = \mathbf{1}^T = g_{ij}\mathbf{g}^i \otimes \mathbf{g}^j = \mathbf{a} + \mathbf{n} \otimes \mathbf{n} = \mu^\alpha_\phi \mathbf{a}_\alpha \otimes \mathbf{g}^\phi + \mathbf{n} \otimes \mathbf{n} \\ &\mathbf{g} = \mathbf{g}^T = \mathbf{1} - \zeta\mathbf{b} \end{aligned} \tag{2.1.10}$$

This leads to the relations

$$\mathbf{g}_\phi = \delta^\alpha_\phi \mathbf{g}\mathbf{a}_\alpha \quad , \quad \mathbf{g}_3 = \mathbf{g}\mathbf{n} \tag{2.1.11}$$

which allow to express in coordinate - free notation quantities at $P \in \mathcal{P}$ in terms of those defined at $M \in \mathcal{M}$ and the distance ζ from $\mathcal{M}$.

2.2. Deformation of thin shells under Kirchhoff - Love constraints

Consider a shell S consisting of particles X,Y,.... A one-to-one correspondence between particles $X \in S$ and points $P \in \mathcal{P} \subset E$ in three-dimensional Euclidean space is the configuration of the shell, $P = \kappa(X)$, $\kappa : S \rightarrow \mathcal{P}$. In the Lagrangean description the displacement of a particle $X \in S$ from its position P in the reference configuration to a position $\bar{P} = \bar{\kappa}(X)$ in the deformed one, $\bar{P} = \chi(P)$, $\chi = \bar{\kappa} \circ \kappa^{-1}$, is given by a displacement vector

$$\mathbf{v} = \mathbf{v}(P) = \bar{\mathbf{p}}[\chi(P)] - \mathbf{p}(P) = v_i \mathbf{g}^i \tag{2.2.1}$$

Thus for differentials

$$d\bar{\mathbf{p}} = \bar{\mathbf{p}}_{,i} d\theta^i = \mathbf{F} d\mathbf{p} \tag{2.2.2}$$

where the tensor $\mathbf{F} \in L_2$ has the form

$$\mathbf{F} = \mathbf{F}(P) = \mathbf{1} + \operatorname{grad} \mathbf{v} = \mathbf{g}_i \otimes \mathbf{g}^i + \mathbf{v}_{,i} \otimes \mathbf{g}^i = \bar{\mathbf{g}}_i \otimes \mathbf{g}^i \tag{2.2.3}$$

This tensor used by Truesdell and Noll [1] is known as the spatial deformation gradient tensor. It fully describes the state of deformation in a neighbourhood of the particle $X \in S$ during the shell deformation from the reference to deformed configurations.

In general, material fibres, normal to the reference surface $\mathcal{M}$, may after deformation become neither straight nor normal to the surface $\bar{\mathcal{M}} = \chi(\mathcal{M})$. In the general case the deformation function χ may be expanded into Taylor series in the vicinity of the shell middle surface. The linear part of the expansion describes exactly the shell deformation at its middle surface. Describing shell deformation only by the linear part of the expansion (or, equivalently, assuming the linear distribution of displacements in the shell space) various relations of the Reissner-type non-linear theory of shells may be discussed [7,16]. In particular, the assumption of the linear distribution of displacements enabled to develope the exact theory of finite rotations in shells [7,8].

In this work we are interested in constructing various relations for the first - approximation geometrically non-linear theory of thin isotropic elastic shells. In such a case the stress state in the shell space is approximately plane and parallel to the shell middle surface [17]. The shell strain energy function depends explicitly only upon the stretching and bending of the shell middle surface. Its dependence upon the transverse strains appear only implicitly, through the modified shell elasticity tensor [10]. In order to simplify all intermediate transformations, we initially impose here the Kirchhoff - Love constraints on the shell deformation. The change of the shell thickness during deformation will, however, be taken into account in the constitutive equations.

According to the Kirchhoff - Love constraints, material fibres that are normal to the reference shell middle surface M remain, after the shell deformation, normal to the deformed surface $\bar{M}$ and do not change their length. Under these constraints the shell deformation gradient tensor $\underset{\sim}{G} \in L_2$ takes at M the following form

$$\underset{\sim}{G} = \underset{\sim}{F}(P')\Big|_{\zeta = 0} = \bar{\underset{\sim}{a}}_\alpha \otimes \underset{\sim}{a}^\alpha + \bar{\underset{\sim}{n}} \otimes \underset{\sim}{n} \tag{2.2.4}$$

Under (2.2.4) and (2.1.10) for the spatial quantities we obtain

$$\underset{\sim}{F} = \bar{\underset{\sim}{g}}\underset{\sim}{G}\underset{\sim}{g}^{-1} = (\underset{\sim}{G} - \zeta\bar{\underset{\sim}{b}}\underset{\sim}{G})\underset{\sim}{g}^{-1} \quad , \quad \underset{\sim}{v} = \underset{\sim}{u} + \zeta\underset{\sim}{\beta} \tag{2.2.5}$$

where under K - L constraints $\underset{\sim}{\beta}$ depends upon $\underset{\sim}{u}$ according to

$$\underset{\sim}{\beta} = (\underset{\sim}{G} - \underset{\sim}{1})\underset{\sim}{n} = \bar{\underset{\sim}{n}} - \underset{\sim}{n} = n_\alpha \underset{\sim}{a}^\alpha + (n - 1)\underset{\sim}{n} \tag{2.2.6}$$

In the Lagrangean description of strain we use the Green strain tensor $\underset{\sim}{E} \in L_2$ defined by

$$\underset{\sim}{E} = \underset{\sim}{E}(P) = \frac{1}{2}(\underset{\sim}{F}^T\underset{\sim}{F} - \underset{\sim}{1}) = E_{ij}\underset{\sim}{g}^i \otimes \underset{\sim}{g}^j \tag{2.2.7}$$

In viev of (2.2.5) this tensor may be expressed in terms of the surface strain measures according to

$$\underset{\sim}{E} = \underset{\sim}{g}^{-1}(\underset{\sim}{\gamma} + \zeta\underset{\sim}{\kappa} + \zeta^2\underset{\sim}{\nu})\underset{\sim}{g}^{-1} \tag{2.2.8}$$

where

$$\begin{aligned} \underset{\sim}{\gamma} &= \frac{1}{2}(\underset{\sim}{G}^T\underset{\sim}{G} - \underset{\sim}{1}) = \gamma_{\alpha\beta}\underset{\sim}{a}^\alpha \otimes \underset{\sim}{a}^\beta \\ \underset{\sim}{\kappa} &= -(\underset{\sim}{G}^T\bar{\underset{\sim}{b}}\underset{\sim}{G} - \underset{\sim}{b}) = \kappa_{\alpha\beta}\underset{\sim}{a}^\alpha \otimes \underset{\sim}{a}^\beta \\ \underset{\sim}{\nu} &= \frac{1}{2}(\underset{\sim}{G}^T\bar{\underset{\sim}{b}}^2\underset{\sim}{G} - \underset{\sim}{b}^2) = \nu_{\alpha\beta}\underset{\sim}{a}^\alpha \otimes \underset{\sim}{a}^\beta = \frac{1}{2}[(\underset{\sim}{b} - \underset{\sim}{\kappa})(\underset{\sim}{1} + 2\underset{\sim}{\gamma})^{-1}(\underset{\sim}{b} - \underset{\sim}{\kappa}) - \underset{\sim}{b}^2] \end{aligned} \tag{2.2.9}$$

In view of (2.2.5) the deformation gradient tensor $\underset{\sim}{G}$ provides complete information about the shell non-linear deformation compatible with K - L constraints. Since $\underset{\sim}{G}$ is nonsingular then, according to the polar decomposition theorem [1], it can be represented uniquely by two following formulae

$$\underset{\sim}{G} = \underset{\sim}{R}\underset{\sim}{U} = \underset{\sim}{V}\underset{\sim}{R} \quad , \quad \underset{\sim}{G}^{-1} = \underset{\sim}{U}^{-1}\underset{\sim}{R}^T = \underset{\sim}{R}^T\underset{\sim}{V}^{-1} \tag{2.2.10}$$

Here $\underset{\sim}{U}$ and $\underset{\sim}{V}$ are the right and left stretch tensors, respectively, and $\underset{\sim}{R}$ is the finite rotation tensor. The tensors $\underset{\sim}{U}$ and $\underset{\sim}{V}$ are positive definite and symmetric, while $\underset{\sim}{R}$ is the proper orthogonal tensor. In terms of $\underset{\sim}{G}$ we have

$$\underset{\sim}{U} = \sqrt{\underset{\sim}{G}^T\underset{\sim}{G}} \quad , \quad \underset{\sim}{V} = \sqrt{\underset{\sim}{G}\underset{\sim}{G}^T} \quad , \quad \underset{\sim}{U} = \underset{\sim}{R}^T\underset{\sim}{V}\underset{\sim}{R} \tag{2.2.11}$$

$$\underset{\sim}{R} = \underset{\sim}{G}\underset{\sim}{U}^{-1} = \underset{\sim}{V}^{-1}\underset{\sim}{G} \quad , \quad \underset{\sim}{R}^T = \underset{\sim}{R}^{-1} \quad , \quad \det\underset{\sim}{R} = +1 \tag{2.2.12}$$

By the formulae (2.2.1), (2.2.5) and (2.2.10) the deformation of a neighbourhood about a point of the shell middle surface is decomposed into a rigid - body translation, a pure stretch along principal directions of strain and a rigid - body rotations of the principal directions. Decomposition of $\underset{\sim}{G}$ in terms of $\underset{\sim}{U}$ in (2.2.10) is compatible with the Lagrangean description, in terms of $\underset{\sim}{V}$ it is compatible with the Eulerian description.

From (2.2.4) and (2.2.10) we obtain

$$\bar{\mathbf{a}}_\alpha = \mathbf{G}\mathbf{a}_\alpha = \mathbf{R}\overset{\vee}{\mathbf{a}}_\alpha = \mathbf{V}\overset{*}{\mathbf{a}}_\alpha \quad , \quad \bar{\mathbf{n}} = \mathbf{G}\mathbf{n} = \mathbf{R}\mathbf{n}$$
$$\bar{\mathbf{a}}^\alpha = (\mathbf{G}^{-1})^T\mathbf{a}^\alpha = \mathbf{R}\overset{\vee}{\mathbf{a}}{}^\alpha = \mathbf{V}^{-1}\overset{*}{\mathbf{a}}{}^\alpha \tag{2.2.13}$$

where

$$\overset{\vee}{\mathbf{a}}_\alpha = \mathbf{U}\mathbf{a}_\alpha \quad , \quad \overset{\vee}{\mathbf{a}}{}^\alpha = \mathbf{U}^{-1}\mathbf{a}^\alpha \quad , \quad \overset{*}{\mathbf{a}}_\alpha = \mathbf{R}\mathbf{a}_\alpha \quad , \quad \overset{*}{\mathbf{a}}{}^\alpha = \mathbf{R}\mathbf{a}^\alpha \tag{2.2.14}$$

The intermediate Lagrangean basis $\overset{\vee}{\mathbf{a}}_\alpha$, $\mathbf{n}$ is obtained by stretching the reference basis $\mathbf{a}_\alpha$, $\mathbf{n}$ along the principal directions of the stretch tensor $\mathbf{U}$. The intermediate Eulerian basis $\overset{*}{\mathbf{a}}_\alpha$, $\bar{\mathbf{n}}$ is obtained by a rigid - body rotation of the reference basis $\mathbf{a}_\alpha$, $\mathbf{n}$ by means of the finite rotation tensor $\mathbf{R}$. The basis $\overset{\vee}{\mathbf{a}}_\alpha$, $\mathbf{n}$ was used by Novozhilov and Shamina [6] and by the author [7,9]. The basis $\overset{*}{\mathbf{a}}_\alpha$, $\bar{\mathbf{n}}$ was used by Simmonds and Danielson [4,5].

Using (2.2.13) and (2.2.14) we may construct exact formulae for various tensors

$$\mathbf{U} = \overset{\vee}{\mathbf{a}}_\alpha \otimes \mathbf{a}^\alpha + \mathbf{n} \otimes \mathbf{n} \quad , \quad \mathbf{U}^{-1} = \mathbf{a}_\alpha \otimes \overset{\vee}{\mathbf{a}}{}^\alpha + \mathbf{n} \otimes \mathbf{n}$$
$$\mathbf{V} = \bar{\mathbf{a}}_\alpha \otimes \overset{*}{\mathbf{a}}{}^\alpha + \bar{\mathbf{n}} \otimes \bar{\mathbf{n}} \quad , \quad \mathbf{V}^{-1} = \overset{*}{\mathbf{a}}_\alpha \otimes \bar{\mathbf{a}}^\alpha + \bar{\mathbf{n}} \otimes \bar{\mathbf{n}} \tag{2.2.15}$$
$$\mathbf{R} = \bar{\mathbf{a}}_\alpha \otimes \overset{\vee}{\mathbf{a}}{}^\alpha + \bar{\mathbf{n}} \otimes \mathbf{n} = \overset{*}{\mathbf{a}}_\alpha \otimes \mathbf{a}^\alpha + \bar{\mathbf{n}} \otimes \mathbf{n}$$

Since $\mathbf{U} = \mathbf{U}(M)$ is positive definite and symmetric it has [18] three real and positive eigenvalues U_r in three orthogonal principal directions defined by a triad of unit vectors $\mathbf{f}_r$, which satisfy the set of equations

$$\mathbf{U}\mathbf{f}_r - U_r\mathbf{f}_r = \mathbf{0} \quad , \quad \mathbf{f}_r \cdot \mathbf{f}_s = \delta_{rs} \tag{2.2.16}$$

Under K - L constraints the vector $\mathbf{n}$ coincides with one of the principal directions, say $\mathbf{f}_3 \equiv \mathbf{n}$, with eigenvalue equal to +1. Thus $\mathbf{U}$ can be presented in the following diagonal form

$$\mathbf{U} = U_1\mathbf{f}_1 \otimes \mathbf{f}_1 + U_2\mathbf{f}_2 \otimes \mathbf{f}_2 + \mathbf{n} \otimes \mathbf{n} \tag{2.2.17}$$

According to $(2.2.9)_1$, (2.2.10) and (2.2.17) we have

$$\chi = \frac{1}{2} (\underset{\sim}{U}^2 - \underset{\sim}{1}) = \gamma_1 \mathfrak{k}_1 \ \text{\o} \ \mathfrak{k}_1 + \gamma_2 \mathfrak{k}_2 \ \text{\o} \ \mathfrak{k}_2$$

$$\gamma_\rho = \frac{1}{2} (U_\rho^2 - 1) \quad , \quad \gamma_3 = 0 \tag{2.2.18}$$

In what follows we shall frequently use a modified Lagrangean strain tensor $\overset{v}{\chi}$ defined by

$$\overset{v}{\chi} = \underset{\sim}{U} - \underset{\sim}{1} = \sqrt{\underset{\sim}{1} + 2\chi} - \underset{\sim}{1} = \overset{v}{\gamma}_{\alpha\beta} \underset{\sim}{a}^\alpha \ \text{\o} \ \underset{\sim}{a}^\beta \tag{2.2.19}$$

This tensor depends upon displacements through the non-rational relations. On the other hand, numerous geometric relations, which are non-rational in terms of χ , become polynomials when expressed in terms of $\overset{v}{\chi}$. For example, using (2.2.14), (2.2.19) and (2.2.18) we obtain

$$\overset{v}{\underset{\sim}{a}}_\alpha = \underset{\sim}{a}_\alpha + \overset{v}{\gamma}_{\alpha\beta} \underset{\sim}{a}^\beta \quad , \quad \overset{v}{\underset{\sim}{a}}^\alpha = \sqrt{\frac{a}{\overset{v}{a}}} \ \epsilon^{\alpha\beta} \epsilon_{\lambda\mu} (\delta^\mu_\beta + \overset{v}{\gamma}^\mu_\beta) \underset{\sim}{a}^\lambda$$

$$\overset{-}{a}_{\alpha\beta} = (\delta^\lambda_\alpha + \overset{v}{\gamma}^\lambda_\alpha)(\delta^\mu_\beta + \overset{v}{\gamma}^\mu_\beta) a_{\lambda\mu} \quad , \quad 2\gamma_{\alpha\beta} = 2\overset{v}{\gamma}_{\alpha\beta} + \overset{v}{\gamma}^\lambda_\alpha \overset{v}{\gamma}_{\lambda\beta} \tag{2.2.20}$$

Similar polynomial relations for $\overline{\epsilon}_{\alpha\beta}$, $\overline{\epsilon}^{\alpha\beta}$, $\sqrt{\frac{a}{\overline{a}}}$, $\overline{a}^{\alpha\beta}$ expressed in terms of $\overset{v}{\gamma}_{\alpha\beta}$ and the reference surface geometry are derived in [7,9] , together with some inverse formulae.

2.3. Finite rotation tensor and vector

It follows from the first of $(2.2.15)_3$ with (2.1.1) and (2.2.20) that we obtain the following general formulae for the finite rotation tensor in terms of displacements

$$\underset{\sim}{R} = \overline{a}^{\alpha\beta} (\underset{\sim}{a}_\alpha + \underset{\sim}{u}_{,\alpha}) \ \text{\o} \ (\underset{\sim}{a}_\beta + \overset{v}{\gamma}_{\beta\lambda} \underset{\sim}{a}^\lambda) + (n_\alpha \underset{\sim}{a}^\alpha + n \underset{\sim}{n}) \ \text{\o} \ \underset{\sim}{n} \tag{2.3.1}$$

According to the spectral decomposition theorem [15,18] the proper orthogonal tensor $\underset{\sim}{R}$ has only one real eigenvalue equal to +1 and two complex conjugate eigenvalues $\cos\omega \pm i\sin\omega$. Let $\underset{\sim}{e}_1$ be a unit vector of the corresponding first principal direction, satisfying $\underset{\sim}{R}\underset{\sim}{e}_1 = \underset{\sim}{e}_1$. Taking arbitrarily a unit vector $\underset{\sim}{e}_2 \perp \underset{\sim}{e}_1$ we obtain the third unit vector defined by $\underset{\sim}{e}_3 = \underset{\sim}{e}_1 \times \underset{\sim}{e}_2$. In this orthonormal basis $\underset{\sim}{e}_r$ the tensor $\underset{\sim}{R}$ takes the form [7,9]

$$\underset{\sim}{R} = \underset{\sim}{e}_1 \otimes \underset{\sim}{e}_1 + \cos\omega(\underset{\sim}{e}_2 \otimes \underset{\sim}{e}_2 + \underset{\sim}{e}_3 \otimes \underset{\sim}{e}_3) - \sin\omega(\underset{\sim}{e}_2 \otimes \underset{\sim}{e}_3 - \underset{\sim}{e}_3 \otimes \underset{\sim}{e}_2) \tag{2.3.2}$$

The direction defined by $\underset{\sim}{e}_1$ is called the axis of rotation and the angle ω , $|\omega| < \pi$, is called the angle of rotation about the axis of rotation.

Let us decompose a vector $\underset{\sim}{w} \in W$ in the basis $\underset{\sim}{e}_r$ to obtain

$$\underset{\sim}{w} = w_1\underset{\sim}{e}_1 + w_p(\cos\alpha\,\underset{\sim}{e}_2 + \sin\alpha\,\underset{\sim}{e}_3) \tag{2.3.3}$$

If the vector $\underset{\sim}{w}$ is acted on by the tensor $\underset{\sim}{R}$ then we obtain a new vector $\overset{*}{\underset{\sim}{w}}$, which in the basis $\underset{\sim}{e}_r$ becomes

$$\overset{*}{\underset{\sim}{w}} = \underset{\sim}{R}\underset{\sim}{w} = w_1\underset{\sim}{e}_1 + w_p[\cos(\alpha + \omega)\underset{\sim}{e}_2 + \sin(\alpha + \omega)\underset{\sim}{e}_3] \tag{2.3.4}$$

It is evident from (2.3.4) that the tensor $\underset{\sim}{R}$ rotates the vector $\underset{\sim}{w}$ through the angle ω about the axis of rotation defined by the vector $\underset{\sim}{e}_1$.

In what follows it is convenient to describe the rotational part of shell deformation by means of an equivalent finite rotation vector $\underset{\sim}{\Omega}$ used extensively in analytic mechanics of a rigid - body motion [3]. The direction of $\underset{\sim}{\Omega}$ is defined by $\underset{\sim}{e}_1 \equiv \underset{\sim}{e}$ and the length of it is taken here to be equal to $|\sin\omega|$. Therefore, for $|\omega| < \pi$ we define here the finite rotation vector by the formulae

$$\underset{\sim}{\Omega} = \sin\omega\,\underset{\sim}{e} \tag{2.3.5}$$

Note that $\underset{\sim}{\Omega}$ as defined here is not a vector in the usual sense. In particular, the rules of superposition of finite rotation vectors as discussed, for example, in [3] are different from the usual addition rules of the linear vector space W .

It is easy to show [9] that the rotated vector $\overset{*}{\underset{\sim}{w}}$ is calculated by means of $\underset{\sim}{\Omega}$ according to the following formulae

$$\overset{*}{\mathbf{v}} = \mathbf{v} + \boldsymbol{\Omega} \times \mathbf{v} + \frac{1}{2\cos^2\omega/2}\, \boldsymbol{\Omega} \times (\boldsymbol{\Omega} \times \mathbf{v}) =$$
$$= \cos\omega\, \mathbf{v} + \boldsymbol{\Omega} \times \mathbf{v} + \frac{1}{2\cos^2\omega/2}\, (\boldsymbol{\Omega} \cdot \mathbf{v}) \tag{2.3.6}$$

In particular, from (2.2.13) it follows that

$$\bar{\mathbf{a}}_\alpha = \check{\mathbf{a}}_\alpha + \boldsymbol{\Omega} \times \check{\mathbf{a}}_\alpha + \frac{1}{2\cos^2\omega/2}\, \boldsymbol{\Omega} \times (\boldsymbol{\Omega} \times \check{\mathbf{a}}_\alpha)$$
$$\bar{\mathbf{n}} = \mathbf{n} + \boldsymbol{\Omega} \times \mathbf{n} + \frac{1}{2\cos^2\omega/2}\, \boldsymbol{\Omega} \times (\boldsymbol{\Omega} \times \mathbf{n}) \tag{2.3.7}$$
$$\overset{*}{\mathbf{a}}_\alpha = \mathbf{a}_\alpha + \boldsymbol{\Omega} \times \mathbf{a}_\alpha + \frac{1}{2\cos^2\omega/2}\, \boldsymbol{\Omega} \times (\boldsymbol{\Omega} \times \mathbf{a}_\alpha)$$

The vector $\boldsymbol{\Omega}$ is uniquely defined by the tensor $\mathbf{R}$. Recalling that in Cartesian frame

$$\mathbf{r} = x^k \mathbf{i}_k \quad , \quad \mathbf{a}_\alpha = x^k{}_{,\alpha} \mathbf{i}_k \quad , \quad \mathbf{n} = \frac{1}{2}\, \epsilon^{\alpha\beta} x^k{}_{,\alpha} x^l{}_{,\beta} e_{klm} \mathbf{i}^m \tag{2.3.8}$$

the tensor $\mathbf{R}$ given by (2.3.1) may easily be expressed as $\mathbf{R} = R_{kl} \mathbf{i}^k \otimes \mathbf{i}^l$ where R_{kl}, depend only upon $\mathbf{u}$ and the geometry of M . Then for $\boldsymbol{\Omega}$ we have

$$\boldsymbol{\Omega} = -\frac{1}{2}\, e^{klm} R_{kl} \mathbf{i}_m \quad , \quad \cos\omega = \frac{1}{2}\,(\mathrm{tr}\mathbf{R} - 1) = \frac{1}{2}(R_{11}+R_{22}+R_{33}-1) \tag{2.3.9}$$

which defines the vector $\boldsymbol{\Omega}$ by means of its components with respect to the Cartesian basis $\mathbf{i}_k$. Its components with respect to the reference surface basis may be found by using (2.3.8).

It is possible to express $\boldsymbol{\Omega}$ in terms of $\mathbf{u}$ in equivalent alternative forms, directly with respect to the surface basis. If we multiply (2.3.7) by $\check{\mathbf{a}}_\alpha$ or $\mathbf{n}$ and use (2.3.6) we obtain

$$\bar{\mathbf{a}}_\alpha \cdot \check{\mathbf{a}}_\beta = \cos\omega\, \bar{a}_{\alpha\beta} + \boldsymbol{\Omega} \cdot (\check{\mathbf{a}}_\alpha \times \check{\mathbf{a}}_\beta) + \frac{1}{2\cos^2\omega/2}\, (\boldsymbol{\Omega} \cdot \check{\mathbf{a}}_\alpha)(\boldsymbol{\Omega} \cdot \check{\mathbf{a}}_\beta)$$
$$\bar{\mathbf{n}} \cdot \check{\mathbf{a}}_\alpha = \boldsymbol{\Omega} \cdot (\mathbf{n} \times \check{\mathbf{a}}_\alpha) + \frac{1}{2\cos^2\omega/2}\, (\boldsymbol{\Omega} \cdot \mathbf{n})(\boldsymbol{\Omega} \cdot \check{\mathbf{a}}_\alpha)$$
$$\bar{\mathbf{a}}_\alpha \cdot \mathbf{n} = \boldsymbol{\Omega} \cdot (\check{\mathbf{a}}_\alpha \times \mathbf{n}) + \frac{1}{2\cos^2\omega/2}\, (\boldsymbol{\Omega} \cdot \check{\mathbf{a}}_\alpha)(\boldsymbol{\Omega} \cdot \mathbf{n})$$

which leads to

$$\frac{1}{2}\bar{\epsilon}^{\alpha\beta}\bar{\mathbf{a}}_{\alpha}\cdot\check{\mathbf{a}}_{\beta}=\boldsymbol{\Omega}\cdot\mathbf{n}\quad,\quad\frac{1}{2}\bar{\epsilon}^{\alpha\beta}(\bar{\mathbf{n}}\cdot\check{\mathbf{a}}_{\alpha}-\bar{\mathbf{a}}_{\alpha}\cdot\mathbf{n})=\boldsymbol{\Omega}\cdot\check{\mathbf{a}}^{\beta}$$

Now the general expression for $\boldsymbol{\Omega}$ takes the form

$$\boldsymbol{\Omega}=\frac{1}{2}\bar{\epsilon}^{\alpha\beta}[(\bar{\mathbf{n}}\cdot\check{\mathbf{a}}_{\alpha}-\bar{\mathbf{a}}_{\alpha}\cdot\mathbf{n})\check{\mathbf{a}}_{\beta}+(\bar{\mathbf{a}}_{\alpha}\cdot\check{\mathbf{a}}_{\beta})\mathbf{n}] \tag{2.3.10a}$$

An equivalent alternative formula for $\boldsymbol{\Omega}$ has also been found in [7,9] to be

$$\boldsymbol{\Omega}=\frac{1}{2}(\check{\mathbf{a}}_{\alpha}\times\bar{\mathbf{a}}^{\alpha}+\mathbf{n}\times\bar{\mathbf{n}}) \tag{2.3.10b}$$

When expressed in the common reference basis $\mathbf{a}_{\alpha}$, $\mathbf{n}$ both forms (2.3.10) lead, after some transformation, to the following formula

$$2\boldsymbol{\Omega}=\epsilon_{\lambda\mu}[n^{\lambda}-\bar{a}^{\alpha\beta}(\delta_{\alpha}^{\lambda}+\check{\gamma}_{\alpha}^{\lambda})\phi_{\beta}]\mathbf{a}^{\mu}+\epsilon_{\lambda\mu}\bar{a}^{\alpha\beta}(\delta_{\alpha}^{\lambda}+\check{\gamma}_{\alpha}^{\lambda})l^{\mu}_{\cdot\beta}\mathbf{n} \tag{2.3.11}$$

Many geometric relations, which are expressed in terms of displacements, can also be presented in terms of $\check{\gamma}_{\alpha\beta}$ and $\boldsymbol{\Omega}$. For example, from (2.1.2), (2.1.3) and (2.3.7) it follows that

$$l_{\alpha\beta}=(\delta_{\beta}^{\lambda}+\check{\gamma}_{\beta}^{\lambda})[a_{\lambda\alpha}+\epsilon_{\lambda\alpha}(\boldsymbol{\Omega}\cdot\mathbf{n})-\frac{1}{2\cos^{2}\omega/2}(\boldsymbol{\Omega}\times\mathbf{a}_{\lambda})(\boldsymbol{\Omega}\times\mathbf{a}_{\alpha})]$$

$$\phi_{\beta}=(\delta_{\beta}^{\lambda}+\check{\gamma}_{\beta}^{\lambda})[\epsilon_{\alpha\lambda}(\boldsymbol{\Omega}\cdot\mathbf{a}^{\alpha})-\frac{1}{2\cos^{2}\omega/2}(\boldsymbol{\Omega}\times\mathbf{n})(\boldsymbol{\Omega}\times\mathbf{a}_{\lambda})]$$

$$n_{\alpha}=\epsilon_{\alpha\lambda}(\boldsymbol{\Omega}\cdot\mathbf{a}^{\lambda})-\frac{1}{2\cos^{2}\omega/2}(\boldsymbol{\Omega}\times\mathbf{a}_{\alpha})(\boldsymbol{\Omega}\times\mathbf{n}) \tag{2.3.12}$$

$$n=1-\frac{1}{2\cos^{2}\omega/2}(\boldsymbol{\Omega}\times\mathbf{n})(\boldsymbol{\Omega}\times\mathbf{n})$$

$$\mathbf{u}_{,\beta}=\check{\gamma}_{\beta}^{\alpha}\mathbf{a}_{\alpha}+(\delta_{\beta}^{\lambda}+\check{\gamma}_{\beta}^{\lambda})[\boldsymbol{\Omega}\times\check{\mathbf{a}}_{\lambda}+\frac{1}{2\cos^{2}\omega/2}\boldsymbol{\Omega}\times(\boldsymbol{\Omega}\times\check{\mathbf{a}}_{\lambda})]$$

Differentiation of $\boldsymbol{\Omega}$ and $\mathbf{R}$ along surface convected coordinate lines follows from the rules given in [2,20] for a three - dimensional continua. In particular, at the shell middle surface where $\zeta = 0$ we obtain [7] the following formula for differentiation of the finite rotation vector

$$\boldsymbol{\Omega}_{,\beta} = \cos\omega\, \mathbf{k}_\beta + \frac{1}{2}\boldsymbol{\Omega}\times\mathbf{k}_\beta - \frac{1}{4\cos^2\omega/2}\boldsymbol{\Omega}\times(\boldsymbol{\Omega}\times\mathbf{k}_\beta) \tag{2.3.13}$$

Here $\mathbf{k}_\beta$ may be called the vector of change of curvature along the coordinate line θ^α = const. , since multiplying (2.3.13) by $\boldsymbol{\Omega}$ we obtain

$$\mathbf{k}_\beta \cdot \boldsymbol{\Omega} = \omega_{,\beta}\sin\omega \quad , \quad \mathbf{k}_\beta \cdot \mathbf{e} = \omega_{,\beta} \tag{2.3.14}$$

Differentiating $(2.3.14)_1$ with respect to θ^α and using (2.3.13) we arrive after some transformations at

$$\epsilon^{\alpha\beta}(\mathbf{k}_{\beta|\alpha} + \frac{1}{2}\mathbf{k}_\alpha \times \mathbf{k}_\beta) = \mathbf{0} \tag{2.3.15}$$

These are integrability conditions of differential equations (2.3.13).

If we invert (2.3.13) the following formula for $\mathbf{k}_\beta$ in terms of $\boldsymbol{\Omega}$ may be derived

$$\mathbf{k}_\beta = \boldsymbol{\Omega}_{,\beta} + \frac{1}{2\cos^2\omega/2}\boldsymbol{\Omega}_{,\beta}\times\boldsymbol{\Omega} + \omega_{,\beta}\,\mathrm{tg}\,\omega/2\,\boldsymbol{\Omega} \tag{2.3.16}$$

It is possible to express $\mathbf{k}_\beta$ in terms of the surface strain measures. Appropriate transformations given by the author [7,8] lead to

$$\mathbf{k}_\beta = \bar{\epsilon}^{\lambda\mu}[(\kappa_{\beta\lambda} + b^\kappa_\beta \overset{\vee}{\gamma}_{\kappa\lambda})\mathbf{a}_\mu + (\overset{\vee}{\gamma}_{\beta\mu|\lambda} - \frac{1}{2}\overset{\vee}{\gamma}{}^\kappa_\mu \overset{\vee}{\gamma}_{\kappa\lambda|\beta})\mathbf{n}] \tag{2.3.17}$$

When (2.3.17) and $(2.2.20)_1$ are introduced into (2.3.15) we obtain three compatibility conditions expressed in terms of $\overset{\vee}{\gamma}_{\alpha\beta}$ and $\kappa_{\alpha\beta}$. They assure the existence of displacement field $\mathbf{u}$ compatible with these strain measures. If (2.3.16) is introduced into (2.3.15), we obtain three compatibility conditions expressed entirely in terms of $\boldsymbol{\Omega}$.

By solving (2.3.17) with respect to symmetric $\kappa_{\alpha\beta}$ we obtain

$$\kappa_{\alpha\beta} = \frac{1}{2}(\bar{\epsilon}_{\alpha\lambda}\mathbf{k}_\beta + \bar{\epsilon}_{\beta\lambda}\mathbf{k}_\alpha)\cdot\overset{\vee}{\mathbf{a}}{}^\lambda - \frac{1}{2}(b^\lambda_\alpha \overset{\vee}{\gamma}_{\lambda\beta} + b^\lambda_\beta \overset{\vee}{\gamma}_{\lambda\alpha}) \tag{2.3.18}$$

This relation together with (2.3.16) gives also general formula for $\kappa_{\alpha\beta}$ in terms of $\boldsymbol{\Omega}$ and $\overset{\vee}{\gamma}_{\alpha\beta}$.

The rotation of shell material fibres coinciding with principal directions of strain are described completely by $\mathbf{R}$ or $\boldsymbol{\Omega}$. Other shell fibres may suffer a rotation also during the pure stretch along principal directions of strain. Sometimes it is convenient to replace these two rotations by one equivalent total rotation. This approach is used below to describe the total rotation of the shell boundary element.

3. DEFORMATION OF SHELL BOUNDARY

Deformation of a shell near the lateral boundary is essentialy three - dimensional. Therefore, only by using three - dimensional analysis we are able to obtain the complete information as to the stress and strain state in this shell region. However, our goal is to derive adequate boundary conditions for interior shell equations within the first - approximation theory of thin isotropic elastic shells. Only for this limited goal it is justified to discuss deformation of the shell boundary under the Kirchhoff - Love constraints.

3.1. Total rotation of a boundary

Let C be a boundary curve at M defined by $\theta^\alpha = \theta^\alpha(s)$, where s is the length parameter along C. We assume here, that in the reference shell configuration the lateral boundary surface is rectilinear and orthogonal to M along C. The position of any $P \in \partial P$ is given by

$$\underset{\sim}{p} = \underset{\sim}{p}(s,\zeta) = \underset{\sim}{r}(s) + \zeta\underset{\sim}{n}(s) \tag{3.1.1}$$

With each point M of C we associate vectors: $\underset{\sim}{t} = \dfrac{d\underset{\sim}{r}}{ds}$, the unit tangent to C, and $\underset{\sim}{\nu} = \underset{\sim}{t} \times \underset{\sim}{n}$, the outward unit normal.

After the shell deformation compatible with K - L constraints C transforms into $\bar{C}$ and ∂P into $\partial\bar{P}$, which still remains orthogonal to $\bar{M}$, and for any $\bar{P} \in \partial\bar{P}$ we have

$$\bar{\mathbf{p}} = \bar{\mathbf{p}}(s,\zeta) = \bar{\mathbf{r}}(s) + \zeta\bar{\mathbf{n}}(s) \tag{3.1.2}$$

During the shell deformation the orthonormal triad $\mathbf{v}$, $\mathbf{t}$, $\mathbf{n}$ transforms into an orthogonal triad $\bar{\mathbf{a}}_\nu$, $\bar{\mathbf{a}}_t$, $\bar{\mathbf{n}}$, where

$$\begin{aligned} \bar{\mathbf{a}}_t &= \frac{d\bar{\mathbf{r}}}{ds} = t^\alpha \bar{\mathbf{a}}_\alpha = \mathbf{t} + \frac{d\mathbf{u}}{ds} \\ \bar{\mathbf{a}}_\nu &= \bar{\mathbf{a}}_t \times \bar{\mathbf{n}} = \bar{\epsilon}_{\lambda\alpha} t^\alpha \bar{\mathbf{a}}^\lambda = \sqrt{\frac{\bar{a}}{a}}\,\nu_\alpha \bar{\mathbf{a}}^\alpha = (\mathbf{t} + \frac{d\mathbf{u}}{ds}) \times (\mathbf{n} + \boldsymbol{\beta}) \end{aligned} \tag{3.1.3}$$

$$|\bar{\mathbf{a}}_t| = |\bar{\mathbf{a}}_\nu| = \bar{a}_t = \sqrt{1 + 2\gamma_{tt}} \quad , \quad \gamma_{tt} = \gamma_{\alpha\beta} t^\alpha t^\beta \tag{3.1.4}$$

According to the polar decomposition theorem (2.2.10) the boundary deformation can also be decomposed into a rigid - body translation, a pure strain performed by means of $\mathbf{U}$, $\boldsymbol{\gamma}$ or $\check{\boldsymbol{\gamma}}$ and a rigid - body rotation performed by means of $\mathbf{R}$ or $\boldsymbol{\Omega}$. Therefore, there should be such intermediate vectors $\check{\mathbf{a}}_t$, $\check{\mathbf{a}}_\nu$ for which

$$\begin{aligned} \bar{\mathbf{a}}_t &= \check{\mathbf{a}}_t + \boldsymbol{\Omega} \times \check{\mathbf{a}}_t + \frac{1}{2\cos^2\omega/2}\,\boldsymbol{\Omega} \times (\boldsymbol{\Omega} \times \check{\mathbf{a}}_t) \\ \bar{\mathbf{a}}_\nu &= \check{\mathbf{a}}_\nu + \boldsymbol{\Omega} \times \check{\mathbf{a}}_\nu + \frac{1}{2\cos^2\omega/2}\,\boldsymbol{\Omega} \times (\boldsymbol{\Omega} \times \check{\mathbf{a}}_\nu) \end{aligned} \tag{3.1.5}$$

In view of (2.2.20) these vectors are calculated according to

$$\check{\mathbf{a}}_t = \check{\gamma}_{\nu t}\mathbf{v} + (1 + \check{\gamma}_{tt})\mathbf{t} \quad , \quad \check{\mathbf{a}}_\nu = (1 + \check{\gamma}_{tt})\mathbf{v} - \check{\gamma}_{\nu t}\mathbf{t} \tag{3.1.6}$$

where

$$\check{\gamma}_{\nu t} = \check{\gamma}_{\alpha\beta}\nu^\alpha t^\beta \quad , \quad \check{\gamma}_{tt} = \check{\gamma}_{\alpha\beta} t^\alpha t^\beta \tag{3.1.7}$$

are physical components of $\check{\boldsymbol{\gamma}}$ at C .

The directions defined by $\mathbf{v}$ and $\mathbf{t}$ do not coincide, in general, with principal directions of strain at $M \in C$ defined by $\mathbf{c}_1$ and $\mathbf{c}_2$ in (2.2.17). Since during the pure strain only the principal directions are stretched without rotation, the vectors $\mathbf{v}$ and $\mathbf{t}$ not only change their lengths but, in general, suffer rotation as well.

Let us denote by $\overset{\vee}{\underset{\sim}{\Omega}}_t$ the finite rotation vector of the boundary caused by the pure strain. Applying $(2.3.10)_2$ we obtain

$$\overset{\vee}{\underset{\sim}{\Omega}}_t = \sin \overset{\vee}{\omega}_t \underset{\sim}{n} \quad , \quad \sin \overset{\vee}{\omega}_t = - \frac{\overset{\vee}{\gamma}_{\nu t}}{\bar{a}_t}$$

$$\cos \overset{\vee}{\omega}_t = \frac{1}{\bar{a}_t}(1 + \overset{\vee}{\gamma}_{tt}) \quad , \quad 2\cos^2 \overset{\vee}{\omega}_t/2 = \frac{1}{\bar{a}_t}(\bar{a}_t + 1 + \overset{\vee}{\gamma}_{tt}) \tag{3.1.8}$$

and transformation of $\underset{\sim}{\nu}$ and $\underset{\sim}{t}$ into $\overset{\vee}{\underset{\sim}{a}}_\nu$ and $\overset{\vee}{\underset{\sim}{a}}_t$ may be expressed in the form

$$\overset{\vee}{\underset{\sim}{a}}_\nu = \bar{a}_t[\underset{\sim}{\nu} + \overset{\vee}{\underset{\sim}{\Omega}}_t \times \underset{\sim}{\nu} + \frac{1}{2\cos^2 \overset{\vee}{\omega}_t/2} \overset{\vee}{\underset{\sim}{\Omega}}_t \times (\overset{\vee}{\underset{\sim}{\Omega}}_t \times \underset{\sim}{\nu})]$$

$$\overset{\vee}{\underset{\sim}{a}}_t = \bar{a}_t[\underset{\sim}{t} + \overset{\vee}{\underset{\sim}{\Omega}}_t \times \underset{\sim}{t} + \frac{1}{2\cos^2 \overset{\vee}{\omega}_t/2} \overset{\vee}{\underset{\sim}{\Omega}}_t \times (\overset{\vee}{\underset{\sim}{\Omega}}_t \times \underset{\sim}{t})] \tag{3.1.9}$$

It is seen from (3.1.9) and (3.1.5) that transformation of $\underset{\sim}{\nu}$ and $\underset{\sim}{t}$ into $\bar{\underset{\sim}{a}}_\nu$ and $\bar{\underset{\sim}{a}}_t$ consists of extension by the factor $\bar{a}_t$ and two successive rotations performed by $\overset{\vee}{\underset{\sim}{\Omega}}_t$ and $\underset{\sim}{\Omega}$. In what follows it is convenient to replace these two rotations by a single equivalent rotation performed by an equivalent total rotation vector $\underset{\sim}{\Omega}_t$. Adopting the superposition rule given, for example, in [3] to our definition (2.3.5) of the finite rotation vector we obtain

$$\underset{\sim}{\Omega}_t = (1 - \frac{\overset{\vee}{\underset{\sim}{\Omega}}_t \cdot \underset{\sim}{\Omega}}{4\cos^2 \overset{\vee}{\omega}_t/2 \cos^2 \omega/2})[\cos^2\omega/2 \overset{\vee}{\underset{\sim}{\Omega}}_t + \cos^2 \overset{\vee}{\omega}_t/2 \underset{\sim}{\Omega} + \frac{1}{2}\underset{\sim}{\Omega} \times \overset{\vee}{\underset{\sim}{\Omega}}_t] \tag{3.1.10}$$

$$\underset{\sim}{\Omega}_t = \sin \omega_t \underset{\sim}{e}_t \tag{3.1.11}$$

Now the transformation of $\underset{\sim}{\nu}$, $\underset{\sim}{t}$, $\underset{\sim}{n}$ into $\bar{\underset{\sim}{a}}_\nu$, $\bar{\underset{\sim}{a}}_t$, $\bar{\underset{\sim}{n}}$ becomes

$$\bar{\underset{\sim}{a}}_\nu = \bar{a}_t[\underset{\sim}{\nu} + \underset{\sim}{\Omega}_t \times \underset{\sim}{\nu} + \frac{1}{2\cos^2\omega_t/2} \underset{\sim}{\Omega}_t \times (\underset{\sim}{\Omega}_t \times \underset{\sim}{\nu})]$$

$$\bar{\underset{\sim}{a}}_t = \bar{a}_t[\underset{\sim}{t} + \underset{\sim}{\Omega}_t \times \underset{\sim}{t} + \frac{1}{2\cos^2\omega_t/2} \underset{\sim}{\Omega}_t \times (\underset{\sim}{\Omega}_t \times \underset{\sim}{t})] \tag{3.1.12}$$

$$\bar{\underset{\sim}{n}} = \underset{\sim}{n} + \underset{\sim}{\Omega}_t \times \underset{\sim}{n} + \frac{1}{2\cos^2\omega_t/2} \underset{\sim}{\Omega}_t \times (\underset{\sim}{\Omega}_t \times \underset{\sim}{n})$$

Note the formal similarity of (3.1.12) and (3.1.9).

An alternative equivalent relation for $\underset{\sim}{\Omega}_t$ can be found directly, by using the cross product as in (2.3.10b) to obtain

$$2\underset{\sim}{\Omega}_t = \frac{1}{\bar{a}_t}(\underset{\sim}{\nu} \times \bar{\underset{\sim}{a}}_\nu + \underset{\sim}{t} \times \bar{\underset{\sim}{a}}_t) + \underset{\sim}{n} \times \bar{\underset{\sim}{n}} \qquad (3.1.13)$$

which gives much simpler formula for $\underset{\sim}{\Omega}_t$ in terms of $\underset{\sim}{u}$.

3.2. Differentiation along the boundary

Differentiating $\underset{\sim}{\nu}$, $\underset{\sim}{t}$, $\underset{\sim}{n}$ along C we obtain [9]

$$\frac{d\underset{\sim}{\nu}}{ds} = \underset{\sim}{\omega}_t \times \underset{\sim}{\nu} , \quad \frac{d\underset{\sim}{t}}{ds} = \underset{\sim}{\omega}_t \times \underset{\sim}{t} , \quad \frac{d\underset{\sim}{n}}{ds} = \underset{\sim}{\omega}_t \times \underset{\sim}{n}$$
$$\underset{\sim}{\omega}_t = \sigma_t \underset{\sim}{\nu} + \tau_t \underset{\sim}{t} + \kappa_t \underset{\sim}{n} \qquad (3.2.1)$$

where in terms of the surface geometry

$$\sigma_t = t^\alpha t^\beta b_{\alpha\beta} , \quad \tau_t = - \nu^\alpha t^\beta b_{\alpha\beta}$$
$$\kappa_t = t_\alpha \nu^\alpha{}_{|\beta} t^\beta = - \nu_\alpha t^\alpha{}_{|\beta} t^\beta \qquad (3.2.2)$$

Here σ_t is the normal curvature, τ_t is the geodesic torsion and κ_t is the geodesic curvature of C, respectively.

Similar rules hold when differentiating $\underset{\sim}{\nu}$, $\underset{\sim}{t}$, $\underset{\sim}{n}$ in direction of the outward unit normal

$$\frac{d\underset{\sim}{\nu}}{ds_\nu} = - \underset{\sim}{\omega}_\nu \times \underset{\sim}{\nu} , \quad \frac{d\underset{\sim}{t}}{ds_\nu} = - \underset{\sim}{\omega}_\nu \times \underset{\sim}{t} , \quad \frac{d\underset{\sim}{n}}{ds_\nu} = - \underset{\sim}{\omega}_\nu \times \underset{\sim}{n}$$
$$\underset{\sim}{\omega}_\nu = \tau_\nu \underset{\sim}{\nu} + \sigma_\nu \underset{\sim}{t} + \kappa_\nu \underset{\sim}{n} \qquad (3.2.3)$$

where

$$\sigma_\nu = \nu^\alpha \nu^\beta b_{\alpha\beta} , \quad \tau_\nu = - t^\alpha \nu^\beta b_{\alpha\beta} = \tau_t$$
$$\kappa_\nu = \nu_\alpha t^\alpha{}_{|\beta} \nu^\beta = - t_\alpha \nu^\alpha{}_{|\beta} \nu^\beta \qquad (3.2.4)$$

Let us define unit vectors at the deformed boundary $\bar{C}$

$$\bar{\nu} = \frac{1}{\bar{a}_t}\bar{a}_\nu \quad , \quad \bar{t} = \frac{1}{\bar{a}_t}\bar{a}_t \tag{3.2.5}$$

Differentiating the orthonormal triad $\bar{\nu}$, $\bar{t}$, $\bar{n}$ along C we obtain

$$\frac{d\bar{\nu}}{ds} = \bar{\omega}_t \times \bar{\nu} \quad , \quad \frac{d\bar{t}}{ds} = \bar{\omega}_t \times \bar{t} \quad , \quad \frac{d\bar{n}}{ds} = \bar{\omega}_t \times \bar{n}$$
$$\bar{\omega}_t = \bar{a}_t(\bar{\sigma}_t\bar{\nu} + \bar{\tau}_t\bar{t} + \bar{\kappa}_t\bar{n}) \tag{3.2.6}$$

where $\bar{\sigma}_t$, $\bar{\tau}_t$ and $\bar{\kappa}_t$ are the normal curvature, the geodesic torsion and the geodesic curvature of $\bar{C}$, respectively

Using (3.1.12) and (3.2.5) the following identity may be derived

$$\bar{n}\cdot\nu - \bar{\nu}\cdot n = 2\Omega_t\cdot t \tag{3.2.7}$$

which differentiated with respect to s gives

$$2\frac{d\Omega_t}{ds}\cdot t = \bar{\omega}_t\cdot(\bar{n}\times\nu - \bar{\nu}\times n) -$$
$$- \omega_t\cdot(n\times\bar{\nu} - \nu\times\bar{n}) - 2(\Omega_t\times\omega_t)\cdot t \tag{3.2.8}$$

Since the vector $\bar{\omega}_t$ is defined in deformed shell configuration we use the polar decomposition theorem (2.2.10) in order to express it in terms of a vector $\check{\omega}_t$ according to

$$\bar{\omega}_t = \check{\omega}_t + \Omega_t\times\check{\omega}_t + \frac{1}{2\cos^2\omega_t/2}\Omega_t\times(\Omega_t\times\check{\omega}_t)$$
$$\check{\omega}_t = \bar{a}_t(\bar{\sigma}_t\nu + \bar{\tau}_t t + \bar{\kappa}_t n) \tag{3.2.9}$$

Using (3.2.6), (3.2.1) and (3.2.9), together with some vector and trigonometric identities, from (3.2.8) we arrive at the following formula for differentiation of the total finite rotation vector

$$\frac{d\Omega_t}{ds} = \cos\omega_t k_t + \frac{1}{2}\Omega_t\times k_t - \frac{1}{4\cos^2\omega_t/2}\Omega_t\times(\Omega_t\times k_t) \tag{3.2.10}$$

where

$$\underset{\sim}{k}_t = \overset{\vee}{\underset{\sim}{\omega}}_t - \underset{\sim}{\omega}_t = - k_{tt}\underset{\sim}{\nu} + k_{\nu t}\underset{\sim}{t} - k_{nt}\underset{\sim}{n} \tag{3.2.11}$$

$$- k_{tt} = \bar{a}_t\bar{\sigma}_t - \sigma_t \quad , \quad k_{\nu t} = \bar{a}_t\bar{\tau}_t - \tau_t \quad , \quad - k_{nt} = \bar{a}_t\bar{\kappa}_t - \kappa_t \tag{3.2.12}$$

Here $\underset{\sim}{k}_t$ defined by (3.2.11) is called the vector of change of curvature of the shell boundary contour.

Note that all differentiation formulae depend only upon $\underset{\sim}{k}_t$ and γ_{tt}. If these parameters are known then $\overset{\vee}{\underset{\sim}{\omega}}_t$ follows from (3.2.11), $\bar{\underset{\sim}{\omega}}_t$ follows from (3.2.9) and derivatives of $\bar{\underset{\sim}{\nu}}$, $\bar{\underset{\sim}{t}}$, $\bar{\underset{\sim}{n}}$ become definite as well, according to (3.2.6).

In order to calculate $\underset{\sim}{k}_t$ let us differentiate $\bar{\underset{\sim}{a}}_t$ and $\bar{\underset{\sim}{n}}$ with respect to s to obtain

$$\begin{aligned} \frac{d}{ds}\bar{\underset{\sim}{a}}_t &= (t^{\alpha}{}_{|\beta} + \bar{a}^{\alpha\nu}\gamma_{\nu\lambda\beta}t^{\lambda})t^{\beta}\bar{\underset{\sim}{a}}_{\alpha} + (\sigma_t - \kappa_{tt})\bar{\underset{\sim}{n}} \\ \frac{d}{ds}\bar{\underset{\sim}{n}} &= - (b_{\alpha\beta} - \kappa_{\alpha\beta})t^{\beta}\bar{\underset{\sim}{a}}^{\alpha} \end{aligned} \tag{3.2.13}$$

From (3.2.5) we also have

$$\frac{d\bar{\underset{\sim}{t}}}{ds} = \frac{1}{\bar{a}_t}\frac{d\bar{\underset{\sim}{a}}_t}{ds} - \frac{1}{\bar{a}_t^2}\frac{d\gamma_{tt}}{ds}\bar{\underset{\sim}{t}} \tag{3.2.14}$$

It follows now from (3.2.6), (3.2.13) and (3.2.14) that

$$\begin{aligned} \bar{a}_t\bar{\sigma}_t &= \frac{1}{\bar{a}_t}(\sigma_t - \kappa_{tt}) \quad , \quad \bar{a}_t\bar{\tau}_t = - \frac{1}{\bar{a}_t}\sqrt{\frac{\bar{a}}{a}}\,\nu_{\kappa}\bar{a}^{\kappa\alpha}(b_{\alpha\beta} - \kappa_{\alpha\beta})t^{\beta} \\ \bar{a}_t\bar{\kappa}_t &= + \frac{1}{\bar{a}_t^2}\sqrt{\frac{\bar{a}}{a}}\,(\kappa_t - \nu_{\kappa}\bar{a}^{\kappa\lambda}\gamma_{\lambda\alpha\beta}t^{\alpha}t^{\beta}) \end{aligned} \tag{3.2.15}$$

When all tensor components in (3.2.16) are expressed in terms of physical components [9] from (3.2.12) we obtain

$$\begin{aligned} k_{tt} &= \sigma_t(1 - \frac{1}{\bar{a}_t}) + \frac{\kappa_{tt}}{\bar{a}_t} \\ k_{\nu t} &= \sqrt{\frac{a}{\bar{a}}}\,[\bar{a}_t(\tau_t + \kappa_{\nu t}) + \frac{2\gamma_{\nu t}}{\bar{a}_t^2}(\sigma_t - \kappa_{tt})] - \tau_t \end{aligned} \tag{3.2.16}$$

$$k_{nt} = \kappa_t\left(1 - \frac{1}{\bar{a}_t^2}\sqrt{\frac{\bar{a}}{a}}\right) - \frac{2\gamma_{\nu t}}{\bar{a}_t^2}\sqrt{\frac{a}{\bar{a}}}\left(\frac{d\gamma_{tt}}{ds} + 2\kappa_t\gamma_{\nu t}\right) + \tag{3.2.16}$$
$$+\sqrt{\frac{a}{\bar{a}}}\left[2\frac{d\gamma_{\nu t}}{ds} - \frac{d\gamma_{tt}}{ds_\nu} + 2\kappa_\nu\gamma_{\nu t} + 2\kappa_t(\gamma_{\nu\nu} - \gamma_{tt})\right]$$

Using the results of [7] an equivalent expression for $k_{\nu t}$ is derived

$$k_{\nu t} = \tau_t\left(\frac{1}{\bar{a}_t}\sqrt{\frac{\bar{a}}{a}} - 1\right) + \frac{1}{\bar{a}_t}\sqrt{\frac{\bar{a}}{a}}\,[\bar{a}_t^2(\kappa_{\nu t} + 2\sigma_t\gamma_{\nu t} - 2\tau_t\gamma_{\nu\nu}) - \tag{3.2.17}$$
$$- 2\gamma_{\nu t}(\kappa_{tt} + 2\sigma_t\gamma_{tt} - 2\tau_t\gamma_{\nu t})]$$

Therefore, $\underset{\sim}{k}_t$ has been expressed entirely in terms of physical components of the surface strain measures at the boundary.

3.3. Geometric boundary conditions

According to (3.1.2), (2.2.1) and $(2.2.5)_2$ the deformed lateral boundary surface $\partial\bar{P}$ is uniquely defined by assuming two vector functions

$$\underset{\sim}{u}(s) = \underset{\sim}{A}(s) \quad , \quad \underset{\sim}{\beta}(s) = \underset{\sim}{B}(s) \quad \text{at } C \tag{3.3.1}$$

If we express $\underset{\sim}{\beta}$ by components according to

$$\underset{\sim}{\beta} = \beta_\nu\bar{\underset{\sim}{a}}_\nu + \beta_t\bar{\underset{\sim}{a}}_t + \beta\bar{\underset{\sim}{n}} \tag{3.3.2}$$

then under K - L constraints

$$\beta_\nu = -\frac{1}{\bar{a}_t^2}\sqrt{\frac{\bar{a}}{a}}\,\nu_\alpha\bar{a}^{\alpha\beta}(\underset{\sim}{u}_{,\beta}\cdot\underset{\sim}{n})$$
$$\beta_t = -\frac{1}{\bar{a}_t^2}\frac{d\underset{\sim}{u}}{ds}\cdot\underset{\sim}{n} \quad , \quad \beta = 1 - \sqrt{1 - \bar{a}_t^2(\beta_\nu^2 + \beta_t^2)} \tag{3.3.3}$$

It is seen that only $\underset{\sim}{u}$ and β_ν appear as independent parameters in the non-linear K - L type theory of shells.

The conditions of the type

$$\underset{\sim}{u}(s) = \underset{\sim}{A}(s) \quad , \quad \beta_\nu(s) = b(s) \quad \text{at} \quad C \tag{3.3.4}$$

are called displacement boundary conditions of the K - L non-linear shell theory.

Let us differentiate (3.1.2) to obtain

$$\frac{\partial \bar{\underset{\sim}{p}}}{\partial s} = \bar{\underset{\sim}{a}}_t + \zeta \frac{d}{ds} \bar{\underset{\sim}{n}} \quad , \quad \frac{\partial \bar{\underset{\sim}{p}}}{\partial \zeta} = \bar{\underset{\sim}{n}} \quad , \quad \frac{d\bar{\underset{\sim}{r}}}{ds} = \bar{\underset{\sim}{a}}_t \tag{3.3.5}$$

These differential equations define the same boundary surface $\partial\bar{P}$ implicitly, with accuracy up to a constant translation in space. In order to obtain $\partial\bar{P}$ explicitly the equation (3.3.5) should be solved.

The right-hand sides of (3.3.5) are given if values of $\bar{\underset{\sim}{a}}_t$ and $\bar{\underset{\sim}{n}}$ are assumed. In (3.1.12) the vectors $\bar{\underset{\sim}{a}}_t$ and $\bar{\underset{\sim}{n}}$ were expressed in terms of $\underset{\sim}{\Omega}_t$ and γ_{tt}. Therefore, in order to establish differential equations (3.3.5) it is enough to assume only $\underset{\sim}{\Omega}_t$ and γ_{tt} at C.

Conditions of the type

$$\underset{\sim}{\Omega}_t(s) = \underset{\sim}{m}(s) \quad , \quad \gamma_{tt}(s) = l(s) \quad \text{at} \quad C \tag{3.3.6}$$

are called kinematical boundary conditions of the K - L non-linear theory of shells.

For values of the assumed vectors we have

$$\begin{aligned} \underset{\sim}{m} &= \frac{1}{2\sqrt{1+2l}} \{\underset{\sim}{\nu} \times [\underset{\sim}{t} \times \underset{\sim}{B} + \frac{d\underset{\sim}{A}}{ds} \times (\underset{\sim}{n} + \underset{\sim}{B})] + \underset{\sim}{t} \times \frac{d\underset{\sim}{A}}{ds}\} + \underset{\sim}{n} \times \underset{\sim}{B} \\ l &= \underset{\sim}{t} \cdot \frac{d\underset{\sim}{A}}{ds} + \frac{1}{2} \frac{d\underset{\sim}{A}}{ds} \cdot \frac{d\underset{\sim}{A}}{ds} \end{aligned} \tag{3.3.7}$$

Let us differentiate again (3.3.5) to obtain

$$\frac{\partial^2 \bar{\underset{\sim}{p}}}{\partial s^2} = \frac{d}{ds} \bar{\underset{\sim}{a}}_t + \zeta \frac{d^2}{ds^2} \bar{\underset{\sim}{n}} \quad , \quad \frac{\partial^2 \bar{\underset{\sim}{p}}}{\partial s \partial \zeta} = \frac{d}{ds} \bar{\underset{\sim}{n}} \quad , \quad \frac{d^2 \bar{\underset{\sim}{r}}}{ds^2} = \frac{d}{ds} \bar{\underset{\sim}{a}}_t \tag{3.3.8}$$

These differential equations also define the same boundary surface $\partial\bar{P}$ implicitly, with accuracy up to a translation linearly varying with s (that means, up to a rigid - body motion in space). The right-hand sides of (3.3.8) are established if values of $\frac{d}{ds}\bar{a}_t$ and $\frac{d}{ds}\bar{n}$ are assumed. However, according to (3.2.6), (3.2.9) and (3.2.11) these vectors may be expressed in terms of k_t and γ_{tt}, which are sufficient for establishing the equations (3.3.8).

Conditions of the type

$$k_t(s) = q(s) \quad , \quad \gamma_{tt}(s) = 1(s) \quad \text{at } C \tag{3.3.9}$$

are called deformational boundary conditions of the K - L non-linear theory of shells. For q here we have the relation

$$q = \frac{dm}{ds} - \frac{1}{1 + \sqrt{1 - m \cdot m}}\left[m \frac{d}{ds}\sqrt{1 - m \cdot m} + m \times \frac{dm}{ds}\right] \tag{3.3.10}$$

When values of u and β_ν are known along C, then Ω_t, γ_{tt} and k_t are easily calculated by using (3.3.7) and (3.3.10). If only k_t is known in advance, in order to obtain Ω_t the differential equation (3.2.10) should be solved. Note that the structure of (3.2.10) is analogous to the one describing the motion of a rigid body about a fixed point[3] and the methods of solution developed in analytic mechanics[21] may be of assistance in calculating Ω_t from the known k_t.

If values of Ω_t and γ_{tt} are known along C, then β_ν follows from (2.3.7) and (3.3.2). In order to obtain u at C the following differential equation should be solved

$$\frac{du}{ds} = \overset{\vee}{\gamma}_{tt} t + \bar{a}_t\left[\Omega_t \times t + \frac{1}{2\cos^2\omega_t/2}\,\Omega_t \times (\Omega_t \times t)\right] \tag{3.3.11}$$

Most shell problems are solved in terms of displacements and then displacement boundary conditions are used. The kinematical boundary conditions are adequate for the shell problems formulated by means of the finite rotation vector[4]. Particularly interesting seem to be deformational boundary conditions, since they are expressed entirely in terms of the shell strain measures. This allows us to formulate the shell problems directly in terms of strain (or stress) measures.

4. BASIC SHELL EQUATIONS

The two-dimensional equilibrium equations and natural boundary conditions, in terms of symmetric internal force and moment resultants, may be obtained most easily by using the virtual work principle [11]. Here one should clearly distinguish between the Eulerian and Lagrangean descriptions [7,9]. When expressed in vector form the Eulerian and Lagrangean equilibrium equations and natural boundary conditions are related through the deformation gradient tensor $\underset{\sim}{G}$. By decomposing $\underset{\sim}{G}$ according to (2.2.10) we may construct various component forms of shell equations with respect to bases $\underset{\sim}{a}_\alpha$, $\underset{\sim}{n}$ or $\check{\underset{\sim}{a}}_\alpha$, $\underset{\sim}{n}$ or $\overset{*}{\underset{\sim}{a}}_\alpha$, $\bar{\underset{\sim}{n}}$ or $\bar{\underset{\sim}{a}}_\alpha$, $\bar{\underset{\sim}{n}}$, respectively. Some of these equations and their consistent simplification under small elastic strains and under additionally restricted rotations will be discussed below.

4.1. Equilibrium equations and static boundary conditions

Let a shell with simply connected middle surface be in equilibrium, under the surface force $\bar{\underset{\sim}{p}} = \bar{p}^\alpha \bar{\underset{\sim}{a}}_\alpha + \bar{p}\bar{\underset{\sim}{n}}$, per unit area of deformed surface $\bar{M}$, and under the boundary force $\bar{\underset{\sim}{F}} = \bar{F}^\alpha \bar{\underset{\sim}{a}}_\alpha + \bar{F}\bar{\underset{\sim}{n}} = \bar{F}_\nu \bar{\underset{\sim}{\nu}} + \bar{F}_t \bar{\underset{\sim}{t}} + \bar{F}\bar{\underset{\sim}{n}}$ and the boundary couple $\bar{\underset{\sim}{K}} = \bar{\epsilon}_{\lambda\mu} \bar{K}^\lambda \bar{\underset{\sim}{a}}^\mu = -\bar{K}_t \bar{\underset{\sim}{\nu}} + \bar{K}_\nu \bar{\underset{\sim}{t}}$, per unit length of deformed boundary $\bar{C}$. For completeness, we could also introduce here the surface moments $\bar{\underset{\sim}{m}}$ and the normal component of the boundary couple $\bar{K} = \bar{\underset{\sim}{K}} \cdot \bar{\underset{\sim}{n}}$. Since for thin shells, within the first approximation theory, these loadings are of secondary importance, we assume here at once $\bar{\underset{\sim}{m}} \equiv \underset{\sim}{0}$ and $\bar{K} \equiv 0$.

For any additional virtual displacement field $\delta\bar{\underset{\sim}{u}} = \delta\bar{u}_\alpha \bar{\underset{\sim}{a}}^\alpha + \delta\bar{w}\bar{\underset{\sim}{n}}$, subject to geometrical constraints, there should be symmetric Eulerian stress and couple resultant tensors

$$\bar{\underset{\sim}{N}} = \bar{N}^{\alpha\beta}\bar{\underset{\sim}{a}}_{\alpha} \otimes \bar{\underset{\sim}{a}}_{\beta} \quad , \quad \bar{\underset{\sim}{M}} = \bar{M}^{\alpha\beta}\bar{\underset{\sim}{a}}_{\alpha} \otimes \bar{\underset{\sim}{a}}_{\beta} \tag{4.1.1}$$

such that the Eulerian virtual work principle IVW = EVW takes the form

$$\iint_{\bar{M}}(\bar{N}^{\alpha\beta}\delta\bar{\gamma}_{\alpha\beta} + \bar{M}^{\alpha\beta}\delta\bar{\kappa}_{\alpha\beta})d\bar{A} = \iint_{\bar{M}}\bar{\underset{\sim}{p}}\cdot\delta\bar{\underset{\sim}{u}}d\bar{A} + \int_{\bar{C}}(\bar{\underset{\sim}{F}}\cdot\delta\bar{\underset{\sim}{u}} + \bar{\underset{\sim}{K}}\cdot\delta\bar{\underset{\sim}{\Omega}}_{t})d\bar{s} \tag{4.1.2}$$

where in deformed surface geometry

$$\begin{aligned}
\delta\bar{\gamma}_{\alpha\beta} &= \frac{1}{2}(\delta\bar{u}_{\alpha\|\beta} + \delta\bar{u}_{\beta\|\alpha}) - \bar{b}_{\alpha\beta}\delta\bar{w} \\
\delta\bar{\kappa}_{\alpha\beta} &= -\delta\bar{w}_{\|\alpha\beta} - \bar{b}^{\lambda}_{\alpha}\delta\bar{u}_{\lambda\|\beta} - \bar{b}^{\lambda}_{\beta}\delta\bar{u}_{\lambda\|\alpha} - \bar{b}^{\lambda}_{\alpha\|\beta}\delta\bar{u}_{\lambda} + \bar{b}^{\lambda}_{\alpha}\bar{b}_{\lambda\beta}\delta\bar{w} \\
\delta\bar{\underset{\sim}{\Omega}}_{t} &= \bar{\epsilon}^{\beta\alpha}(\delta\bar{\phi}_{\alpha}\bar{\underset{\sim}{a}}_{\beta} + \frac{1}{2}\delta\bar{\omega}_{\beta\alpha}\bar{\underset{\sim}{n}}) - \delta\bar{\theta}_{\nu t}\bar{\underset{\sim}{n}}
\end{aligned} \tag{4.1.3}$$

and $\bar{s}$ is the length parameter along $\bar{C}$.

By applying the Stockes' theorem the virtual work principle can be transformed [9] into

$$\begin{aligned}
-\iint_{\bar{M}}\bar{\underset{\sim}{N}}^{\beta}{}_{\|\beta}\cdot\delta\bar{\underset{\sim}{u}}d\bar{A} + J_i &= \iint_{\bar{M}}\bar{\underset{\sim}{p}}\cdot\delta\bar{\underset{\sim}{u}}d\bar{A} + J_e \\
J_i &= \int_{\bar{C}}(\bar{\underset{\sim}{P}}_{\nu}\cdot\delta\bar{\underset{\sim}{u}} + \bar{M}_{\nu\nu}\delta\bar{\beta}_{\nu})d\bar{s} + \sum_{n}\Delta\bar{M}_{t\nu}\bar{\underset{\sim}{n}}\cdot\delta\bar{\underset{\sim}{u}} \\
J_e &= \int_{\bar{C}}(\bar{\underset{\sim}{R}}\cdot\delta\bar{\underset{\sim}{u}} + \bar{K}_{\nu}\delta\bar{\beta}_{\nu})d\bar{s} + \sum_{n}\Delta\bar{K}_{t}\bar{\underset{\sim}{n}}\cdot\delta\bar{\underset{\sim}{u}}
\end{aligned} \tag{4.1.4}$$

where

$$\begin{aligned}
\bar{\underset{\sim}{N}}^{\beta} &= \bar{Q}^{\alpha\beta}\bar{\underset{\sim}{a}}_{\alpha} + \bar{Q}^{\beta}\bar{\underset{\sim}{n}} \quad , \quad \delta\bar{\beta}_{\nu} = (\delta\bar{\underset{\sim}{\Omega}}_{t} \times \bar{\underset{\sim}{n}})\cdot\bar{\underset{\sim}{\nu}} = \delta\bar{\underset{\sim}{\Omega}}_{t}\cdot\bar{\underset{\sim}{t}} \\
\bar{Q}^{\alpha\beta} &= \bar{N}^{\alpha\beta} - \bar{b}^{\alpha}_{\lambda}\bar{M}^{\lambda\beta} \quad , \quad \bar{Q}^{\beta} = \bar{M}^{\alpha\beta}{}_{\|\alpha} \\
\bar{\underset{\sim}{P}}_{\nu} &= \bar{\underset{\sim}{N}}^{\beta}\bar{\nu}_{\beta} + \frac{d}{d\bar{s}}(\bar{M}_{t\nu}\bar{\underset{\sim}{n}}) \quad , \quad \bar{\underset{\sim}{R}} = \bar{\underset{\sim}{F}} + \frac{d}{d\bar{s}}(\bar{K}_{t}\bar{\underset{\sim}{n}})
\end{aligned} \tag{4.1.5}$$

$$\begin{aligned}
\sum_{n}\Delta\bar{M}_{t\nu}\bar{\underset{\sim}{n}}\cdot\delta\bar{\underset{\sim}{u}} &= \sum_{M_n}[\bar{M}_{t\nu}(\bar{s}_n + 0) - \bar{M}_{t\nu}(\bar{s}_n - 0)]\bar{\underset{\sim}{n}}(\bar{s}_n)\cdot\delta\bar{\underset{\sim}{u}}(\bar{s}_n) \\
\sum_{n}\Delta\bar{K}_{t}\bar{\underset{\sim}{n}}\cdot\delta\bar{\underset{\sim}{u}} &= \sum_{M_n}[\bar{K}_{t}(\bar{s}_n + 0) - \bar{K}_{t}(\bar{s}_n - 0)]\bar{\underset{\sim}{n}}(\bar{s}_n)\cdot\delta\bar{\underset{\sim}{u}}(\bar{s}_n)
\end{aligned} \tag{4.1.6}$$

and $\bar{M}_n$, n = 1,2,...,N are corners of $\bar{C}$ labelled by $\bar{s} = \bar{s}_n$.

The local form of (4.1.4) gives the Eulerian equilibrium equations and the Eulerian static boundary conditions

$$\bar{\mathbf{N}}^{\beta}\|_{\beta} + \bar{\mathbf{p}} = \mathbf{0} \quad \text{in} \quad \bar{M}$$

$$\bar{\mathbf{P}}_{\nu} = \bar{\mathbf{R}} \quad , \quad \bar{M}_{\nu\nu} = \bar{K}_{\nu} \quad \text{on} \quad \bar{C} \tag{4.1.7}$$

$$\Delta\bar{M}_{t\nu}\bar{\mathbf{n}} = \Delta\bar{K}_{t}\bar{\mathbf{n}} \quad \text{at each} \quad \bar{M}_n$$

Usually the undeformed reference configuration is the only known in advance. It is desirable then to express all shell equations in terms of quantities defined in and/or refered to the known geometry of the reference middle surface M .

Transformation rules between deformed and reference geometry of the shell middle surface are [9,12]

$$d\bar{s} = \bar{a}_t ds \quad , \quad d\bar{A} = \sqrt{\frac{\bar{a}}{a}}\, dA$$

$$\bar{\nu}_{\beta} d\bar{s} = \sqrt{\frac{\bar{a}}{a}}\, \nu_{\beta} ds \quad , \quad \bar{\nu}^{\beta} d\bar{s} = \sqrt{\frac{a}{\bar{a}}}\, (\delta^{\beta}_{\alpha} + 2\epsilon_{\alpha\lambda}\epsilon^{\beta\mu}\gamma^{\lambda}_{\mu})\nu^{\alpha} ds \tag{4.1.8}$$

$$\bar{t}_{\beta} d\bar{s} = (\delta^{\alpha}_{\beta} + 2\gamma^{\alpha}_{\beta})t_{\alpha} ds \quad , \quad \bar{t}^{\beta} d\bar{s} = t^{\beta} ds$$

Remind also transformation rules for covariant differentiation of a vector v^{β} and a symmetric second-order tensor $T^{\alpha\beta} = T^{\beta\alpha}$ to be [9]

$$\left(\sqrt{\frac{a}{\bar{a}}}\, v^{\beta}\right)\Big\|_{\beta} = \sqrt{\frac{a}{\bar{a}}}\, v^{\beta}\big|_{\beta}$$

$$\left(\sqrt{\frac{a}{\bar{a}}}\, T^{\alpha\beta}\right)\Big\|_{\beta} = \sqrt{\frac{a}{\bar{a}}}\, [T^{\alpha\beta}|_{\beta} + \bar{a}^{\alpha\kappa}(2\gamma_{\kappa\lambda|\mu} - \gamma_{\lambda\mu|\kappa})T^{\lambda\mu}] \tag{4.1.9}$$

Let us introduce symmetric Lagrangean stress and couple resultant tensors $\mathbf{N} = N^{\alpha\beta}\mathbf{a}_{\alpha} \otimes \mathbf{a}_{\beta}$ and $\mathbf{M} = M^{\alpha\beta}\mathbf{a}_{\alpha} \otimes \mathbf{a}_{\beta}$ by the relations

$$\bar{\mathbf{N}} = \sqrt{\frac{a}{\bar{a}}}\, \mathbf{G}\mathbf{N}\mathbf{G}^T \quad , \quad \bar{\mathbf{M}} = \sqrt{\frac{a}{\bar{a}}}\, \mathbf{G}\mathbf{M}\mathbf{G}^T$$

$$\bar{N}^{\alpha\beta} = \sqrt{\frac{a}{\bar{a}}}\, N^{\alpha\beta} \quad , \quad \bar{M}^{\alpha\beta} = \sqrt{\frac{a}{\bar{a}}}\, M^{\alpha\beta} \tag{4.1.10}$$

The structure of transformations (4.1.10) is analogous to the one relating the Cauchy stress tensor to the second Piola - Kirchhoff stress tensor in the non-linear continuum mechanics [15,18]. Therefore, the Lagrangean quantities $\underset{\sim}{N}$ and $\underset{\sim}{M}$ may also be called the second Piola - Kirchhoff stress and couple resultant tensors, respectively.

Introducing (2.2.13) and (4.1.8) to (4.1.10) into (4.1.4) we obtain

$$
\begin{aligned}
&- \iint_M (\underset{\sim}{G}\underset{\sim}{N}^\beta)\big|_\beta \cdot \delta\underset{\sim}{u}\,dA + J_i = \iint_M \underset{\sim}{p}\cdot\delta\underset{\sim}{u}\,dA + J_e \\
&J_i = \int_C (\underset{\sim}{P}_\nu\cdot\delta\underset{\sim}{u} + \bar{M}_{\nu\nu}\delta\underset{\sim}{\Omega}_t\cdot\underset{\sim}{\bar{a}}_t)ds + \sum_n \Delta\bar{M}_{t\nu}\underset{\sim}{\bar{n}}\cdot\delta\underset{\sim}{u} \qquad (4.1.11)\\
&J_e = \int_C (\underset{\sim}{R}\cdot\delta\underset{\sim}{u} + \bar{K}_\nu\delta\underset{\sim}{\Omega}_t\cdot\underset{\sim}{\bar{a}}_t)ds + \sum_n \Delta\bar{K}_t\underset{\sim}{\bar{n}}\cdot\delta\underset{\sim}{u}
\end{aligned}
$$

Here the surface force $\underset{\sim}{p}$ is per unit area of M and the boundary force $\underset{\sim}{F}$ and boundary couple $\underset{\sim}{K}$ are per unit length of C . They are supposed to be given through their components with respect to the reference surface geometry

$$
\begin{aligned}
&\underset{\sim}{p} = \sqrt{\frac{\bar{a}}{a}}\,\underset{\sim}{\bar{p}} = p^\alpha\underset{\sim}{a}_\alpha + p\underset{\sim}{n} \\
&\underset{\sim}{F} = \bar{a}_t\underset{\sim}{\bar{F}} = F^\alpha\underset{\sim}{a}_\alpha + F\underset{\sim}{n} = F_\nu\underset{\sim}{\nu} + F_t\underset{\sim}{t} + F\underset{\sim}{n} \qquad (4.1.12)\\
&\underset{\sim}{K} = \bar{a}_t\underset{\sim}{\bar{K}} = \epsilon_{\alpha\beta}K^\alpha\underset{\sim}{a}^\beta + K\underset{\sim}{n} = - K_t\underset{\sim}{\nu} + K_\nu\underset{\sim}{t} + K\underset{\sim}{n}
\end{aligned}
$$

Since we have assumed $\bar{K} \equiv 0$ then K^α and K cannot be independent. After transformation we obtain

$$
\begin{aligned}
&K = -\frac{1}{n}\epsilon_{\lambda\mu}K^\lambda n^\mu \qquad (4.1.13)\\
&\hat{K}^\alpha = \bar{\epsilon}^{\alpha\beta}\underset{\sim}{K}\cdot\underset{\sim}{\bar{a}}_\beta = \sqrt{\frac{a}{\bar{a}}}\,\epsilon^{\alpha\beta}\epsilon_{\lambda\mu}K^\lambda(1^\mu_{\cdot\beta} - \frac{1}{n}n^\mu\phi_\beta)
\end{aligned}
$$

For the remaining Lagrangean quantities appearing in (4.1.11) we obtain the following relations

$$\underset{\sim}{N}^{\beta} = Q^{\alpha\beta}\underset{\sim}{a}_{\alpha} + Q^{\beta}\underset{\sim}{n} = \sqrt{\frac{\bar{a}}{a}}\,\underset{\sim}{G}^{-1}\bar{\underset{\sim}{N}}^{\beta}$$

$$Q^{\alpha\beta} = N^{\alpha\beta} - \bar{b}^{\alpha}_{\lambda}M^{\lambda\beta} \quad , \quad Q^{\beta} = M^{\alpha\beta}{}_{|\alpha} + \bar{a}^{\beta\kappa}(2\gamma_{\kappa\lambda|\mu} - \gamma_{\lambda\mu|\kappa})M^{\lambda\mu} \qquad (4.1.14)$$

$$\underset{\sim}{P}_{\nu} = \underset{\approx}{G}\underset{\sim}{N}^{\beta}\nu_{\beta} + \frac{d}{ds}(\bar{M}_{t\nu}\bar{\underset{\sim}{n}}) \quad , \quad \underset{\sim}{R} = \underset{\sim}{F} + \frac{d}{ds}(\bar{K}_{t}\bar{\underset{\sim}{n}})$$

$$\bar{M}_{t\nu} = \frac{1}{\bar{a}_t^2}M^{\alpha\beta}(\delta^{\lambda}_{\alpha} + 2\gamma^{\lambda}_{\alpha})t_{\lambda}\nu_{\beta} \quad , \quad \bar{K}_{t} = \frac{1}{\bar{a}_t^2}\hat{K}^{\alpha}(\delta^{\lambda}_{\alpha} + 2\gamma^{\lambda}_{\alpha})t_{\lambda}$$

$$\bar{M}_{\nu\nu} = \frac{1}{\bar{a}_t^2}\sqrt{\frac{\bar{a}}{a}}M^{\alpha\beta}\nu_{\alpha}\nu_{\beta} \quad , \quad \bar{K}_{\nu} = \frac{1}{\bar{a}_t^2}\sqrt{\frac{\bar{a}}{a}}\hat{K}^{\alpha}\nu_{\alpha} \qquad (4.1.15)$$

The relations (4.1.11) are the transformed form of the following Lagrangean virtual work principle

$$\iint_{M}(N^{\alpha\beta}\delta\gamma_{\alpha\beta} + M^{\alpha\beta}\delta\kappa_{\alpha\beta})dA = \iint_{M}\underset{\sim}{p}\cdot\delta\underset{\sim}{u}dA + \int_{C}(\underset{\sim}{F}\cdot\delta\underset{\sim}{u} + \underset{\sim}{K}\cdot\delta\underset{\sim}{\Omega}_{t})ds \qquad (4.1.16)$$

where now

$$\delta\underset{\sim}{u} = \delta u_{\alpha}\underset{\sim}{a}^{\alpha} + \delta w\underset{\sim}{n} \quad , \quad \delta\gamma_{\alpha\beta} = \frac{1}{2}(l^{\lambda}_{\cdot\alpha}\delta l_{\lambda\beta} + l^{\lambda}_{\cdot\beta}\delta l_{\lambda\alpha} + \phi_{\alpha}\delta\phi_{\beta} + \phi_{\beta}\delta\phi_{\alpha})$$

$$\delta\kappa_{\alpha\beta} = -\,n(\delta\phi_{\alpha|\beta} + b^{\lambda}_{\beta}\delta l_{\lambda\alpha}) - (\phi_{\alpha|\beta} + b^{\lambda}_{\beta}l_{\lambda\alpha})\delta n - \qquad (4.1.17)$$

$$-\,n_{\lambda}(\delta l^{\lambda}_{\cdot\alpha|\beta} - b^{\lambda}_{\beta}\delta\phi_{\alpha}) - (l^{\lambda}_{\cdot\alpha|\beta} - b^{\lambda}_{\beta}\phi_{\alpha})\delta n$$

The local form of (4.1.11) or (4.1.16) gives the Lagrangean equilibrium equations and the Lagrangean static boundary conditions

$$(\underset{\approx}{G}\underset{\sim}{N}^{\beta})_{|\beta} + \underset{\sim}{p} = \underset{\sim}{0} \quad \text{in } M$$

$$\underset{\sim}{P}_{\nu} = \underset{\sim}{R} \quad , \quad M^{\alpha\beta}\nu_{\alpha}\nu_{\beta} = \hat{K}^{\alpha}\nu_{\alpha} \quad \text{on } C \qquad (4.1.18)$$

$$\Delta\bar{M}_{t\nu}\underset{\approx}{G}\underset{\sim}{n} = \Delta\bar{K}_{t}\underset{\approx}{G}\underset{\sim}{n} \quad \text{at each } M_n$$

By expressing vector equilibrium equations $(4.1.18)_1$ in different shell bases various component forms of it in terms of $N^{\alpha\beta}$ and $M^{\alpha\beta}$ may be derived [7,9]. Let us present some of them here.

When $(4.1.18)_1$ is expressed by components in the base $\underset{\sim}{a}_\alpha$, $\underset{\sim}{n}$ with the help of (2.2.13) and (2.2.1) the following component form of equilibrium equations is obtained [12,25]

$$
\begin{aligned}
&(1^{\alpha}_{\cdot\lambda}Q^{\lambda\beta} + n^{\alpha}Q^{\beta})_{|\beta} - b^{\alpha}_{\beta}(\phi_\lambda Q^{\lambda\beta} + nQ^{\beta}) + p^{\alpha} = 0 \\
&(\phi_\lambda Q^{\lambda\beta} + nQ^{\beta})_{|\beta} + b_{\alpha\beta}(1^{\alpha}_{\cdot\lambda}Q^{\lambda\beta} + n^{\alpha}Q^{\beta}) + p = 0
\end{aligned}
\qquad (4.1.19)
$$

If we express $(4.1.7)_1$ in the base $\bar{\underset{\sim}{a}}_\alpha$, $\bar{\underset{\sim}{n}}$ and transform all components according to (4.1.10) and (4.1.9) then

$$
\begin{aligned}
&Q^{\alpha\beta}{}_{|\beta} + \bar{a}^{\alpha\kappa}\gamma_{\kappa\lambda\beta}Q^{\lambda\beta} - \bar{b}^{\lambda}_{\beta}Q^{\beta} + \sqrt{\frac{\bar{a}}{a}}\,\bar{p}^{\alpha} = 0 \\
&Q^{\beta}{}_{|\beta} + \bar{b}_{\alpha\beta}Q^{\alpha\beta} + \sqrt{\frac{\bar{a}}{a}}\,\bar{p} = 0
\end{aligned}
\qquad (4.1.20)
$$

When also the base vectors $\bar{\underset{\sim}{a}}_\alpha$, $\bar{\underset{\sim}{n}}$ are transformed according to (2.2.1) from $(4.1.7)_1$ and (4.1.20) we obtain the component form of equilibrium equations equivalent to (4.1.19) to be

$$
\begin{aligned}
&1^{\alpha}_{\cdot\lambda}(Q^{\lambda\beta}{}_{|\beta} + \bar{a}^{\lambda\kappa}\gamma_{\kappa\mu\beta}Q^{\mu\beta} - \bar{b}^{\lambda}_{\beta}Q^{\beta}) + n^{\alpha}(Q^{\beta}{}_{|\beta} + \bar{b}_{\lambda\beta}Q^{\lambda\beta}) + p^{\alpha} = 0 \\
&\phi_\lambda(Q^{\lambda\beta}{}_{|\beta} + \bar{a}^{\lambda\kappa}\gamma_{\kappa\mu\beta}Q^{\mu\beta} - \bar{b}^{\lambda}_{\beta}Q^{\beta}) + n(Q^{\beta}{}_{|\beta} + \bar{b}_{\lambda\beta}Q^{\lambda\beta}) + p = 0
\end{aligned}
\qquad (4.1.21)
$$

The transformation formula $(4.1.14)_1$ can also be presented in a more extended form

$$
\begin{aligned}
&\bar{\underset{\sim}{N}}^{\beta} = \sqrt{\frac{a}{\bar{a}}}\,\underset{\approx}{R}\overset{\vee}{\underset{\sim}{N}}{}^{\beta} = \sqrt{\frac{\bar{a}}{a}}\,\underset{\approx}{V}\overset{*}{\underset{\sim}{N}}{}^{\beta} \\
&\overset{\vee}{\underset{\sim}{N}}{}^{\beta} = Q^{\alpha\beta}\overset{\vee}{\underset{\sim}{a}}_\alpha + Q^{\beta}\underset{\sim}{n} \quad , \quad \overset{*}{\underset{\sim}{N}}{}^{\beta} = Q^{\alpha\beta}\overset{*}{\underset{\sim}{a}}_\alpha + Q^{\beta}\bar{\underset{\sim}{n}}
\end{aligned}
\qquad (4.1.22)
$$

Therefore the equilibrium equations (4.1.7) and (4.1.18) can be presented as follows

$$
(\underset{\approx}{V}\overset{*}{\underset{\sim}{N}}{}^{\beta})_{|\beta} + \sqrt{\frac{\bar{a}}{a}}\,\bar{\underset{\sim}{p}} = \underset{\sim}{0} \quad , \quad (\underset{\approx}{R}\overset{\vee}{\underset{\sim}{N}}{}^{\beta})_{|\beta} + \underset{\sim}{p} = \underset{\sim}{0} \qquad (4.1.23)
$$

When written in components with respect to $\overset{*}{\underset{\sim}{a}}_\alpha$, $\bar{\underset{\sim}{n}}$ or $\overset{\vee}{\underset{\sim}{a}}_\alpha$, $\underset{\sim}{n}$ several addition component forms of shell equilibrium equations can be obtained, (see [4,7,9])

4.2. Modified static boundary conditions

It follows from the structure of J_i in $(4.1.11)_2$ that in Lagrangean description the effective internal force $\underset{\sim}{P}_\nu$ and the moment $\bar{a}_t\bar{M}_{\nu\nu}$ are static quantities at C which produce virtual work on variations of displacement parameters $\underset{\sim}{u}$ and β_ν . Assuming at C values for $\underset{\sim}{P}_\nu$ and for $\bar{a}_t\bar{M}_{\nu\nu}$ we obtain the basic variant $(4.1.18)_2$ of static boundary conditions energetically compatible with displacement boundary conditions (3.3.4).

Let $\underset{\sim}{F}_\nu$ and $\underset{\sim}{B}_\nu(0)$ be a total force and a total couple, with respect to an origin 0 in space, of all internal stress and couple resultants acting at a part of the boundary $\bar{C}$. In the Lagrangean description these vectors are defined by

$$\underset{\sim}{F}_\nu = \underset{\sim}{F}^o_\nu + \int_{M_o}^{M} \underset{\sim}{P}_\nu ds \quad , \quad \underset{\sim}{B}_\nu(0) = \underset{\sim}{B}^o_\nu(0) + \int_{M_o}^{M} (\bar{M}_{\nu\nu}\bar{\underset{\sim}{a}}_t + \bar{\underset{\sim}{r}} \times \underset{\sim}{P}_\nu)ds \qquad (4.2.1)$$

where $\underset{\sim}{F}^o_\nu$ and $\underset{\sim}{B}^o_\nu(0)$ are initial values of $\underset{\sim}{F}_\nu$ and $\underset{\sim}{B}_\nu(0)$ at $M = M_o$.

The total couple $\underset{\sim}{B}_\nu(\bar{M}) \equiv \underset{\sim}{B}_\nu$, with respect to a current point $\bar{M}$ of $\bar{C}$, is calculated according to

$$\underset{\sim}{B}_\nu = \underset{\sim}{B}_\nu(0) - \bar{\underset{\sim}{r}} \times \underset{\sim}{F}_\nu \qquad (4.2.2)$$

Differentiating $(4.2.1)_1$ and (4.2.2) we obtain

$$\frac{d\underset{\sim}{F}_\nu}{ds} = \underset{\sim}{P}_\nu \quad , \quad \frac{d\underset{\sim}{B}_\nu}{ds} = \bar{M}_{\nu\nu}\bar{\underset{\sim}{a}}_t - \bar{\underset{\sim}{a}}_t \times \underset{\sim}{F}_\nu \qquad (4.2.3)$$

Let us differentiate $\delta\underset{\sim}{u}$ and $\delta\underset{\sim}{\Omega}_t$ with respect to $\bar{s}$ and take into account $(4.1.8)_1$, which leads to

$$\frac{d}{ds}\delta\underset{\sim}{\Omega}_t = \bar{a}_t\delta\underset{\sim}{k}_t \quad , \quad \frac{d}{ds}\delta\underset{\sim}{u} = \delta\bar{\gamma}_{tt}\bar{\underset{\sim}{a}}_t + \delta\underset{\sim}{\Omega}_t \times \bar{\underset{\sim}{a}}_t \tag{4.2.4}$$

Here $\delta\underset{\sim}{k}_t$ is the vector of virtual change of curvature of the shell boundary contour $\bar{C}$ and $\delta\bar{\gamma}_{tt} = \delta\bar{\gamma}_{\alpha\beta}\bar{t}^{\alpha}\bar{t}^{\beta} = \frac{1}{\bar{a}_t^2}\delta\gamma_{tt}$, $\delta\gamma_{tt} = \delta\gamma_{\alpha\beta}t^{\alpha}t^{\beta}$, according to $(4.1.8)_3$.

Now it is possible to transform J_i in (4.1.11) as follows

$$\begin{aligned} J_i &= \int_C [\frac{d}{ds}(\underset{\sim}{F}_\nu\cdot\delta\underset{\sim}{u}) - \underset{\sim}{F}_\nu\cdot\frac{d}{ds}\delta\underset{\sim}{u} + \bar{M}_{\nu\nu}\delta\underset{\sim}{\Omega}_t\cdot\bar{\underset{\sim}{a}}_t]ds + \sum_n \Delta\bar{M}_{t\nu}\bar{\underset{\sim}{n}}\cdot\delta\underset{\sim}{u} = \\ &= \int_C [(\bar{M}_{\nu\nu}\bar{\underset{\sim}{a}}_t - \bar{\underset{\sim}{a}}_t \times \underset{\sim}{F}_\nu)\cdot\delta\underset{\sim}{\Omega}_t - \frac{1}{\bar{a}_t^2}(\bar{\underset{\sim}{a}}_t\cdot\underset{\sim}{F}_\nu)\delta\gamma_{tt}]ds + \\ &+ \sum_n (\Delta\bar{M}_{t\nu}\bar{\underset{\sim}{n}} - \Delta\underset{\sim}{F}_\nu)\cdot\delta\underset{\sim}{u} \end{aligned} \tag{4.2.5}$$

By introducing $(4.2.3)_2$ and $(4.2.4)_1$ into (4.2.5) we also get

$$\begin{aligned} J_i &= -\int_C [\bar{a}_t\underset{\sim}{B}_\nu\cdot\delta\underset{\sim}{k}_t + \frac{1}{\bar{a}_t^2}(\bar{\underset{\sim}{a}}_t\cdot\underset{\sim}{F}_\nu)\delta\gamma_{tt}]ds + \\ &+ \sum_n [(\Delta\bar{M}_{t\nu}\bar{\underset{\sim}{n}} - \Delta\underset{\sim}{F}_\nu)\cdot\delta\underset{\sim}{u} - \Delta\underset{\sim}{B}_\nu\cdot\delta\underset{\sim}{\Omega}_t] \end{aligned} \tag{4.2.6}$$

where

$$\begin{aligned} \sum_n \Delta\underset{\sim}{F}_\nu\cdot\delta\underset{\sim}{u} &= \sum_{M_n} [\underset{\sim}{F}_\nu(s_n + 0) - \underset{\sim}{F}_\nu(s_n - 0)]\cdot\delta\underset{\sim}{u}(s_n) \\ \sum_n \Delta\underset{\sim}{B}_\nu\cdot\delta\underset{\sim}{\Omega}_t &= \sum_{M_n} [\underset{\sim}{B}_\nu(s_n + 0) - \underset{\sim}{B}_\nu(s_n - 0)]\cdot\delta\underset{\sim}{\Omega}_t(s_n) \end{aligned} \tag{4.2.7}$$

The relations (4.2.5) and (4.2.6) show that during virtual deformation some static parameters produce at C a virtual work on variations of geometric parameters $\underset{\sim}{\Omega}_t$, γ_{tt} and $\underset{\sim}{k}_t$, γ_{tt} which establish the deformed lateral boundary surface $\partial\bar{P}$. Therefore, within K - L type non--linear theory of shells to each of displacemental, kinematical and deformational quantity discussed in p.3.3. there corresponds a static quantity according to the following schema

$$\underset{\sim}{u} \Longleftrightarrow \underset{\sim}{P}_\nu \quad , \quad \beta_\nu \Longleftrightarrow \bar{a}_t \bar{M}_{\nu\nu}$$

$$\underset{\sim}{\Omega}_t \Longleftrightarrow \bar{M}_{\nu\nu} \underset{\sim}{\bar{a}}_t - \underset{\sim}{\bar{a}}_t \times \underset{\sim}{F}_\nu \quad , \quad \gamma_{tt} \Longleftrightarrow - \frac{1}{\bar{a}_t^2} (\underset{\sim}{\bar{a}}_t \cdot \underset{\sim}{F}_\nu) \tag{4.2.8}$$

$$\underset{\sim}{k}_t \Longleftrightarrow - \bar{a}_t \underset{\sim}{B}_\nu \quad , \quad \gamma_{tt} \Longleftrightarrow - \frac{1}{\bar{a}_t^2} (\underset{\sim}{\bar{a}}_t \cdot \underset{\sim}{F}_\nu)$$

In exactly the same way we can transform in (4.1.11) the integral J_e of the virtual work due to the external boundary loading. Defining a total force $\underset{\sim}{T}$ and a total couple $\underset{\sim}{C}$, with respect to a current point $\bar{M}$, of the external boundary force and boundary couple

$$\underset{\sim}{T} = \underset{\sim}{T}^o + \int_{M_o}^{M} \underset{\sim}{R} ds \quad , \quad \underset{\sim}{C} = \underset{\sim}{C}(0) - \underset{\sim}{\bar{r}} \times \underset{\sim}{T}$$
$$\underset{\sim}{C}(0) = \underset{\sim}{C}^o(0) + \int_{M_o}^{M} (\bar{K}_\nu \underset{\sim}{\bar{a}}_t + \underset{\sim}{\bar{r}} \times \underset{\sim}{R}) ds \tag{4.2.9}$$

the integral J_e is transformed into

$$J_e = \int_C [(\bar{K}_\nu \underset{\sim}{\bar{a}}_t - \underset{\sim}{\bar{a}}_t \times \underset{\sim}{T}) \cdot \delta \underset{\sim}{\Omega}_t - \frac{1}{\bar{a}_t^2} (\underset{\sim}{\bar{a}}_t \cdot \underset{\sim}{T}) \delta \gamma_{tt}] ds +$$
$$+ \sum_n (\Delta \bar{K}_t \underset{\sim}{\bar{n}} - \Delta \underset{\sim}{T}) \cdot \delta \underset{\sim}{u} \tag{4.2.10}$$

$$J_e = - \int_C [\bar{a}_t \underset{\sim}{C} \cdot \delta \underset{\sim}{k}_t + \frac{1}{\bar{a}_t^2} (\underset{\sim}{\bar{a}}_t \cdot \underset{\sim}{T}) \delta \gamma_{tt}] ds +$$
$$+ \sum_n [(\Delta \bar{K}_t \underset{\sim}{\bar{n}} - \Delta \underset{\sim}{T}) \cdot \delta \underset{\sim}{u} - \Delta \underset{\sim}{C} \cdot \delta \underset{\sim}{\Omega}_t] \tag{4.2.11}$$

Introducing (4.2.5) and (4.2.10) or (4.2.6) and (4.2.11), respectively, into (4.1.11) we obtain the following modified forms of static boundary conditions

$$\bar{M}_{\nu\nu} \underset{\sim}{\bar{a}}_t - \underset{\sim}{\bar{a}}_t \times \underset{\sim}{F}_\nu = \bar{K}_\nu \underset{\sim}{\bar{a}}_t - \underset{\sim}{\bar{a}}_t \times \underset{\sim}{T} \quad , \quad \underset{\sim}{\bar{a}}_t \cdot \underset{\sim}{F}_\nu = \underset{\sim}{\bar{a}}_t \cdot \underset{\sim}{T} \quad \text{on } C$$
$$\Delta \bar{M}_{t\nu} \underset{\sim}{G} \underset{\sim}{n} - \Delta \underset{\sim}{F}_\nu = \Delta \bar{K}_t \underset{\sim}{G} \underset{\sim}{n} - \Delta \underset{\sim}{T} \quad \text{at each } M_n \tag{4.2.12}$$

$$\underset{\sim}{B}_\nu = \underset{\sim}{C} \quad , \quad \bar{\underset{\sim}{a}}_t \cdot \underset{\sim}{F}_\nu = \bar{\underset{\sim}{a}}_t \cdot \underset{\sim}{T} \quad \text{on} \quad C$$
$$(4.2.13)$$
$$\Delta \bar{M}_{t\nu} \underset{\sim}{G}\underset{\sim}{n} - \Delta \underset{\sim}{F}_\nu = \Delta \bar{K}_t \underset{\sim}{G}\underset{\sim}{n} - \Delta \underset{\sim}{T} \quad \text{and} \quad \Delta \underset{\sim}{B}_\nu = \Delta \underset{\sim}{C} \quad \text{at each} \quad M_n$$

These static boundary conditions are energetically compatible with kinematical (3.3.6) and deformational (3.3.9) boundary conditions, respectively.

When the shell problems are solved in displacements the basic variant (4.1.18) of static boundary conditions should be used. The modified static boundary conditions (4.2.12) are adequate for shell problems formulated by means of the finite rotation vector $\underset{\sim}{\Omega}$ and stress resultants $N^{\alpha\beta}$. The modified static boundary conditions (4.2.13) should be used when formulating the shell problems entirely in terms of the shell strain or stress measures.

4.3. Simplified shell relations under small elastic strains

The various shell relations discussed so far have purely geometrical character, which follows from the assumption of K - L constraints on deformation process of the shell. The relations still contain unrestricted strains and unrestricted rotations and do not depend upon the shell material properties. In what follows we shall discuss possible simplification of shell relations in the case of a thin shell composed of an isotropic elastic material under the assumption, that strains are small everywhere in the shell space.

For a shell subjected to forces applied only at its lateral boundaries John [22] obtained exact a priori estimates of stresses and their derivatives in the interior domain of the shell. The common measure of small quantities in [22] is the small parameter θ defined by

$$\theta = \max\left(\frac{h}{d}, \sqrt{\frac{h}{R}}, \sqrt{\eta}\right) \tag{4.3.1}$$

where d is the distance of the shell point from the lateral boundary, R is a large parameter and η is the largest strain in the shell space.

The estimates of [22] may still be used if we admit some smooth surface force. $\underset{\sim}{\bar{p}} = \bar{p}^{\alpha}\underset{\sim}{\bar{a}}_{\alpha} + \bar{p}\underset{\sim}{\bar{n}}$, with small variability, to be applied at the upper and/or the lower boundary of the shell space, provided that it does not disturb the approximately plain stress state in the shell [23] and gives the surface components [7]

$$\bar{p}^{\alpha} = O(E\eta\theta) \quad , \quad \bar{p} = O(E\eta\theta^2) \tag{4.3.2}$$

In this case the small parameter θ may be redefined on physical grounds [24] as follows

$$\theta = \max\left(\frac{h}{L}, \frac{h}{d}, \sqrt{\frac{h}{R}}, \sqrt{\eta}\right) \tag{4.3.3}$$

where L is the smallest wavelength of deformation patterns at $\mathcal{M}$ and R is the smallest principal radius of curvature of $\mathcal{M}$.

Under small strains for the bending theory of shells it is assumed, that strains in the shell space, caused by stretching and bending of the shell middle surface, are of comparable order in the entire shell region. In this case we have the following estimates

$$\begin{gathered}
a_{\alpha\beta} = O(1) \quad , \quad b_{\alpha\beta} = O\left(\frac{1}{R}\right) \\
\gamma_{\alpha\beta} = O(\eta) \quad , \quad h\kappa_{\alpha\beta} = O(\eta) \quad , \quad h^2\nu_{\alpha\beta} = O(\eta\theta^2) \\
\frac{\bar{a}}{a} = 1 + 2\gamma^{\lambda}_{\lambda} + O(\eta^2) = 1 + O(\eta) \\
\bar{a}^{\alpha\beta} = a^{\alpha\beta} - 2\gamma^{\alpha\beta} + O(\eta^2) = a^{\alpha\beta} + O(\eta)
\end{gathered} \tag{4.3.4}$$

Within the first - approximation theory of thin isotropic elastic shells, the strain energy function Σ, per unit area of $\mathcal{M}$, can be consistently approximated by the following expression [7,10,24]

$$\Sigma = \frac{h}{2} H^{\alpha\beta\lambda\mu}\left(\gamma_{\alpha\beta}\gamma_{\lambda\mu} + \frac{h^2}{12}\kappa_{\alpha\beta}\kappa_{\lambda\mu}\right) + O(Eh\eta^2\theta^2) \tag{4.3.5}$$

$$H^{\alpha\beta\lambda\mu} = \frac{E}{2(1+\nu)} (a^{\alpha\lambda}a^{\beta\mu} + a^{\alpha\mu}a^{\beta\lambda} + \frac{2\nu}{1-\nu} a^{\alpha\beta}a^{\lambda\mu})$$

The formula (4.3.5) takes into account the main contributions to the elastic strain energy of a shell caused by stretching and bending of the shell middle surface as well as by the transverse strains. The last contribution is taken into account by using in (4.3.5) the modified elasticity tensor $H^{\alpha\beta\lambda\mu}$.

Appropriate constitutive equations of the first - approximation theory compatible with (4.3.5) take the form

$$N^{\alpha\beta} = \frac{\partial \Sigma}{\partial \gamma_{\alpha\beta}} = C[(1-\nu)\gamma^{\alpha\beta} + \nu a^{\alpha\beta}\gamma^{\lambda}_{\lambda}] + O(Eh\eta\theta^2)$$

$$M^{\alpha\beta} = \frac{\partial \Sigma}{\partial \kappa_{\alpha\beta}} = D[(1-\nu)\kappa^{\alpha\beta} + \nu a^{\alpha\beta}\kappa^{\lambda}_{\lambda}] + O(Eh^2\eta\theta^2) \qquad (4.3.6)$$

and their inverse

$$\gamma_{\alpha\beta} = A[(1+\nu)N_{\alpha\beta} - \nu a_{\alpha\beta}N^{\lambda}_{\lambda}] + O(\eta\theta^2)$$

$$\kappa_{\alpha\beta} = B[(1+\nu)M_{\alpha\beta} - \nu a_{\alpha\beta}M^{\lambda}_{\lambda}] + O(\frac{\eta\theta^2}{h}) \qquad (4.3.7)$$

where

$$C = \frac{Eh}{1-\nu^2} \; , \quad D = \frac{Eh^3}{12(1-\nu^2)} \; , \quad A = \frac{1}{Eh} \; , \quad B = \frac{12}{Eh^3} \qquad (4.3.8)$$

Under small strains $\gamma_{\alpha\beta}$ is still given by (2.1.4) while for $\kappa_{\alpha\beta}$ in (2.1.5) we should put

$$n = [1 + \theta^{\kappa}_{\kappa} + \frac{1}{2}(\theta^{\kappa}_{\kappa})^2 - \frac{1}{2}\theta^{\kappa}_{\mu}\theta^{\mu}_{\kappa} + \phi^2][1 - \gamma^{\lambda}_{\lambda} + O(\eta^2)]$$

$$n_{\mu} = [-(1 + \theta^{\kappa}_{\kappa})\phi_{\mu} + \phi^{\lambda}(\theta_{\lambda\mu} - \omega_{\lambda\mu})][1 + O(\eta)] \qquad (4.3.9)$$

$$\gamma^{\lambda}_{\lambda} = \theta^{\lambda}_{\lambda} + \frac{1}{2}\theta^{\lambda}_{\mu}\theta^{\mu}_{\lambda} + \frac{1}{2}\phi^{\lambda}\phi_{\lambda} + \phi^2$$

Note that in n here we have to take into account more accurate estimate for $\sqrt{\frac{\bar{a}}{a}}$, since the leading term of the product $n(\phi_{\alpha|\beta} + b^{\lambda}_{\beta}1_{\lambda\alpha})$

cancels with $-b_{\alpha\beta}$ in (2.1.5). As a result, in geometrically non-linear shell problems (with unrestricted rotations) $\gamma_{\alpha\beta}$ are quadratic polynomials and $\kappa_{\alpha\beta}$ are polynomials of fifth degree in displacements u_α, w and their surface gradients.

Within small strains the finite rotation vector $\underset{\sim}{\Omega}$, expressed exactly by (2.3.11), reduces to

$$\underset{\sim}{\Omega} = \{\epsilon^{\beta\alpha}[\phi_\alpha(1 + \frac{1}{2}\theta^\kappa_\kappa) - \frac{1}{2}\phi^\lambda(\theta_{\lambda\alpha} - \omega_{\lambda\alpha})]\underset{\sim}{a}_\beta + \phi\underset{\sim}{n}\}[1 + O(\eta)] \quad (4.3.10)$$

The total finite rotation vector $\underset{\sim}{\Omega}_t$ of the boundary, defined exactly by (3.1.13), may be simplified according to

$$\underset{\sim}{\Omega}_t = \frac{1}{2}(\underset{\sim}{\nu} \times \underset{\sim}{\bar{a}}_\nu + \underset{\sim}{t} \times \underset{\sim}{\bar{a}}_t + \underset{\sim}{n} \times \underset{\sim}{\bar{n}})[1 + O(\eta)] \quad (4.3.11)$$

where

$$\underset{\sim}{\bar{a}}_t = (\theta_{\nu t} - \phi)\underset{\sim}{\nu} + (1 + \theta_{tt})\underset{\sim}{t} + \phi_t\underset{\sim}{n}\ , \quad \underset{\sim}{\bar{a}}_\nu = \underset{\sim}{\bar{a}}_t \times \underset{\sim}{\bar{n}}$$

$$\underset{\sim}{\bar{n}} = \{[-\phi_\nu(1 + \theta_{tt}) + \phi_t(\theta_{\nu t} + \phi)]\underset{\sim}{\nu} + [\phi_\nu(\theta_{\nu t} - \phi) - \phi_t(1 + \theta_{\nu\nu})]\underset{\sim}{t} + [1 + \frac{1}{2}(\phi_\nu^2 + \phi_t^2) + \frac{1}{2}(\theta^\kappa_\kappa)^2 - \theta^\lambda_\kappa\theta^\kappa_\lambda]\underset{\sim}{n}\}[1 + O(\eta)] \quad (4.3.12)$$

$$\phi_t = \frac{dw}{ds} - \tau_t u_\nu + \sigma_t u_t\ , \quad \phi = -\frac{du_\nu}{ds} + \kappa_t u_t - \tau_t w + \theta_{\nu t}$$

$$\theta_{\nu\nu} = \frac{du_\nu}{ds_\nu} + \kappa_\nu u_t - \sigma_\nu w\ , \quad \theta_{tt} = \frac{du_t}{ds} + \kappa_t u_\nu - \sigma_t w \quad (4.3.13)$$

$$\theta_{\nu t} = \frac{1}{2}\left(\frac{du_t}{ds_\nu} + \frac{du_\nu}{ds} - \kappa_t u_t - \kappa_\nu u_\nu\right) + \tau_t w$$

Note that here $\underset{\sim}{\Omega}$ is quadratic and $\underset{\sim}{\Omega}_t$ is cubic with respect to displacements and their surface gradients.

When introduced into the Lagrangean virtual work principle (4.1.16) the relations (2.1.4), (2.1.5) with (4.3.9) and (4.3.11) with (4.3.12) would give us the reduced set of Lagrangean shell equations of the type (4.1.18). The set becomes extremely complex when expressed in terms of displacements and we do not elaborate it here. Anyway, in such general case the only reasonable way to obtain a solution at once in terms of displacements is to apply numerical methods directly to a functional based on reduced form of (4.1.16) and not to the resulting set of shell equations.

The error indicated in the constitutive equations (4.3.6) and (4.3.7) allows to make essention reductions also in basic shell equations of the type (4.1.20). Let us present here the reduction procedure leading to the set of bending shell equations [9] formulated by means of the shell strain measures $\gamma_{\alpha\beta}$ and $\kappa_{\alpha\beta}$.

Within the error indicated in (4.3.6) for $Q^{\alpha\beta}$ and Q^{β} in (4.1.20) we obtain the following estimates

$$
\begin{aligned}
Q^{\alpha\beta} &= C[(1-\nu)\gamma^{\alpha\beta} + \nu a^{\alpha\beta}\gamma^{\lambda}_{\lambda}] + O(Eh\eta\theta^2) \\
Q^{\beta} &= D[(1-\nu)\kappa^{\alpha\beta}{}_{|\alpha} + \nu a^{\alpha\beta}\kappa^{\lambda}_{\lambda|\alpha}] + O(Eh^2\frac{\eta\theta^2}{\lambda})
\end{aligned}
\tag{4.3.14}
$$

where the parameter λ , used here to estimate surface derivatives , is defined by

$$
\lambda = \frac{h}{\theta} = \min(L , d , \sqrt{hR} , \frac{h}{\sqrt{\eta}}) \tag{4.3.15}
$$

Using (4.3.4) for terms appearing in the compatibility conditions (2.1.6) we have the estimates

$$
\bar{a}^{\kappa\nu}b_{\kappa\lambda}\gamma_{\nu\beta\mu} = O(\frac{\eta\theta^2}{h\lambda}) \quad , \quad \bar{a}^{\kappa\nu}\kappa_{\kappa\lambda}\gamma_{\nu\beta\mu} = O(\frac{\eta\theta^2}{h\lambda}) \tag{4.3.16}
$$

$$
K\gamma^{\kappa}_{\kappa} = O(\frac{\eta\theta^2}{\lambda^2}) \quad , \quad \bar{a}^{\kappa\nu}\gamma_{\kappa\alpha\mu}\gamma_{\nu\beta\lambda} = O(\frac{\eta\theta^2}{\lambda^2}) \tag{4.3.17}
$$

and the compatibility conditions reduce to the form

$$
\begin{aligned}
&\kappa^{\beta}_{\alpha|\beta} - \kappa^{\beta}_{\beta|\alpha} = O(\frac{\eta\theta^2}{h\lambda}) \\
&\gamma^{\beta}_{\alpha}{}^{|\alpha}_{|\beta} - \gamma^{\alpha}_{\alpha}{}^{|\beta}_{|\beta} - (b^{\beta}_{\kappa}\kappa^{\alpha}_{\beta} - b^{\alpha}_{\alpha}\kappa^{\beta}_{\beta}) + \frac{1}{2}(\kappa^{\beta}_{\alpha}\kappa^{\alpha}_{\beta} - \kappa^{\alpha}_{\alpha}\kappa^{\beta}_{\beta}) = O(\frac{\eta\theta^2}{\lambda^2})
\end{aligned}
\tag{4.3.18}
$$

The conditions $(4.3.18)_1$ allow to obtain sharper estimate for Q^{β}

$$
Q^{\beta} = D\kappa^{\lambda}_{\lambda}{}^{|\beta} + O(Eh^2\frac{\eta\theta^2}{\lambda}) \tag{4.3.19}
$$

When $(4.3.14)_1$ and (4.3.19) are introduced into the equilibrium equations (4.1.20) they reduce within the same error to

$$C[(1-\nu)\gamma^{\beta}_{\alpha|\beta} + \nu\gamma^{\beta}_{\beta|\alpha}] + \bar{p}_{\alpha} = O(Eh\frac{\eta\theta^2}{\lambda})$$
$$D\kappa^{\alpha|\beta}_{\alpha|\beta} + C(b^{\alpha}_{\beta} - \kappa^{\alpha}_{\beta})[(1-\nu)\gamma^{\beta}_{\alpha} + \nu\delta^{\beta}_{\alpha}\gamma^{\lambda}_{\lambda}] + \bar{p} = O(Eh^2\frac{\eta\theta^2}{\lambda^2}) \tag{4.3.20}$$

Under small strains and bending shell theory components (3.2.16) of $\underset{\sim}{k}_t$ can be reduced to the linear form

$$k_{tt} = \kappa_{tt} + O(\frac{\eta\theta^2}{h}) \quad , \quad k_{\nu t} = \kappa_{\nu t} + O(\frac{\eta\theta^2}{h})$$
$$k_{nt} = 2\frac{d\gamma_{\nu t}}{ds} - \frac{d\gamma_{tt}}{ds_{\nu}} + 2\kappa_{\nu}\gamma_{\nu t} + \kappa_t(\gamma_{\nu\nu} - \gamma_{tt}) + O(\frac{\eta\theta^3}{h}) \tag{4.3.21}$$

The appropriate static boundary conditions may be obtained by consistent reduction of quantities appearing in (4.2.13). In this case

$$\underset{\sim}{P}_{\nu} = (P_{\nu\nu}\bar{\underset{\sim}{\nu}} + P_{t\nu}\bar{\underset{\sim}{t}} + P_{n\nu}\bar{\underset{\sim}{n}})[1 + O(\eta)]$$
$$P_{\nu\nu} = C(\gamma_{\nu\nu} + \nu\gamma_{tt}) + O(Eh\eta\theta^2) \quad , \quad P_{t\nu} = C(1-\nu)\gamma_{\nu t} + O(Eh\eta\theta^2)$$
$$P_{n\nu} = D[\frac{d\kappa_{\nu\nu}}{ds_{\nu}} + \frac{d\kappa_{tt}}{ds_{\nu}} + 2(1-\nu)\frac{d\kappa_{t\nu}}{ds}] + \tag{4.3.22}$$
$$+ D(1-\nu)[\kappa_t(\kappa_{\nu\nu} - \kappa_{tt}) + 2\kappa_{\nu}\kappa_{t\nu}] + O(Eh\eta\theta^3)$$
$$\bar{M}_{\nu\nu} = M_{\nu\nu} + O(Eh^2\eta\theta^2) \quad , \quad \bar{M}_{t\nu} = M_{t\nu} + O(Eh^2\eta\theta^2)$$

Therefore, for the static parameters (4.2.13) to be assumed on C we obtain the following simplified formula

$$\underset{\sim}{B}_{\nu} = \underset{\sim}{B}^{o}_{\nu} + \int_{M_o}^{M} [D(\kappa_{\nu\nu} + \nu\kappa_{tt})\bar{\underset{\sim}{t}} + \bar{\underset{\sim}{r}} \times \underset{\sim}{P}_{\nu}]ds - \bar{\underset{\sim}{r}} \times \int_{M_o}^{M} \underset{\sim}{P}_{\nu}ds$$
$$\bar{\underset{\sim}{a}}_t \cdot \underset{\sim}{F}_{\nu} = \bar{\underset{\sim}{t}} \cdot (\underset{\sim}{F}^{o}_{\nu} + \int_{M_o}^{M} \underset{\sim}{P}_{\nu}ds) \tag{4.3.23}$$

where $\underset{\sim}{P}_{\nu}$ is given by (4.3.22).

The resulting set of six bending shell equations (4.3.18) and (4.3.20) is remarkably simple . Four of them are linear and two are quadratic with respect to the shell strain measures $\gamma_{\alpha\beta}$ and $\kappa_{\alpha\beta}$. Also all the formulae (4.3.21) and (4.3.22) for boundary quantities are linear in the strain measures.

The set of shell equations may be used for solving bending shell problems directly in terms of the strain measures, without having had to calculate displacements first. The resulting strain measures give us also the stress and couple resultants and, therefore, the stress distribution in the shell space. For some shell problems this may end the solution. The displacement field may be obtained, if necessary, by additional integration of the strain - displacement relations.

4.4. Canonical intrinsic shell equations

In many shell problems there may be some regions, in which small strains caused by membrane forces may happen to be of essentially different order (higher or smaller by the factor θ^2) from those caused by moments. Within these regions the reduced bending shell equations $(4.3.18)_1$ and $(4.3.20)_1$ may not be accurate enough, since they contain terms of only one kind: changes of curvatures and membrane strains, respectively.

The refinement of $(4.3.18)_1$ and $(4.3.20)_1$ can be carried out by selecting stress resultants $N^{\alpha\beta}$ and changes of curvatures $\kappa_{\alpha\beta}$ as two independent variables of shell equations. The refinement procedure was originaly suggested by Danielson [26] in terms of slightly different basic variables and with surface force $\bar{p}$ taken into account. It was also applied by Koiter and Simmonds [24], in the absence of surface forces. Here following our earlier results [9] we present the refined shell equations with all surface forces (4.3.2) taken into account.

With $N^{\alpha\beta}$ and $\kappa_{\alpha\beta}$ as independent variables we are able to obtain much better estimate for $Q^{\alpha\beta}$ which, according to $(4.1.14)_2$, (4.3.6) and (4.3.4) becomes

$$Q^{\alpha\beta} = N^{\alpha\beta} - D(b^{\alpha}_{\kappa} - \kappa^{\alpha}_{\kappa})[(1 - \nu)\kappa^{\kappa\beta} + \nu a^{\kappa\beta}\kappa^{\lambda}_{\lambda}] + O(Eh\eta\theta^4) \qquad (4.4.1)$$

This estimate introduces an error $O(Eh\frac{\eta\theta^4}{\lambda})$ into the equilibrium equations $(4.1.20)_1$. Within the same error the second term of $(4.1.20)_1$ is reduced with the help of (4.3.4), (4.4.1) and (4.3.7) as follows

$$\begin{aligned}\bar{a}^{\alpha\kappa}\gamma_{\kappa\lambda\beta}Q^{\lambda\beta} &= (2\gamma^{\alpha}_{\lambda}|_{\beta} - \gamma_{\lambda\beta}|^{\alpha})N^{\lambda\beta} + O(Eh\frac{\eta\theta^4}{\lambda}) = \\ &= 2A[(1+\nu)N^{\alpha}_{\lambda} - \nu\delta^{\alpha}_{\lambda}N^{\kappa}_{\kappa}]|_{\beta}N^{\lambda\beta} - \\ &- A[(1+\nu)N_{\lambda\beta} - \nu a_{\lambda\beta}N^{\kappa}_{\kappa}]|^{\alpha}N^{\lambda\beta} + O(Eh\frac{\eta\theta^4}{\lambda}) = \\ &= 2A[N^{\alpha}_{\lambda}N^{\lambda\beta} + \nu(N^{\alpha}_{\lambda}N^{\lambda\beta} - N^{\alpha\beta}N^{\lambda}_{\lambda})]|_{\beta} - \\ &- \frac{1}{2}[(1+\nu)N_{\lambda\beta}N^{\lambda\beta} - \nu N^{\lambda}_{\lambda}N^{\beta}_{\beta}]|^{\alpha} + \\ &+ 2A[(1+\nu)N^{\alpha}_{\lambda}\bar{p}^{\lambda} - \nu N^{\lambda}_{\lambda}\bar{p}^{\alpha}] + O(Eh\frac{\eta\theta^4}{\lambda})\end{aligned} \tag{4.4.2}$$

Using the identities

$$N^{\alpha}_{\lambda} = \delta^{\alpha}_{\lambda}N^{\kappa}_{\kappa} - \epsilon^{\alpha\mu}\epsilon_{\lambda\nu}N^{\nu}_{\mu}\ , \quad N^{\lambda\beta} = a^{\beta\lambda}N^{\rho}_{\rho} - \epsilon^{\beta\gamma}\epsilon^{\lambda\kappa}N_{\gamma\kappa} \tag{4.4.3}$$

we obtain the relation

$$\begin{aligned}(N^{\alpha}_{\lambda}N^{\lambda\beta} - N^{\alpha\beta}N^{\lambda}_{\lambda})|_{\beta} &= a^{\alpha\beta}(N_{\lambda\kappa}N^{\lambda\kappa} - N^{\lambda}_{\lambda}N^{\kappa}_{\kappa})|_{\beta} - (N^{\alpha}_{\lambda}N^{\lambda\beta} - N^{\alpha\beta}N^{\lambda}_{\lambda})|_{\beta} = \\ &= \frac{1}{2}(N_{\lambda\beta}N^{\lambda\beta} - N^{\lambda}_{\lambda}N^{\beta}_{\beta})|^{\alpha}\end{aligned} \tag{4.4.4}$$

which is also an identity for any continuously differentiable symmetric tensor components.

With the help of (4.4.4) the relation (4.4.2) takes now the form

$$\begin{aligned}\bar{a}^{\alpha\kappa}\gamma_{\kappa\lambda\beta}Q^{\lambda\beta} &= 2A(N^{\alpha}_{\lambda}N^{\lambda\beta})|_{\beta} - \frac{1}{2}A[(1-\nu)N_{\lambda\beta}N^{\lambda\beta} + \nu N^{\lambda}_{\lambda}N^{\beta}_{\beta}]|^{\alpha} + \\ &+ 2A[(1+\nu)N^{\alpha}_{\lambda}\bar{p}^{\lambda} - \nu N^{\lambda}_{\lambda}\bar{p}^{\alpha}] + O(Eh\frac{\eta\theta^4}{\lambda})\end{aligned} \tag{4.4.5}$$

For the third term in $(4.1.20)_1$, by taking into account (4.3.19) and (4.3.4), we obtain the following estimate

$$-\bar{b}^{\alpha}_{\beta}Q^{\beta} = -D(b^{\alpha}_{\beta} - \kappa^{\alpha}_{\beta})\kappa^{\lambda}_{\lambda}|^{\beta} + O(Eh\frac{\eta\theta^4}{\lambda}) \tag{4.4.6}$$

Taking into account (4.4.1), (4.4.5) and (4.4.6) with $(4.3.20)_2$ the refined equilibrium equations take the following form

$$
\begin{aligned}
&N^{\beta}_{\alpha|\beta} - D[(1-\nu)b^{\lambda}_{\alpha}\kappa^{\beta}_{\lambda} + \nu b^{\beta}_{\alpha}\kappa^{\lambda}_{\lambda}]_{|\beta} - Db^{\beta}_{\alpha}\kappa^{\lambda}_{\lambda|\beta} + D[(1-\nu)\kappa^{\lambda}_{\alpha}\kappa^{\beta}_{\lambda} + \\
&+ \nu\kappa^{\beta}_{\alpha}\kappa^{\lambda}_{\lambda}]_{|\beta} + D\kappa^{\beta}_{\alpha}\kappa^{\lambda}_{\lambda|\beta} + 2A(N^{\lambda}_{\alpha}N^{\beta}_{\lambda})_{|\beta} - \frac{1}{2}A[(1-\nu)N^{\beta}_{\lambda}N^{\lambda}_{\beta} + \nu N^{\lambda}_{\lambda}N^{\beta}_{\beta}]_{|\alpha} + \\
&+ A[2(1+\nu)N^{\lambda}_{\alpha}\bar{p}_{\lambda} + (1-3\nu)N^{\lambda}_{\lambda}\bar{p}_{\alpha}] + \bar{p}_{\alpha} = O(Eh\frac{\eta\theta^4}{\lambda}) \qquad (4.4.7)\\
&D\kappa^{\alpha}_{\alpha}{}^{|\beta}_{|\beta} + (b^{\alpha}_{\beta} - \kappa^{\alpha}_{\beta})N^{\beta}_{\alpha} + \bar{p} = O(Eh^2\frac{\eta\theta^2}{\lambda^2})
\end{aligned}
$$

In exactly the same way the reduction of compatibility conditions $(2.1.6)_1$ may be carried out within a smaller error than in $(4.3.18)_1$. Let us replace $\bar{a}^{\kappa\nu}$ by $a^{\kappa\nu}$ in the second term of $(2.1.6)_1$. This, according to (4.3.4), introduces an error $O(\frac{\eta\theta^4}{h\lambda})$. Then using (2.1.7), (2.1.4) and $(4.3.7)_1$ this term can be estimated by

$$
\begin{aligned}
\epsilon_{\alpha\beta}\epsilon^{\lambda\mu}\bar{b}^{\kappa}_{\lambda}\gamma^{\ \beta}_{\kappa\cdot\mu} &= (b^{\kappa}_{\alpha} - \kappa^{\kappa}_{\alpha})(2\gamma^{\beta}_{\kappa|\beta} - \gamma^{\beta}_{\beta|\kappa}) - (b^{\kappa}_{\beta} - \kappa^{\kappa}_{\beta})\gamma^{\beta}_{\kappa|\alpha} + O(\frac{\eta\theta^4}{h\lambda}) = \\
&= -2A(1+\nu)(b^{\beta}_{\alpha} - \kappa^{\beta}_{\alpha})\bar{p}_{\beta} + A\nu(b^{\beta}_{\beta} - \kappa^{\beta}_{\beta})N^{\lambda}_{\lambda|\alpha} - \qquad (4.4.8)\\
&\quad - A(1+\nu)[(b^{\kappa}_{\beta} - \kappa^{\kappa}_{\beta})N^{\beta}_{\kappa|\alpha} + (b^{\kappa}_{\alpha} - \kappa^{\kappa}_{\alpha})N^{\lambda}_{\lambda|\kappa}] + O(\frac{\eta\theta^4}{h\lambda})
\end{aligned}
$$

Taking into account (4.4.8), $(4.3.20)_1$ and $(4.3.7)_1$ the refined compatibility conditions take the following form

$$
\begin{aligned}
&\kappa^{\beta}_{\alpha|\beta} - \kappa^{\beta}_{\beta|\alpha} - A(1+\nu)[b^{\lambda}_{\beta}N^{\beta}_{\lambda|\alpha} + b^{\beta}_{\alpha}N^{\lambda}_{\lambda|\beta}] + A\nu b^{\beta}_{\beta}N^{\lambda}_{\lambda|\alpha} + \\
&+ A(1+\nu)[\kappa^{\lambda}_{\beta}N^{\beta}_{\lambda|\alpha} + \kappa^{\beta}_{\alpha}N^{\lambda}_{\lambda|\beta}] - A\nu\kappa^{\beta}_{\beta}N^{\lambda}_{\lambda|\alpha} - \\
&- 2A(1+\nu)b^{\beta}_{\alpha}\bar{p}_{\beta} + 2A(1+\nu)\kappa^{\beta}_{\alpha}\bar{p}_{\beta} = O(\frac{\eta\theta^4}{h\lambda}) \qquad (4.4.9)\\
&AN^{\alpha}_{\alpha}{}^{|\beta}_{|\beta} + (b^{\alpha}_{\beta}\kappa^{\beta}_{\alpha} - b^{\alpha}_{\alpha}\kappa^{\beta}_{\beta}) - \frac{1}{2}(\kappa^{\alpha}_{\beta}\kappa^{\beta}_{\alpha} - \kappa^{\alpha}_{\alpha}\kappa^{\beta}_{\beta}) + A(1+\nu)\bar{p}^{\alpha}{}_{|\alpha} = O(\frac{\eta\theta^2}{\lambda^2})
\end{aligned}
$$

The relations (4.4.7) and (4.4.9) are canonical form of intrinsic equations in the first - approximation non-linear theory of thin elastic shells [24]. The appropriate deformational and static boundary conditions expressed in terms of $N^{\alpha\beta}$ and $\kappa_{\alpha\beta}$ follow from the more accurate reduction of (3.2.16) and (4.2.13), respectively.

The canonical intrinsic shell equations (4.4.7) and (4.4.9) describe accurately the behaviour of a thin shell in the whole internal region independently of the strain state in the shell. For some special shell problems these equations may be simplified. Some simplified forms of the equations were discussed by the author [25,27] based on restrictions assumed for the bending-to-membrane strain ratio.

5. GEOMETRICALLY NON-LINEAR THEORY OF SHELLS WITH RESTRICTED ROTATIONS

5.1. Classification of rotations

For many engineering purposes it is hardly necessary to allow the rotations of any magnitude. Some shell structures would become unserviceable if really unrestricted rotations were permitted to occure. Therefore it is certainly worthwhile to discuss the possible reduction of geometrically non-linear shell relations resulting from consistently restricted rotations.

The well known classifications of the approximate variants of shell equations were proposed by Mushtari and Galimov [28] and Koiter [11]. In [28] restrictions on components of the linearized rotation vector $\underset{\sim}{\Phi}$ were used to make distinction between three approximate variants of shell equations with "small, medium or large bending". In [11] four approximate variants with "infinitesimal, small finite, moderate or large deflections" were clearly defined by putting various restrictions on displacement gradients and components of $\underset{\sim}{\Phi}$. Note that in the classifications the word "rotation" does not appear at all, since neither $\underset{\sim}{\Phi}$ nor displacement gradients themselves are the finite rotations of the shell material elements.

By the polar decomposition (2.2.10) finite strains and finite rotations have clearly been separated. Within the geometrically non - linear theory we have restricted strains to be small everywhere in the shell. It seems therefore natural to restrict now in a consistent manner either the finite rotation tensor or the equivalent finite rotation vector. Since $\underset{\sim}{R}$ and $\underset{\sim}{\Omega}$ are uniquely described in (2.3.2) and (2.3.5) by the angle of rotation ω and the unit vector $\underset{\sim}{e}$ describing a direction of the rotation axis the both parameters may be restricted independently.

Within geometrically non-linear theory of thin isotropic and elastic shells there exist a small parameter θ defined in (4.3.3). This parameter can be used to introduce the following classification of rotations [7]:

$$\begin{aligned} &\omega \le O(\theta^2) \text{ - small rotations} \\ &\omega = O(\theta) \text{ - moderate rotations} \\ &\omega = O(\sqrt{\theta}) \text{ - large rotations} \\ &\omega \ge O(1) \text{ - finite rotations} \end{aligned} \tag{5.1.1}$$

Since for $|\omega| < \pi/2$ we have $O(|\underset{\sim}{\Omega}|) = O(\sin\omega) = O(\omega)$, the classification proposed here restricts only the magnitude $|\underset{\sim}{\Omega}|$ of the finite rotation vector, leaving the direction of the rotation axis to be arbitrary. It is known, however, that many shell structures are manufactured to be quite rigid in the direction tangent to the reference surface M even if they are allowed to be flexible in the direction orthogonal to M. For this reason, when discussing simplified variants of shell equations, there may be of interest to put restrictions of different order on appropriate components $\underset{\sim}{\Omega}\cdot\underset{\sim}{a}_\alpha$ or $\underset{\sim}{\Omega}\cdot\underset{\sim}{n}$. The name "small, moderate, large or finite rotation" may then be associated with the particular component of the finite rotation vector.

As an example, let us discuss here the simplest case of the non-linear theory of shells admiting only small rotations. In this case we have the following estimates

$$\begin{aligned} &|\underset{\sim}{\Omega}| = O(\theta^2) \quad , \quad \underset{\sim}{\Omega}\cdot\underset{\sim}{a}_\alpha = O(\theta^2) \quad , \quad \underset{\sim}{\Omega}\cdot\underset{\sim}{n} = O(\theta^2) \\ &\phi_\alpha = O(\theta^2) \quad , \quad \phi = O(\theta^2) \quad , \quad \theta_{\alpha\beta} = O(\theta^2) \end{aligned} \tag{5.1.2}$$

The shell strain measures become

$$\gamma_{\alpha\beta} = \theta_{\alpha\beta} + O(\eta\theta^2) \tag{5.1.3}$$

$$\kappa_{\alpha\beta} = -\frac{1}{2}[\phi_{\alpha|\beta} + \phi_{\beta|\alpha} + b^{\lambda}_{\alpha}(\theta_{\lambda\beta} - \omega_{\lambda\beta}) + b^{\lambda}_{\beta}(\theta_{\lambda\alpha} - \omega_{\lambda\alpha})] + O(\frac{\eta\theta^2}{\lambda})$$

The finite rotation vectors $\underset{\sim}{\Omega}$ and $\underset{\sim}{\Omega}_t$ can be approximated by

$$\underset{\sim}{\Omega} = \underset{\sim}{\phi}[1 + O(\theta^2)] \quad , \quad \underset{\sim}{\Omega}_t = (\underset{\sim}{\phi} - \theta_{\alpha\beta}\nu^{\alpha}t^{\beta}\underset{\sim}{n})[1 + O(\theta^2)]$$
$$\underset{\sim}{\phi} = \epsilon^{\beta\alpha}\phi_{\alpha}\underset{\sim}{a}_{\beta} + \phi\underset{\sim}{n} \tag{5.1.4}$$

where $\underset{\sim}{\phi}$ is the linearized rotation vector.

When introduced into the Lagrangean virtual work principle (4.1.16) these relations lead to the Lagrangean equilibrium equations and static boundary conditions

$$\underset{\sim}{N}^{\beta}|_{\beta} + \underset{\sim}{p} = \underset{\sim}{0} \quad \text{in } M$$
$$\underset{\sim}{N}^{\beta}\nu_{\beta} + \frac{d}{ds}(M_{t\nu}\underset{\sim}{n}) = \underset{\sim}{F} + \frac{d}{ds}(K_t\underset{\sim}{n}) \quad , \quad M_{\nu\nu} = K_{\nu} \quad \text{on } C \tag{5.1.5}$$
$$\Delta M_{t\nu}\underset{\sim}{n} = \Delta K_t\underset{\sim}{n} \quad \text{at each } M_n$$

$$\underset{\sim}{N}^{\beta} = (N^{\alpha\beta} - b^{\alpha}_{\lambda}M^{\lambda\beta})\underset{\sim}{a}_{\alpha} + M^{\alpha\beta}|_{\alpha}\underset{\sim}{n} \tag{5.1.6}$$

Therefore, within small rotations the geometrically non-linear theory of shells reduces to the classical linear theory of shells [29,30,31]. Note that the linearized rotation vector $\underset{\sim}{\phi}$ enters many relations of the linear theory of shells. It was used, in particular, to construct multivalued solutions for shells with multi-connected regions [32,33].

5.2. Moderate rotation theory of shells

When rotations are assumed to be moderate then, using (4.3.10) and (2.1.4), we have the estimates

$$|\underset{\sim}{\Omega}| = O(\theta) \quad , \quad \underset{\sim}{\Omega}\cdot\underset{\sim}{a}_{\alpha} = O(\theta) \quad , \quad \underset{\sim}{\Omega}\cdot\underset{\sim}{n} = O(\theta)$$
$$\phi_{\alpha} = O(\theta) \quad , \quad \phi = O(\theta) \quad , \quad \theta_{\alpha\beta} = O(\theta^2) \tag{5.2.1}$$

It follows from the error indicated in the shell strain energy function (4.3.5) that $\gamma_{\alpha\beta}$ may be simplified by omitting terms $O(\eta\theta^2)$ while in $\kappa_{\alpha\beta}$ we can omit terms $O(\frac{\eta\theta}{\lambda})$. Therefore, the shell strain measures take here the following approximate forms

$$\gamma_{\alpha\beta} = \theta_{\alpha\beta} + \frac{1}{2}\phi_\alpha\phi_\beta + \underline{\frac{1}{2}a_{\alpha\beta}\phi^2 - \frac{1}{2}(\theta^\lambda_\alpha\omega_{\lambda\beta} + \theta^\lambda_\beta\omega_{\lambda\alpha})} + O(\eta\theta^2)$$
$$\kappa_{\alpha\beta} = -\frac{1}{2}[\phi_{\alpha|\beta} + \phi_{\beta|\alpha} + \underline{b^\lambda_\alpha(\theta_{\lambda\beta} - \omega_{\lambda\beta}) + b^\lambda_\beta(\theta_{\lambda\alpha} - \omega_{\lambda\alpha})}] + O(\frac{\eta\theta}{\lambda}) \qquad (5.2.2)$$

Since $\frac{1}{2}(b^\lambda_\alpha\theta_{\lambda\beta} + b^\lambda_\beta\theta_{\lambda\alpha}) = O(\frac{\eta\theta}{\lambda})$ we could also omit these terms in $(5.2.2)_2$ within the accuracy of the first - approximation theory. These terms are linear in displacements and it is rather a convention to keep them here together with definition of $\kappa_{\alpha\beta}$ as a shell strain measure.

The finite rotation vectors are approximated by

$$\underset{\sim}{\Omega} = (\epsilon^{\beta\alpha}\phi_\alpha + \underline{\frac{1}{2}\phi^\beta\phi})\underset{\sim}{a}_\beta + \phi\underset{\sim}{n} + O(\eta\theta)$$

$$\underset{\sim}{\Omega}_t = \underset{\sim}{\Omega} - \gamma_{\nu t}\underset{\sim}{n} + O(\theta^3) = \qquad (5.2.3)$$
$$= (\phi_t + \underline{\frac{1}{2}\phi_\nu\phi})\underset{\sim}{\nu} + (-\phi_\nu + \underline{\frac{1}{2}\phi_t\phi})\underset{\sim}{t} + (\phi - \theta_{\nu t} - \frac{1}{2}\phi_\nu\phi_t)\underset{\sim}{n} + O(\eta\theta)$$

If we introduce (5.2.2) into the left - hand side of (4.1.16) after transformation we obtain the following formula for the internal virtual work

$$IVW = -\iint_{M}(\underset{\approx}{GN}^\beta)_{|\beta}\cdot\delta\underset{\sim}{u}\,dA + J_i$$
$$J_i = \int_C \{[\underset{\approx}{GN}^\beta\nu_\beta + \frac{d}{ds}(M_{t\nu}\underset{\sim}{n})]\cdot\delta\underset{\sim}{u} - M_{\nu\nu}\delta\phi_\nu\}ds + \sum_n \Delta M_{t\nu}\delta w \qquad (5.2.4)$$

where $\delta\phi_\nu = -\delta\beta_\nu$ and

$$\underset{\approx}{GN}^\beta = [N^{\alpha\beta} - \underline{b^\alpha_\lambda M^{\lambda\beta} - \frac{1}{2}\omega^{\alpha\beta}N^\lambda_\lambda - \frac{1}{2}(\omega^{\alpha\lambda}N^\beta_\lambda + \omega^{\beta\lambda}N^\alpha_\lambda)} +$$
$$+ \underline{\frac{1}{2}(\theta^{\alpha\lambda}N^\beta_\lambda - \theta^{\beta\lambda}N^\alpha_\lambda)}]\underset{\sim}{a}_\alpha + (\phi_\alpha N^{\alpha\beta} + M^{\alpha\beta}{}_{|\beta})\underset{\sim}{n} \qquad (5.2.5)$$

Note that within the shell theory admiting moderate rotations the moment $M_{t\nu}$ appears in $(5.2.4)_2$ multiplied by $\mathbf{n}$ and not by $\bar{\mathbf{n}}$ as in the general case $(4.1.14)_3$. This results from linearity of $\kappa_{\alpha\beta}$ in (5.2.2) and is compatible with accuracy of the shell strain energy (4.3.5). Although the energy arguments cannot be applied directly to the external boundary couple $\mathbf{K}$, it would seem to be inconsistent to use a better approximation for $\mathbf{K}$ in EVW than that resulting for $\mathbf{M}^{\beta} = \epsilon_{\alpha\lambda} M^{\alpha\beta} \mathbf{a}^{\lambda}$ at the boundary in IVW. Since according to $(4.1.13)_1$, (5.2.1) and (4.3.9) there is $K = K^{\alpha} \cdot O(\theta)$ the dominant terms in $\mathbf{K}$ and $\boldsymbol{\Omega}_t$ are

$$\begin{aligned} \mathbf{K} &= (-K_t \boldsymbol{\nu} + K_{\nu} \mathbf{t})[1 + O(\theta)] \\ \boldsymbol{\Omega}_t &= (\phi_t \boldsymbol{\nu} - \phi_{\nu} \mathbf{t} + \phi \mathbf{n})[1 + O(\theta)] \end{aligned} \tag{5.2.6}$$

Introducing (5.2.2) and (5.2.6) into (4.1.16) we obtain the Lagrangean equilibrium equations and the Lagrangean static boundary conditions

$$\begin{aligned} &(\mathbf{G}\mathbf{N}^{\beta})_{|\beta} + \mathbf{p} = \mathbf{0} \quad \text{in } M \\ &\mathbf{G}\mathbf{N}^{\beta}\nu_{\beta} + \frac{d}{ds}(M_{t\nu}\mathbf{n}) = \mathbf{F} + \frac{d}{ds}(K_t \mathbf{n}) \;, \quad M_{\nu\nu} = K_{\nu} \quad \text{on } C \\ &\Delta M_{t\nu}\mathbf{n} = \Delta K_t \mathbf{n} \quad \text{at each } M_n \end{aligned} \tag{5.2.7}$$

The component form of $(5.2.7)_1$ with (5.2.5) reads

$$\begin{aligned} &\underline{[N^{\alpha\beta} - b^{\alpha}_{\lambda} M^{\lambda\beta} - \tfrac{1}{2}\omega^{\alpha\beta} N^{\lambda}_{\lambda} - \tfrac{1}{2}(\omega^{\alpha\lambda} N^{\beta}_{\lambda} + \omega^{\beta\lambda} N^{\alpha}_{\lambda}) +} \\ &\quad \underline{+ \tfrac{1}{2}(\theta^{\alpha\lambda} N^{\beta}_{\lambda} - \theta^{\beta\lambda} N^{\alpha}_{\lambda})]}_{|\beta} - b^{\alpha}_{\beta}(\phi_{\lambda} N^{\lambda\beta} + M^{\lambda\beta}{}_{|\lambda}) + p^{\alpha} = 0 \\ &(\phi_{\alpha} N^{\alpha\beta} + M^{\alpha\beta}{}_{|\alpha})_{|\beta} + b_{\alpha\beta}\underline{[N^{\alpha\beta} - b^{\alpha}_{\lambda} M^{\lambda\beta} - \tfrac{1}{2}(\omega^{\alpha\lambda} N^{\beta}_{\lambda} + \omega^{\beta\lambda} N^{\alpha}_{\lambda})]} + p = 0 \end{aligned} \tag{5.2.8}$$

The equations of moderate rotation theory of shells are quite complex, partialy due to the presence of the last two terms (underlined by dots) in the expression $(5.2.2)_1$ for $\gamma_{\alpha\beta}$. Since these terms are $O(\eta\theta)$ by neglecting them in $(5.2.2)_1$ we would introduce an additional error $O(Eh\eta^2\theta)$ into the strain energy (4.3.5), which sometimes may cause some decrease in accuracy of the solution. At the expence of indicated loss in accuracy, we still may neglect [11,9] these terms in (5.2.2). As a result terms underlined by dots will not appear in (5.2.5) and (5.2.8).

In many engineering structures only rotations about a tangent to the shell middle surface are allowed to be moderate, while rotations about a normal are supposed to be small. In such a case we have

$$\begin{aligned} &\Omega\cdot a_\alpha = O(\theta) \quad , \quad \Omega\cdot n = O(\theta^2) \\ &\phi_\alpha = O(\theta) \quad , \quad \phi = O(\theta^2) \quad , \quad \theta_{\alpha\beta} = O(\theta^2) \end{aligned} \tag{5.2.9}$$

and all the relations (5.2.2) to (5.2.8) can be considerably simplified, without any loss in accuracy, by omitting there all terms underlined by a solid line.

Further possible simplifications of shell equations, within moderate rotation theory of shells, were discussed by Koiter [11] and the author [9].

5.3. Large rotation theory of shells

When rotations are allowed to be large then from (4.3.10) and (2.1.4) we have

$$\begin{aligned} &|\Omega| = O(\sqrt{\theta}) \quad , \quad \Omega\cdot a_\alpha = O(\sqrt{\theta}) \quad , \quad \Omega\cdot n = O(\sqrt{\theta}) \\ &\phi_\alpha = O(\sqrt{\theta}) \quad , \quad \phi = O(\sqrt{\theta}) \quad , \quad \theta_{\alpha\beta} = O(\theta) \\ &\Omega = \epsilon^{\beta\alpha}[\phi_\alpha(1 + \tfrac{1}{2}\theta^\kappa_\kappa) - \tfrac{1}{2}\phi^\lambda(\theta_{\lambda\alpha} - \omega_{\lambda\alpha})]a_\beta + \phi n + O(\eta\sqrt{\theta}) \end{aligned} \tag{5.3.1}$$

The strain tensor cannot be simplified here. In order to reduce the tensor of change of curvature we use the relations [9]

$$\begin{aligned} &\omega_{\lambda\alpha|\beta} = \theta_{\alpha\beta|\lambda} - \theta_{\lambda\beta|\alpha} + b_{\alpha\beta}\phi_\lambda - b_{\lambda\beta}\phi_\alpha \\ &\phi_\lambda\theta^\kappa_\kappa - \phi^\mu\theta_{\mu\lambda} = -\tfrac{1}{2}\phi_\lambda\phi^2 + O(\theta^2) \end{aligned} \tag{5.3.2}$$

The relation $(5.3.2)_1$ is an identity, while $(5.3.2)_2$ is satisfied within the large rotation theory of shells.

When (4.3.9), (5.3.2) together with (5.3.1) are introduced into (2.1.5) then within $O(\frac{\eta\theta}{\lambda})$ we obtain for $\kappa_{\alpha\beta}$ the following formula

$$\begin{aligned}\kappa_{\alpha\beta} = & -\frac{1}{2}[\phi_{\alpha|\beta} + \phi_{\beta|\alpha} + b^{\lambda}_{\alpha}(\theta_{\lambda\beta} - \omega_{\lambda\beta}) + b^{\lambda}_{\beta}(\theta_{\lambda\alpha} - \omega_{\lambda\alpha})] - \\ & -\frac{1}{2} b_{\alpha\beta}\phi^{\lambda}\phi_{\lambda} + \frac{1}{2}(\frac{1}{2}\phi^{\lambda}\phi_{\lambda} + \underline{\theta^{\lambda}_{\kappa}\theta^{\kappa}_{\lambda} - \frac{1}{2}\theta^{\kappa}_{\kappa}\theta^{\lambda}_{\lambda}})(\phi_{\alpha|\beta} + \phi_{\beta|\alpha}) - \\ & -\underline{\frac{1}{4}\phi^{\lambda}\phi_{\lambda}(b^{\lambda}_{\alpha}\omega_{\lambda\beta} - b^{\lambda}_{\beta}\omega_{\lambda\alpha})} + (\phi^{\lambda} + \phi_{\kappa}\omega^{\kappa\lambda} - \underline{\frac{1}{2}\phi^{\lambda}\phi^{2}})\theta_{\lambda\alpha\beta} + O(\frac{\eta\theta}{\lambda})\end{aligned} \tag{5.3.3}$$

$$\theta_{\lambda\alpha\beta} = \theta_{\lambda\alpha|\beta} + \theta_{\lambda\beta|\alpha} - \theta_{\alpha\beta|\lambda}$$

Using the relation (4.3.11) we also reduce the total finite rotation vector to the form

$$\begin{aligned}2\underset{\sim}{\Omega}_{t} = & (2\phi_{t} + \phi_{\nu}\phi - \phi_{\nu}\theta_{\nu t} + \phi_{t}\theta_{\nu\nu})\underset{\sim}{\nu} + \\ & + [-2\phi_{\nu} + \phi_{t}\phi - 3\phi_{\nu}\theta_{tt} + 3\phi_{t}\theta_{\nu t} + \underline{2\phi_{\nu}\phi\theta_{\nu t} + \phi_{t}\phi(\theta_{tt} - \theta_{\nu\nu})} - \\ & - \underline{\phi_{\nu}\phi^{2}}]\underset{\sim}{t} + [2\phi - 2\theta_{\nu t} - \phi_{\nu}\phi_{t}(1 + \theta_{tt}) - \frac{1}{2}(\phi^{2}_{\nu} - \phi^{2}_{t})\phi + \\ & + \frac{1}{2}(\phi^{2}_{\nu} + 3\phi^{2}_{t})\theta_{\nu t}]\underset{\sim}{n} + O(\theta^{2}\sqrt{\theta})\end{aligned} \tag{5.3.4}$$

When introduced into (4.1.16) the strain measures (2.1.4) and (5.3.3) lead to the Lagrangean equilibrium equations $(4.1.18)_1$, where

$$\begin{aligned}\underset{\sim}{GN}^{\beta} = & \{1^{\alpha}_{\cdot\lambda}N^{\lambda\beta} - b^{\alpha}_{\lambda}M^{\lambda\beta} + \underline{\frac{1}{4}\phi^{\kappa}\phi_{\kappa}(b^{\alpha}_{\lambda}M^{\lambda\beta} - b^{\beta}_{\lambda}M^{\alpha\lambda})} + \\ & + (2\theta^{\alpha\beta} - a^{\alpha\beta}\theta^{\kappa}_{\kappa})\phi_{\lambda|\mu}M^{\lambda\mu} + \underline{\frac{1}{2}\omega^{\alpha\beta}\phi^{\kappa}\theta_{\kappa\lambda\mu}M^{\lambda\mu}} - \\ & - \frac{1}{2}(\phi^{\alpha}\theta^{\beta}_{\cdot\lambda\mu} - \phi^{\beta}\theta^{\alpha}_{\cdot\lambda\mu})M^{\lambda\mu} - [(1 - \underline{\frac{1}{2}\phi^{2}})(\phi^{\alpha}M^{\lambda\beta} + \phi^{\beta}M^{\alpha\lambda} - \phi^{\lambda}M^{\alpha\beta}) + \\ & + \phi_{\kappa}(\omega^{\kappa\alpha}M^{\lambda\beta} + \omega^{\kappa\beta}M^{\alpha\lambda} - \omega^{\kappa\lambda}M^{\alpha\beta})]_{|\lambda}\}\underset{\sim}{a}_{\alpha} + \\ & + \{\phi_{\lambda}N^{\lambda\beta} + [(1 - \frac{1}{2}\phi^{\kappa}\phi_{\kappa} + \frac{1}{2}\theta^{\kappa}_{\kappa}\theta^{\mu}_{\mu} - \theta^{\kappa}_{\mu}\theta^{\mu}_{\kappa})M^{\lambda\beta}]_{|\lambda} +\end{aligned} \tag{5.3.5}$$

$$+ [\phi^{\beta}(- b_{\lambda\mu} + \phi_{\lambda|\mu} - \underline{b^{\kappa}_{\lambda}\omega_{\kappa\mu}}) +$$

$$+ (1 - \underline{\tfrac{1}{2}\phi^2})\theta^{\beta}_{\cdot\lambda\mu} + \omega^{\beta\kappa}\theta_{\kappa\lambda\mu}]M^{\lambda\mu}\}\underset{\sim}{n}$$

It is obvious how to obtain the component form of $(4.1.18)_1$ with (5.3.5), although it is quite complex.

At the expence of a possible loss in accuracy we may neglect in (5.3.3) some terms $O(\frac{\eta\sqrt{\theta}}{\lambda})$ underlined by dots. As a result, terms underlined by dots will not appear in (5.3.5).

In some shell structures only rotations about a tangent may be large while rotations about a normal are at most moderate. In such a case, we assume $\underset{\sim}{\Omega}\cdot\underset{\sim}{n} = O(\theta)$ and $\phi = O(\theta)$ in (5.3.1) and relations (5.3.3), (5.3.4) and (5.3.5) can be simplified, without any loss in accuracy, by omitting there all terms underlined by a solid line.

In many engineering shell structures the rotations about a normal are always small even if rotations about a tangent are allowed to be large. In such a case

$$|\underset{\sim}{\Omega}| = O(\sqrt{\theta}) \quad , \quad \underset{\sim}{\Omega}\cdot\underset{\sim}{a}_{\alpha} = O(\sqrt{\theta}) \quad , \quad \underset{\sim}{\Omega}\cdot\underset{\sim}{n} = O(\theta^2)$$

$$\phi_{\alpha} = O(\sqrt{\theta}) \quad , \quad \phi = O(\theta^2) \quad , \quad \theta_{\alpha\beta} = O(\theta) \tag{5.3.6}$$

$$\underset{\sim}{\Omega} = \epsilon^{\beta\alpha}[\phi_{\alpha}(1 + \tfrac{1}{2}\theta^{\kappa}_{\kappa}) - \tfrac{1}{2}\phi^{\lambda}\theta_{\lambda\alpha}]\underset{\sim}{a}_{\beta} + \phi\underset{\sim}{n} + O(\eta\sqrt{\theta})$$

The shell strain measures can now be approximated by

$$\gamma_{\alpha\beta} = \theta_{\alpha\beta} + \tfrac{1}{2}\phi_{\alpha}\phi_{\beta} + \tfrac{1}{2}\theta^{\lambda}_{\alpha}\theta_{\lambda\beta} - \tfrac{1}{2}(\theta^{\lambda}_{\alpha}\omega_{\lambda\beta} + \theta^{\lambda}_{\beta}\omega_{\lambda\alpha}) + O(\eta\theta^2)$$

$$\begin{aligned}\kappa_{\alpha\beta} = &- \tfrac{1}{2}[\phi_{\alpha|\beta} + \phi_{\beta|\alpha} + b^{\lambda}_{\alpha}(\theta_{\lambda\beta} - \omega_{\lambda\beta}) + b^{\lambda}_{\beta}(\theta_{\lambda\alpha} - \omega_{\lambda\alpha})] - \\ &- \tfrac{1}{2}b_{\alpha\beta}\phi^{\lambda}\phi_{\lambda} + \tfrac{1}{2}[\tfrac{1}{2}\phi^{\lambda}\phi_{\lambda} + \underline{\theta^{\lambda}_{\kappa}\theta^{\kappa}_{\lambda} - \tfrac{1}{2}\theta^{\kappa}_{\kappa}\theta^{\lambda}_{\lambda}}](\phi_{\alpha|\beta} + \phi_{\beta|\alpha}) + \\ &+ \phi^{\lambda}\theta_{\lambda\alpha\beta} + O(\tfrac{\eta\theta}{\lambda})\end{aligned} \tag{5.3.7}$$

These strain measures lead to the vector equilibrium equation (4.1.18)$_1$, where now

$$\underset{\sim}{GN}^{\beta} = [N^{\alpha\beta} + \frac{1}{2}(\theta^{\alpha\lambda}N^{\beta}_{\lambda} + \theta^{\beta\lambda}N^{\alpha}_{\lambda}) + \frac{1}{2}(\theta^{\alpha\lambda}N^{\beta}_{\lambda} - \theta^{\beta\lambda}N^{\alpha}_{\lambda}) - \frac{1}{2}(\omega^{\alpha\lambda}N^{\beta}_{\lambda} + \omega^{\beta\lambda}N^{\alpha}_{\lambda}) - b^{\alpha}_{\lambda}M^{\lambda\beta} + \underline{(2\theta^{\alpha\beta} - a^{\alpha\beta}\theta^{\kappa}_{\kappa})\phi_{\lambda|\mu}}M^{\lambda\mu} - (\phi^{\alpha}M^{\lambda\beta} + \phi^{\beta}M^{\alpha\lambda} - \phi^{\lambda}M^{\alpha\beta})_{|\lambda}]\underset{\sim}{a}_{\alpha} + \{\phi_{\lambda}N^{\lambda\beta} + [(1 - \frac{1}{2}\phi^{\kappa}\phi_{\kappa} + \underline{\frac{1}{2}\theta^{\kappa}_{\kappa}\theta^{\mu}_{\mu} - \theta^{\kappa}_{\mu}\theta^{\mu}_{\kappa}})M^{\lambda\beta}]_{|\lambda} + [\phi^{\beta}(-b_{\lambda\mu} + \phi_{\lambda|\mu}) + \theta^{\beta}_{\cdot\lambda\mu}]M^{\lambda\mu}\}\underset{\sim}{n} \qquad (5.3.8)$$

Again, at the expence of a possible loss in accuracy we may neglect in (5.3.7) terms underlined by dots. As a result, terms underlined by dots will not appear in (5.3.8). Such approximation introduces an additional error $O(Eh\eta^2\theta\sqrt{\theta})$ into the strain energy (4.3.5).

At the expence of a greater error $O(Eh\eta^2\theta)$ in (4.3.5) the strain measures may be approximated by the following formulae

$$\begin{aligned} \gamma_{\alpha\beta} &= \theta_{\alpha\beta} + \frac{1}{2}\phi_{\alpha}\phi_{\beta} + \frac{1}{2}\theta^{\lambda}_{\alpha}\theta_{\lambda\beta} + O(\eta\theta) \\ \kappa_{\alpha\beta} &= -\frac{1}{2}(1 - \frac{1}{2}\phi^{\lambda}\phi_{\lambda})(\phi_{\alpha|\beta} + \phi_{\beta|\alpha}) + \theta^{\lambda}\theta_{\lambda\alpha\beta} + O(\frac{\eta}{\lambda}) \end{aligned} \qquad (5.3.9)$$

which result in equilibrium equation (4.1.18)$_1$ with

$$\underset{\sim}{GN}^{\beta} = [N^{\alpha\beta} + \frac{1}{2}(\theta^{\alpha\lambda}N^{\beta}_{\lambda} + \theta^{\beta\lambda}N^{\alpha}_{\lambda}) - (\phi^{\alpha}M^{\lambda\beta} + \phi^{\beta}M^{\alpha\lambda} - \phi^{\lambda}M^{\alpha\beta})_{|\lambda}]\underset{\sim}{a}_{\alpha} + \{\phi_{\lambda}N^{\lambda\beta} + [(1 - \frac{1}{2}\phi^{\kappa}\phi_{\kappa})M^{\lambda\beta}]_{|\lambda} + (\phi^{\beta}\phi_{\lambda|\mu} + \theta^{\beta}_{\cdot\lambda\mu})M^{\lambda\mu}\}\underset{\sim}{n} \qquad (5.3.10)$$

When expressed in the reference basis (4.1.18)$_1$ with (5.3.8) or (5.3.10) leads to appropriate component forms of equilibrium equations.

REFERENCES

1. Truesdell, C., Noll, W., The non-linear field theories of mechanics, in *Handbuch der Physik*, III/3, Springer-Verlag, Berlin - Heidelberg - New York, 1965.

2. Shamina, V.A., Determination of displacement vector from the components of deformation tensor in non-linear continuum mechanics (in Russian), *Izv. AN SSSR, Mekh. Tv. Tela*, 1, 14-22, 1974.

3. Lurie, A.I., *Analytical mechanics* (in Russian), Nauka, Moscow, 1961.

4. Simmonds, J.G., Danielson, D.A., Nonlinear shell theory with a finite rotation vector, *Proc. Kon. Ned. Ak. Wet.*, Ser.B, 73, 460-478, 1970.

5. Simmonds, J.G., Danielson, D.A., Nonlinear shell theory with finite rotation and stress function vectors, *J. Applied Mech., Trans. ASME*, Ser. E, 39, No 4, 1085-1090, 1972.

6. Novozhilov, V.V., Shamina, V.A., Kinematic boundary conditions in non-linear problems of the theory of elasticity (in Russian), *Izv. AN SSSR Mekh. Tv. Tela*, 5, 63-74, 1975.

7. Pietraszkiewicz, W., *Obroty skończone i opis Lagrange'a w nieliniowej teorii powłok*, Rozprawa habilitacyjna, Biuletyn Instytutu Maszyn Przepływowych PAN, 172(880), Gdańsk, 1976.
English translation: *Finite rotations and Lagrangean description in the non-linear theory of shells*, Polish Scientific Publishers, Warszawa - Poznań, 1979.

8. Pietraszkiewicz, W., Finite rotations in shells, Presented at the III IUTAM Symp. on Shell Theory, Tbilisi, Aug. 22-28, 1978 (to be published in *Proceedings*).

9. Pietraszkiewicz, W., *Introduction to the non-linear theory of shells*, Ruhr - Universität Bochum, Mitt. Inst. für Mech., Nr 10, Bochum, 1977.

10. Pietraszkiewicz, W., Consistent second approximation to the elastic strain energy of a shell, Z. *für Angew. Math. Mech.*, 59, Nr 3/4, 1979 (in print)

11. Koiter, W.T., On the nonlinear theory of thin elastic shells, *Proc. Kon. Ned. Ak. Wet.*, Ser.B, 69, No1, 1-54, 1966.

12. Pietraszkiewicz, W., Lagrangean non-linear theory of shells, *Archives of Mechanics*, 26, No 2, 221-228, 1974.

13. Lichnerowicz, A., *Elements of tensor calculus*, Methuen & Co., London, 1962.

14. Bowen, R.M., Wang, C.C., *Introduction to vectors and tensors*, Plenum Press, New York, 1976.

15. Truesdell, C., *A first course in rational continuum mechanics*, John Hopkins University, Baltimore, Maryland, 1972.

16. Pietraszkiewicz, W., Some relations of the non-linear Reissner theory of shells (in Russian), *Vestnik Leningradskogo Universiteta*, Ser. Mat. Mekh., No 1, 1979 (in print).

17. Koiter, W.T., A consistent first approximation in the general theory of thin elastic shells, in *Theory of Thin Shells*, Proc. IUTAM Symp. Delft, 1959; North-Holland P.Co., Amsterdam, 1960, 12-33.

18. Leigh, D.C., *Nonlinear continuum mechanics*, McGraw-Hill, New York, 1968.

19. Korn, G.A., Korn, T.M., *Mathematical handbook*, 2nd ed., McGraw-Hill, New York, 1968.

20. Shield, R.T., The rotation associated with large strains, *SIAM J. Appl. Math.*, 25, No 3, 483-491, 1973.

21. Gorr, G.A., Kudryashova, L.V., Stepanova, L.A., *Classical problems of rigid-body dynamics, Development and present state* (in Russian), Naukova Dumka, Kiev, 1978.

22. John, F., Estimates for the derivatives of the stresses in a thin

shell and interior shell equations, *Comm. Pure & Appl. Math.*, 18, 235-267, 1965.

23. Koiter, W.T., On the mathematical foundation of shell theory, *Proc. Congr. Int. des Math.*, Nice 1970, tome 3, Gauthier-Villars, Paris 1971, 123-130.

24. Koiter, W.T., Simmonds, J.G., Foundations of shell theory, in *Theoretical and applied mechanics*, Proc. 13th IUTAM Congr., Moscow 1972, Springer-Verlag, Berlin - Heidelberg - New York, 1973, 150-175.

25. Pietraszkiewicz, W., Non-linear theories of thin elastic shells (in Polish), in *Shell structures, theory and applications* (in Polish), 1, Proc. Symp. Kraków, April 25-27, 1974, Orkisz, J., Waszczyszyn, Z., Eds., Polish Scientific Publishers, Warszawa, 1978, 27-50.

26. Danielson, D.A., Simplified intrinsic equations for arbitrary elastic shells, *Int. J. Engng. Sci.*, 8, 251-259, 1970.

27. Pietraszkiewicz, W., Simplified equations for the geometrically non--linear thin elastic shells, *Trans. Inst. Fluid-Flow Mach. Gdańsk*, 75, 165-175, 1978.

28. Mushtari, K.M., Galimov, K.Z., *Non-linear theory of thin elastic shells* (in Russian), Kazan', 1957.
English translation: The Israel Pr. for Sci. Transl., Jerusalem 1961.

29. Naghdi, P.M., Foundations of elastic shell theory, in *Progress in Solid Mechanics*, IV, Amsterdam, 1963.

30. Chernykh, K.F., *Linear theory of shells*, Part 2 (in Russian), Leningrad St. Univ. Press, Leningrad, 1964.
English translation: NASA-TT-F-II 562, 1968.

31. Goldenveizer, A.L., *Theory of thin elastic shells* (in Russian), 2nd ed., Nauka, Moscow 1976.

32. Pietraszkiewicz, W., Multivalued stress functions in the linear theory of shells, *Archiwum Mechaniki Stosowanej*, 20, No 1, 37-45, 1968.

33. Pietraszkiewicz, W., Multivalued solutions for shallow shells, *Archiwum Mechaniki Stosowanej*, 20, No 1, 3-10, 1968.

34. Wempner, G., Finite elements, finite rotations and small strains, *Int. J. Solids & Str.*, 5, 117-153, 1969.

35. Glockner, P.G., Shrivastava, J.P., On the geometry and kinematics of nonlinear deformation of shell space, in *Proc. 11th Midw. Mech. Conf.* Iowa St. Univ., Aug. 18-20, 1969, 331-352.

36. Pietraszkiewicz, W., Three forms of geometrically non-linear bending shell equations, VIIIth Int. Congr. on Appl. of Math. in Engng., Weimar, June 26 ÷ July 2, 1978 (to be publ. in *Trans. Inst. of Fluid-Flow Mach. Gdańsk*)

37. Pietraszkiewicz, W., Some problems of the non-linear theory of shells, (in Polish), in *Shell structures, theory and applications* (in Polish), Proceed. 2nd Polish Conf., Gołuń, Nov. 6-10, 1978, General lectures, Center for Shipbld. Res. Press, Gdańsk, 1978, 119-159.

INELASTIC REPONSE OF THIN SHELLS

Basic Problems and Applications

by

W. Olszak and **A. Sawczuk**

Polish Academy of Sciences, Warsaw

CISM, Udine

Polish Academy of Sciences

Warsaw

INELASTIC RESPONSE OF THIN SHELLS

Basic Problems and Applications

W. Olszak and *A. Sawczuk*

1. Introduction

The applications of shells in modern technology makes the classical theory, based on physical linearity and time independence of the material properties, inadequate for rational design. The inelastic behaviour of shells attracts attention, various forms of the material constitutive equations being used, in order to explain such types of inelastic response as creep, relaxation, or/and plasticity.

For convenience, the inelastic behaviour can be classified into two groups viz: *viscoelasticity* and *plasticity,* depending upon the source of the mechanical energy dissipation. The most simple and tractable types of inelasticity are: *linear visco-elasticity, steady creep, elastic-plastic deformations,* and *plastic flow.*

In the present note, attention is focussed on the statics of *isotropic* inelastic shells within the framework of infinitesimal deformation theories. Hence no distinction is made between the original and the deformed configuration. By necessity, the basic relationships are indicated in a very concise manner. However, at the request of the course participants they have been completed by a comprehensive list of references to original papers presenting either methods of solution or solutions of engineering problems.

Besides, for more general information, also references related to geometrically nonlinear behaviour, mostly regarding the post-yield range, are indicated (cf. Chap. 9).

Finally, some remarks on the evolution of formulating the yield criteria have been added. These are related to phenomena which were originally disregarded but may greatly influence the actual response of structures (cf. Chap. 10).

2. Constitutive Equations

The response of a material to external agencies depends upon the mechanical properties of the solid considered. For the description of any deformation process it is therefore necessary to specify the material *constitutive equation*, i.e. a tensor equation

$$F(\sigma_{ij}, \dot{\sigma}_{ij}, \ddot{\sigma}_{ij}, \ldots, \varepsilon_{ij}, \dot{\varepsilon}_{ij}, \ddot{\varepsilon}_{ij}, \ldots, T, t) = 0 \qquad (i,j = 1,2,3) \tag{1}$$

interrelating the dynamical quantities (stresses) σ_{ij} , the kinematical quantities (strains) ε_{ij} , their time derivatives $\dot{\sigma}_{ij}, \ddot{\sigma}_{ij}, \ldots, \dot{\varepsilon}_{ij}, \ddot{\varepsilon}_{ij}, \ldots,$ the temperature T as well as time t explicitly, if history effects are to be included [1,2] (*). The principal mathematical models introduced in order to describe the material inelasticity can, from the phenomenological point of view, be divided into two groups concerning the primarily *viscous* or primarily *plastic* (time-independent) behaviour.

For a *visco-elastic* material the coefficients appearing in the dynamical and kinematical tensor components of Eq.(1) are material constants. Thus even small stresses produce irreversible time dependent deformation.

The *elastic-plastic* behaviour is essentially different. The corresponding material can sustain a certain amount of stress without showing any irreversible deformation as long as the *yield function*, a scalar function of stress components $\Psi(\sigma_{ij}) - k = 0$, is satisfied (k denoting a material constant). Since for plastic materials, in addition to Eq. (1), the yield relation has to be fulfilled, the corresponding coefficients standing at the tensor components are no longer material constants, but depend on the state of stress. This makes the problems of structural analysis more complicated and requires different mathematical treatment as compared with analogous problems of viscous material response.

Within the theory of visco-elasticity a distinction has to be made between (physically) linear and nonlinear theories depending on whether the tensor components in Eq. (1) appear in linear or nonlinear form.

The general stress-strain law for linear visco-elastic materials in a cartesian reference frame is conventionally written in the form

$$P\sigma_{ij} = 2GQ\varepsilon_{ij} + \frac{1}{3}(3KP - 2GQ)\varepsilon_{kk}\delta_{ij} - 3\alpha KP\delta_{ij}, \tag{2}$$

(*) Numbers in square brackets refer to references listed at the end of the paper.

where P and Q are linear differential operators $P = a_0 + \sum_1^m a_k \partial^k/\partial t^k$, $Q = b_0 + \sum_1^n b_k \partial^k/\partial t^k$; a_k, b_k denote material constants, G and K are the shear and bulk elastic moduli, T and α the temperature and the coefficient of thermal expansion, respectively. The simplest visco-elastic models are those involving the stress and strain tensors and their first time derivatives. *Retarded elasticity* (Kelvin solid) and *creep at a constant rate* (Maxwell solid) are the fundamental combinations of elastic and viscous response [3].

From the view point of applications of visco-elastic analysis to shell problems the model describing creep is of special interest. If, at a constant stress, the stress-strain law for a Maxwell solid is examined, it is found that the strains increase linearly with time. Such a creep mechanism does not find experimental justification, thus more complex models have been proposed by retaining higher order time-derivatives in the operators P and Q from Eq. (2).

More adequate descriptions of creep and relaxation processes may be obtained by applying the *superposition principle* (Boltzmann principle). The state of strain at time t due to effects imposed at times τ is then given by the expression

$$2G\varepsilon_{ij}(t) = \int_{-\infty}^{t} \varphi_1(t-\tau)\,\sigma_{ij}(\tau)\,d\tau + \delta_{ij}\int_{-\infty}^{t} \varphi_2(t-\tau)\,\sigma_{kk}(\tau)\,d\tau\,, \tag{3}$$

φ_1 and φ_2 being the *heredity functions* [1, 2, 4]. Various simplified theories of linear creep have been advanced assuming specific mathematical forms of the kernels φ_1, φ_2, [4–7].

Effective solutions to boundary value problems for linear viscoelastic structures are facilitated by the possibility of removing the time dependence by employing the ***integral transform*** techniques. The *visco-elastic analogy* [7] leads to a practical conclusion that there is no redistribution of stress with time under the action of maintained loads (the magnitude of the stresses and strains being, of course, time-dependent).

The simplest *nonlinear visco-elastic* theories retain in Eq. (1) only stresses, strains and their rates. The following relation for a viscous incompressible material

$$\dot{\varepsilon}_{ij} = \frac{\dot{\sigma}_{ij}}{2G} - \frac{3K-2G}{18KG}\,\dot{\sigma}_{kk}\,\delta_{ij} + f(\sigma_{ij}) \tag{4}$$

represents the *nonlinear creep*. Depending on the rate of increase of (4), three stages of creep can be defined: *primary* (unsteady), *secondary* (steady) and *tertiary* (accelerated) creep, [8–11].

Most analytical studies are related to *steady creep* when Eq. (4) becomes $\dot{\varepsilon}_{ij} = f(\sigma_{ij})\left[\sigma_{ij} - (1/3)\sigma_{kk}\,\delta_{ij}\right]$, $f(\sigma_{ij})$ being a scalar function of stress.

Practical solutions are based on various generalizations of the exponential relations $\dot{\varepsilon} = k\sigma^n$, with k and n being material constants. Since the above relation resembles that of nonlinear elasticity, strain being replaced by strain rate, the *creep-nonlinear elastic analogy* [12] is of value for obtaining solutions.

The phenomenon of *creep buckling* is related to nonlinear viscoelastic response. This is characterized by a deflection increase beyond all bounds in a finite period of time [13, 14]. Thus, in contradistinction to the classical elastic theory, the critical load has to be replaced by a critical time, and the notion of stability has to be appropriately redefined and to be treated accordingly (cf. Chap. 5).

Among the theories dealing with plastic behaviour two groups of theories can be distinguished, depending on whether the deformation laws are written in terms of stresses and strain relationships (*theory of small elasto-plastic deformations*) or else of stresses and strain rates (*"flow theory"*). In both cases, obviously, the yield criterion has to be satisfied for plastic deformations to occur [15, 16].

The constitutive equation of the *deformation theory* for the loading process has the form

$$s_{ij} = 2G(1-\varphi/2G)e_{ij}\,, \qquad \sigma_{ii} = 3K(\varepsilon_{ii} - 3\alpha T)\,, \tag{5}$$

the unloading being purely elastic. In Eq. (5) φ denotes a scalar function of the space coordinates such as $\varphi > 0$ for the loading, and $\varphi = 0$ for the unloading process, [15–17]. The relation (5) induces coaxiality of the stress and strain deviators during the entire process of plastic deformation. This is a fundamental drawback of the theory since it permits only such stress redistributions which leave the principal directions unchanged.

The *flow theory* of elasto-plastic materials employs the relation

$$\dot{\varepsilon}_{ij} = \lambda s_{ij} + \dot{s}_{ij}/2G\,, \qquad \sigma_{ii} = 3K(\varepsilon_{ii} - 3\alpha T)\,, \tag{6}$$

where λ is a scalar function to be evaluated from the yield criterion. For *rigid-perfectly plastic* materials the appropriate flow law is $\dot{\varepsilon}_{ij} = \lambda s_{ij}$. The model of the perfectly plastic solid is employed for the evaluation of the *load carrying capacity* of structures (*limit analysis*) and the corresponding design (*limit design*), [18–24].

3. Assumptions of the Shell Theory

For thin shells the state of stress is specified if the *stress resultants* are known [25], therefore, if the tensors of the membrane forces $N_{\alpha\beta}$, of the moments $M_{\alpha\beta}$, and the vector of the transverse shear Q_α, $(\alpha,\beta = 1,2)$ are known. If the *extensions* and the *curvatures* of the deformed median surface are $\lambda_{\alpha\beta}$ and $\varkappa_{\alpha\beta}$, respectively, the deformations of the shells layers are related by the assumption

$$\varepsilon_{\alpha\beta} = \lambda_{\alpha\beta} + x_3\, \varkappa_{\alpha\beta}\,, \qquad \varepsilon_{\alpha 3} = 0\,, \tag{7}$$

i.e., by the assumption of *straight normals*, x_3 being directed along the outer normal.

The *constitutive equation* (1) for shells has to be written in the form

$$F(N_{\alpha\beta}, \dot{N}_{\alpha\beta}, \ldots, M_{\alpha\beta}, \ldots, \lambda_{\alpha\beta}, \ldots, \varkappa_{\alpha\beta}, \ldots, T, t) = 0\,, \tag{8}$$

since the equilibrium and the kinematical relations are written for the shell median surface. For shells exhibiting plasticity, additional relations representing the yield condition in terms of the stress resultants are required, [26, 27]:

$$F(M_{\alpha\beta}, N_{\alpha\beta}, Q_{\alpha}) = \text{const.} \tag{9}$$

The *energy dissipation density* for shells reduces to the following expression (per unit area of the median surface)

$$D = k(M_{\alpha\beta}\, \dot{\varkappa}_{\alpha\beta} + N_{\alpha\beta}\, \dot{\lambda}_{\alpha\beta}) \tag{10}$$

where k is a constant multiplier depending upon the type of inelastic response, if contributions of shear forces to the dissipation are neglected.

4. Linear Visco-elasticity

Under the assumptions of the shell theory, the constitutive equation (2) for visco-elastic shells of thickness $2H$ takes the form

$$PN_{\alpha\beta} = 4HGQ(\lambda_{\alpha\beta} + A\lambda_{\gamma\gamma}\,\delta_{\alpha\beta})\,, \tag{11}$$

$$PM_{\alpha\beta} = \frac{2}{3}H^3GQ(\varkappa_{\alpha\beta} + A\varkappa_{\gamma\gamma}\,\delta_{\alpha\beta})\,, \tag{12}$$

where $A = (3KP - 2GQ)/(3KP - 4GQ)$. It is seen that membrane forces produce only extensional deformations and that the moments are related to the curvatures only. Thus, the distribution of stress across the shell is similar to that for a linearly elastic shell, the difference being the time dependence of the stress and deformation magnitude. There occurs no redistribution of stress between moments and membrane forces, since, e.g., no term involving $k_{\alpha\beta}$ appears in Eq. (11). If the time dependence of Eqs. (11) and (12) is removed by application of appropriate integral transforms, the problem of a linearly visco-elastic shell is reduced to a corresponding problem in the linear theory of elasticity. Thus, whenever the elastic solution is known, one has only to deal with the question of inverse transform. Such a technique has been used for the solution of certain problems of spherical and cylindrical shells, [28–31].

For simpler types of visco-elasticity direct methods of solution are possible.

If we consider a general form of the creep equation, involving heredity functions as given by Eq. (3), a set of integro-differential equations for the boundary value problem of the shell is obtained. For materials with a *single heredity function,* a number of solutions have been found, [32–36]. Under sustained load, the displacements are proportional to the elastic ones, $W/W_{elastic} = 1+E\varphi(t,\tau)$, $\varphi(t,\tau)$ being the material heredity function and E Young's modulus, [37, 38]. The temperature influences the material properties as well as the stress-strain relations, and attempts have been made to solve thermal problems for shells [39, 40].

5. Steady Creep

The steady creep equations allow to study two important features of nonlinear visco-elasticity, namely: *redistribution of stress* in comparison with the elastic state and *creep buckling.* For plane stress and extensive creep deformation the exponential creep law $\dot\varepsilon = k(\varepsilon)\sigma^n$ takes the form

$$\sigma_{\alpha\beta} = (\dot\varepsilon_{\alpha\beta} + \dot\varepsilon_{\gamma\gamma}\delta_{\alpha\beta})/B^n(\varepsilon_{ij}\varepsilon_{ij})^{\frac{n-1}{2n}} , \tag{13}$$

where B and n are material constants (B may explicitely depend on time). Employing Eq. (13) in the definition of the stress resultants we obtain

$$N_{\alpha\beta} = (\dot\lambda_{\alpha\beta} + \dot\lambda_{\gamma\gamma}\delta_{\alpha\beta})I_1 + (\dot\varkappa_{\alpha\beta} + \dot\varkappa_{\gamma\gamma}\delta_{\alpha\beta})I_2 , \tag{14}$$

$$M_{\alpha\beta} = (\dot\lambda_{\alpha\beta} + \dot\lambda_{\gamma\gamma}\delta_{\alpha\beta})I_2 + (\dot\varkappa_{\alpha\beta} + \dot\varkappa_{\gamma\gamma}\delta_{\alpha\beta})I_3 , \tag{15}$$

where

$$I_i = \frac{1}{B^n}\int_{-H}^{H} \frac{x_3^{i-1}}{(\varepsilon_{ij}\varepsilon_{ij})^{\frac{n-1}{2n}}}\, dx_3 , \quad (i = 1,2,3) . \tag{16}$$

The membrane forces and bending moments are interconnected with the deformation rates by more complicated relations than would follow from the simple superposition of bending and membrane states [26, 27, 41].

The redistribution of stresses in shells also takes place under unsteady creep [42]. The effect of the redistribution is the more pronounced the higher the nonlinearity of the constitutive equation. Deformation fields can be obtained in closed form for some special cases of shell geometry only, [43–52]. Replacing the term B of Eq. (13) by a piecewise linear relationship, one facilitates the procedure of solving the creep problem [53].

A striking feature of nonlinear visco-elastic behaviour is encountered in the case of axial loading with constant load intensity. After a sufficiently long period of time has elapsed, the deflection velocities increase beyond all bounds. Thus buckling takes place at any value of the compressive load at a specific *critical time* [54–57].

Few studies only deal with large deformations of shells [58], primary and accelerated creep up to rupture [59–62], and anisotropy [63].

6. Elasto-plastic Deformations

For a deformation theory of plasticity the stress-strain relation in plane stress has the form

$$\sigma_{\alpha\beta} = \sigma_0(2/3e_{ij}e_{ij})^{1/2}(\varepsilon_{\alpha\beta} + \varepsilon_{\gamma\gamma}\delta_{\alpha\beta}) . \qquad (17)$$

If the material obeys the Huber-von Mises yield criterion, the stress resultants are [64]

$$N_{\alpha\beta} = (\lambda_{\alpha\beta} + \lambda_{\gamma\gamma}\delta_{\alpha\beta})I_1 + (\varkappa_{\alpha\beta} + \varkappa_{\gamma\gamma}\delta_{\alpha\beta})I_2 , \qquad (18)$$

$$M_{\alpha\beta} = (\lambda_{\alpha\beta} + \lambda_{\gamma\gamma}\delta_{\alpha\beta})I_2 + (\varkappa_{\alpha\beta} + \varkappa_{\gamma\gamma}\delta_{\alpha\beta})I_3 , \qquad (19)$$

where

$$I_i = \sigma_0\sqrt{\frac{2}{3}}\int_{-H}^{H}\frac{x_3^{i-1}}{\sqrt{e_{ij}\,e_{ij}}}dx_3 , \quad (i = 1,2,3) . \qquad (20)$$

Thus the bending and membrane actions are not separated. Therefore it is, in general, impossible to split the shell equations into two groups: one dealing only with the membrane, the second only with the bending action. Direct integration of the elasto-plastic shell equations is, as a rule, impossible and approximate methods have to be employed. The "*method of elastic solutions*", a step-by-step procedure, has been used in investigations of the initiation of the plastic deformation process in cylindrical and spherical shells [65–68] as well as in thin-walled beams [69-72]. Approximate solutions are also obtained by *variational methods* [73].

Since the deformation theory of plasticity applies only when the principal axes of the deviators remain fixed during the process of loading, it has to be borne in mind that this can furnish reliable data for early stages of loading only. Elasto-plastic analysis based on the deformation theory is, therefore, seldom applied to shells [74-77].

7. Limit Analysis and Limit Design. Fundamentals and Applications

In the response of an elastic-perfectly plastic structure to a prescribed loading programme, there exists a state when the deformations begin to increase beyond all bounds under a constant load intensity. The corresponding value of the load is called the *ultimate load* or the *collapse load.* Under the collapse load the analysed structure transforms into a mechanism with, at least, one degree of freedom. The plastic motion of the collapse mechanism can be sustained only when the appropriate criterion of yielding is reached, in regions sufficient to allow the continued deformation without any further extension of these regions. The magnitude of the collapse load can be obtained by methods concerning the collapse state. The starting point for the analysis is the rigid-perfectly plastic model of deformation [78–82].

An analysis of the collapse state (*limit analysis*) requires firstly the specification of the yield criterion in terms of the stress resultants. This constitutes a closed convex hypersurface referred to as *yield locus* [64, 83, 84]. The particular forms of the yield loci depend upon the yield conditions. If the Huber-von Mises yield criterion is adopted, then $F = 3\sigma_{\alpha\beta}\sigma_{\alpha\beta} - \sigma_{\alpha\alpha}\sigma_{\gamma\gamma} - 2\sigma_0^2 = 0 \quad (\alpha,\beta = 1,2)$, together with the associated flow law $\dot{\varepsilon}_{\alpha\beta} = \lambda \partial F/\partial\sigma_{\alpha\beta}$, yield the following relations

$$N_{\alpha\beta} = \sigma_0(\dot{\lambda}_{\alpha\beta} + \dot{\lambda}_{\gamma\gamma}\delta_{\alpha\beta})I_1 + \sigma_0(\dot{\varkappa}_{\alpha\beta} + \dot{\varkappa}_{\gamma\gamma}\delta_{\alpha\beta})I_2 , \tag{21}$$

$$M_{\alpha\beta} = \sigma_0(\dot{\lambda}_{\alpha\beta} + \dot{\lambda}_{\gamma\gamma}\delta_{\alpha\beta})I_2 + \sigma_0(\dot{\varkappa}_{\alpha\beta} + \dot{\varkappa}_{\gamma\gamma}\delta_{\alpha\beta})I_3 , \tag{22}$$

where

$$I_i = \sqrt{\frac{2}{3}} \int_{-H}^{H} x_3^{i-1} (\dot{\varepsilon}_{\alpha\beta}\dot{\varepsilon}_{\alpha\beta} + \dot{\varepsilon}_{\gamma\gamma}\dot{\varepsilon}_{\beta\beta})^{-\frac{1}{2}} dx_3 , \qquad (i = 1,2,3). \tag{23}$$

These relations represent a parametric form of the yield hypersurface for uniform shells. In the case of sandwich structures two yield loci are obtained

$$3n_{\alpha\beta}n_{\alpha\beta} + 3m_{\alpha\beta}m_{\alpha\beta} - n_{\alpha\alpha}n_{\beta\beta} - m_{\alpha\alpha}m_{\beta\beta} = 2 \tag{24}$$

$$3m_{\alpha\beta}n_{\alpha\beta} - m_{\alpha\alpha}n_{\beta\beta} = 0 \tag{25}$$

$m_{\alpha\beta}, n_{\alpha\beta}$ being dimensionless stress resultants [64, 85–88].

For piecewise linear yield criteria the yield hypersurfaces are given by a set of intersecting hypersurfaces and hyperplanes [89–92].

Comparisons have been made of various approximations to the yield hypersurfaces, [93–97], and a number of approximated yield loci has been proposed to suit particular requirements [98–106]. For non-homogeneous materials (reinforced concrete) the concept of the yield hypersurface retains its meaning, [107–110].

The search for approximate yield loci, in order to obtain closed form solutions to limit analysis problems, is justified by the possibility of applying the two *fundamental theorems of the limit analysis theory.* The significance of these theorems lies precisely in providing lower and upper bounds to the collapse load.

The *lower bound theorem* states that a rigid plastic structure will not collapse under a loading $\mu_s p_{0i}$ if any stress distribution satisfying the requirements of equilibrium and specified boundary conditions, and not violating the yield locus, is found. The statically admissible load multiplier is homogeneous (of order α , say) in the stresses; thus

$$\mu_s^{\alpha} = \min\left|\frac{K}{F(N^0_{\alpha\beta}, M^0_{\alpha\beta})}\right| \tag{26}$$

represents the best statically admissible load factor, where $N^0_{\alpha\beta}, M^0_{\alpha\beta}$ are the statically admissible stress resultants, and K denotes the material yield modulus.

In contrast, the *upper bound theorem* states that a rigid perfectly plastic structure will collapse under a loading if any collapse mode is found such that the rate at which the external forces do work exceeds or equals the rate of internal dissipation. The best kinematically admissible load factor is

$$\mu_k = \min \int_A (M^*_{\alpha\beta}\dot{\kappa}^*_{\alpha\beta} + N^*_{\alpha\beta}\dot{\lambda}^*_{\alpha\beta})\, dA \Big/ \int_A p_{0i}\dot{u}^*\, dA, \tag{27}$$

where $\dot{u}_i^*$ denotes the virtual velocity field (whence $\dot{\lambda}^*_{\alpha\beta}$, $\dot{\varkappa}^*_{\alpha\beta}$ can be determined from the geometrical relations) and $F(M^*_{\alpha\beta}, N^*_{\alpha\beta}) = M$.

The ***complete solution*** of a limit analysis problem consists in the evaluation of the load factor μ_c, of the extent and velocity of deformation in the plastic region, and of the stress distribution throughout the structure. Thus the complete solution satisfies both fundamental limit analysis theorems, [111].

The existing complete solutions have been found for rotationally symmetric shells, mostly spherical domes [112–114], conical shells, [115–119], and circular cylinders [89, 91, 92, 120–130]. For nonsymmetric shells only a few solutions are known, [131–137].

The theorems on bounds have been applied to many shell problems [138–154]. In application to reinforced concrete shells, the upper bound theorem, supported by experimental evidence, has led to the formulation of the *generalized hinge lines method* [155–161], an engineering approach to the limit analysis problems.

The domain of yield criteria has been enriched by studies concerning yield surfaces for *rib reinforced shells* and *layered structures* as well as by analysis of the *shear force influence* on yielding. Moreover, various ***approximations*** to formulating the yield criteria for shells have been advanced [162–181].

Once the yield criteria are established, it is natural to study the general system of *shell equations* as far as its type and suitable analytical or numerical ***methods of solution*** are concerned [182–195]. Numerous ***solutions*** of limit analysis problems have been found (some of them checked experimentally), mainly for rotationally symmetric *spherical* and *conical shells* [196–210], both in the general plastic analysis and in the generalized yield line theory [282].

An interesting problem in the theory of shells is represented by that of the ***minimum weight design*** (limit design) with the aim to conceive a structure which leads to minimum material consumption for carrying the prescribed loading. (Also other optimum criteria can be formulated). The general theorems of this procedure [211–214] have been applied to simpler problems of shells, [215–218], including anisotropy [219–220] and nonhomogeneity [222–224].

8. Cyclic and Transitory Effects

The theory of limit analysis is already becoming a classical domain of the plastic shell theory at present. It is based on the assumption of the one-parameter dependence of the loading process.

An extension to processes depending on several parameters is of evident importance. Thus in the theory of inelastic structures problems involving *cyclic* and *transitory effects* as well as *thermal agencies* can be grasped and evaluated. In the first instance the behaviour of structures subject to variable loading programmes can be examined, the analysis being based on the *shakedown theorems* [225–227].

The pertinent theory has been generalized to certain types of **strain-hardening**, and recently extended to also include elastic-viscoplastic response. The phenomena of ***shakedown,*** *low cycle fatigue, and incremental collapse* have been found to occur and are being studied [228, 229] (*). They are of importance in the rational design.

9. Nonlinearity; Dynamic Effects

A growing attention is concentrating on problems of ***nonlinear shell theory.*** The relevant basic principles and results are due to the fundamental contributions of W.T. Koiter [230–232, . . .] and to research activities in various centres [233–235, . . .]. For plastic shells, the equations allowing to account for the *geometry changes* and for the *post-yield behaviour* were studied in [236–243], and solutions concerning the post-yield behaviour of cylindrical and spherical shells were obtained [244–251]. *Stable* and *unstable* situations in the post-yield range may occur [252–256]. Attention is focussed on algorithms suitable for a numerical treatment of engineering problems (cf., e.g., [257]).

The analysis of the *dynamic behaviour* of plastic structures has recently been extended to the geometrically nonlinear range [258–263].

10. Evolution of Yield Criteria

It is known that for new materials and for new types of materials great efforts have been made to formulate the appropriate realistic constitutive equations reflecting the actual material response. Less known is that with the advent of some new materials and of newly discovered phenomena which have to be taken into account, also a suitable approach to establishing the appropriate yield criteria should be adopted.

In fact, an interesting evolution in formulating these criteria is taking place.

The known classical approach to define the yield condition is based on the notion of mechanically *isotropic* and *homogeneous* media. If expressed in terms of the stress (or strain) invariants, the simplest possible form (compatible with physical reality) is that of the Huber (1904)-von Mises (1913) yield condition [264, 265]. The other frequently used conditions, those of Coulomb and Tresca, in spite of being historically older (1773 and 1868, respectively) [266, 267], are considerably more complicated if written in terms of stress invariants.

This classical approach has successively been generalized by taking into account more general physical phenomena which were originally disregarded. Thus mechanical *anisotropy*

(*) The French terminology is somewhat different and it seems that the notions of, respectively, "adaptation", "accommodation", and "rochet" will finally be adopted.

and *(macro)nonhomogeneity* were taken into consideration [268–272].

Plastic anisotropy results in changes of the yield hypersurface for shells, thus in changing the collapse loads; certain solutions are available for cylindrical shells [273–283]. The effects of plastic nonhomogeneity on the behaviour of shells have also been investigated [284–287].

When passing to *rheologically reacting* materials, we immediately find two new notions: that of the *static* yield condition and that of the *dynamic* yield condition, these two notions marking in the evolution a remarkable step. The rheological response itself has successively been considered in both its alternative types: (a) that which only sets in after the plastic limit has been exceeded (cf., e.g., [288–291]); or else (b) that – notably more difficult to be accounted for – which accompanies the deformation process from its very beginning [291, 292].

Departure from the assumption of perfect plasticity, when taking into consideration *work hardening* effects according to various theories of hardening [293–295] is of great importance in both the theory of inviscid plasticity and that of rheological approach. As a matter of fact, experimental results clearly prove, that time-dependent phenomena like, e.g., elastic-visco-plastic deformation processes in metals, concretes and soils, occur as a rule in the presence of work-hardening (or work-softening) effects. The literature is rich and abundant (cf., e.g.,[296–299]). They also allow to clear one of the aspects of the post-yield behaviour of shells [300, 301].

Nowadays, *thermal* effects and *combined elasto-visco-plastic* response are attracting more attention [302–304].

The next step consists in investigating the consequences of a *time-variable plasticity condition* which may occur when, e.g., time-variable temperatury fields [305] or artificial irradiation effects [306] are considered; this may for some particular shell structures be of importance. Similar situations may occur if a time-variable humidity content of elasto-visco-plastic soils is being envisaged.

All the above mentioned considerations assume that the (original) investigated elastic deformation fields are uniform (homogeneous). This special assumption, as a rule, does not correspond to physical reality, especially in view of applications in the shell theory. Thus the next step to be taken consists in introducing (*space and time-dependent) nonhomogeneities* of these fields. The ensuing consequences are interesting. They may, under certain conditions, result in a considerable increase of the plastic limit. This fact has for many years already been known from experimental evidence [307, 308].

11. Concluding Remarks

Within the particular theories mentioned in this paper research is in progress; thus new results are pending. Many problems however have even still to be formulated.

Geometry effects, if included in the analysis, lead to the problem of ***stability of inelastic deformation processes.*** The concept of stability will in the future replace the concept of limit analysis.

The thermal and dynamic response of shells made of different inelastic media is so far a virtually open area.

Problems of creep, creep damage and failure, interaction of mechanical and extramechanical agencies, questions of random material properties and loading programmes, become vital in many branches of contemporary and future technologies and utilization of natural resources.

REFERENCES(*)

[1] I.I. Goldenblatt: Some questions of mechanics of deformable media (in Russian), 1955.

[2] A.M. Freudenthal and H. Geiringer: The Mathematical Theories of the I ıel‹ ›tic Continuum, Handbuch der Physik 6, 229–433, Berlin 1958.

[3] R. Bland: The Theory of Linear Viscoelasticity, Oxford 1960.

[4] N.Kh. Arutyunyan: Some Questions of the Creep Theory, (in Russian), Moscow 1952.

[5] Yu.N. Rabotnov: The equilibrium of elastic media with a heredity, (in Russian), Prikl. Mat. Mech. 12 (1948), 53–62.

[6] A.R. Rzhanitsyn: Some Questions of Mechanics of Time-Deformable Systems, (in Russian), Moscow (1949).

[7] T. Alfrey: Mechanical Behaviour of High Polymers, New York 1948.

[8] R.W. Bailey: Creep relationships and their applications to pipes, tubes and cylindrical parts under internal pressure, Proc. Instn. Mech. Engrs., 4 (1951), 164, 425-431.

[9] I. Finnie: Stress analysis in the presence of creep, Appl. Mech. Rev. 13 (1960), 705-712.

[10] L.M. Kachanov: Theory of Creep, (in Russian), Moscow 1960.

[11] F.K.G. Odqvist and J. Hult: Kriechfestigkeit metallischer Werkstoffe, Berlin 1962.

[12] N.J. Hoff: Approximate analysis of structures in the presence of moderately large creep deformations, Quart. Appl. Math. 12 (1954), 49-55.

[13] N.J. Hoff: Theories of creep buckling, Proc. 3rd U.S. Nath. Congress Appl. Mech. (Providence 1958), ASME New York 1959, 29-49.

[14] Yu. N. Rabotnov: The theory of creep and its applications, Proc. 2nd Symp. on Nav. Struc. Mech. (Providence 1960), Oxford, 338-346.

[15] R. Hill: Mathematical Theory of Plasticity, London 1950.

[16] W. Prager and P.G. Hodge: Theory of Perfectly Plastic Solids, New York 1951.

(*) As already mentioned, the following ample reference list including also a number of less known contributions, has been prepared at the request of the course participants. It is meant to facilitate their access to more detailed information.

[17] A.A. Ilyushin: Plasticity, (in Russian), Moscow 1948.

[18] W.T. Koiter: General theorems for elastic-plastic solids, Progress in Solid Mechanics I, Amsterdam 1960, 167-224.

[19] D.C. Drucker: Plasticity, Proc. Ist. Symp. Naval Struct. Mech. (Stanford 1958), Oxford 1959, 407-455.

[20] W. Prager: Introduction to Plasticity, Reading, Mass. 1959.

[21] A.A. Gvozdev: Theory of Limit Equilibrium, (in Russian), Moscow 1949.

[22] W. Prager: The general theory of limit design, Proc. 8th Int. Congr. Appl. Mech. (Istanbul 1952).

[23] P.G. Hodge: Plastic Analysis of Structures, New York 1959.

[24] A. Sawczuk and Th. Jaeger: Grenztragfähigkeits-Theorie der Platten, Berlin 1963.

[25] W. Flügge: Stresses in Shells, Springer Verlag, Berlin 1960.

[26] I.I. Goldenblatt and N.A. Nikolaenko: Creep and Bearing Capacity of Shells, (in Russian), Moscow 1960.

[27] W. Olszak and A. Sawczuk: Inelastic Behaviour in Shells, P. Noordhoff, Groningen 1967.

[28] P.M. Naghdi and W.C. Orthwein: Response of shallow viscoelastic spherical shells to time dependent axisymmetric loads, Quart. Appl. Math. 18 (1960/61), 107-121.

[29] E. Tungl: Durchschlagen einer flachen Kugelschale aus viscoelastischem Material, Österreichisches Ingenieur-Archiv 16 (1926), 280-289.

[30] J.N. Distefano and M.H. Gradowczyk: Creep behaviour of homogeneous anisotropic prismatic shells, Proc. Symp. Non-Classical Shell Problems (Warsaw 1963), 1964.

[31] K. Pister: Axisymmetric deformation of orthotropic visco-elastic cylindrical shells, Proc. Symp. Non-Classical Shell Problems (Warsaw 1963), Amsterdam 1964.

[32] I.I. Goldenblatt and N.A. Nikolaenko: Theory of Creep of Structural Materials and its Applications, (in Russian), Moscow 1960.

[33] M.I. Rozovski: Influence of time factor on strength of a spherical shell subjected to internal pressure, (in Russian), Izv. Akad. Nauk SSSR, OTN, Mekh. mash. (1961) 4, 124-129.

[34] N. Kh. Arutyunyan and M.M. Manukyan: Creep of composite cylindrical tubes , (in Russian), Izv. Akad. Nauk Arm. SSR, Ser. fiz. mat. 10 (1957), 6, 41-58.

[35] N. Kh. Arutyunyan and M.M. Manukyan: Creep of a spherical vessel, Dokl. Akad. Nauk. Arm. SSR 27 (1958), 209-218.

[36] M.I. Estrin: Calculation of shallow axisymmetrical shells made of elasto-viscous materials taking account of large deflections, (in Russian), Voprosy teorii plastitchnosti i protchnosti stroj. konstr. Trudy CNIISK, Moscow 1961, 4, 123-134.

[37] H. Muguruma: Two-dimensional creep deformation of concrete, Proc. Symp. Non-Classical Shell, Problems (Warsaw 1963), Amsterdam 1964.

[38] D.C. Houghton and A. Rothwell: Measured deflections of concrete folded plate and hyperbolic paraboloid shells, Proc. Symp. Non-Classical Shell Problems (Warsaw) 1963, Amsterdam 1964.

[39] B.D. Aggarwala: Thermal stresses in spherical shells of viscoelastic materials, Z. Angew. Math. Mech. 40 (1960) 482-488.

[40] W. Nowacki: Thermal stresses in elastic and viscoelastic shells, Proc. Symp. Non-Classical Shell Problems (Warsaw) 1963, North Holland, Amsterdam 1964.

[41] L.M. Kachanov: Creep of oval tubes with variable wall thickness, (in Russian), Izv. Akad. Nauk SSSR, OTN, 9 (1956), 65-71.

[42] M.P. Bieniek and N.M. Freudenthal: Creep deformation and stresses in pressurized long cylindrical shells, J. Aero Space Sci. 27 (1960), 763-778.

[43] A.E. Gemma: The creep deformation of symmetrically loaded circular cylindrical shells, J. Aero/Space Sci. 27 (1960), 953-954.

[44] A.E. Gemma and J.T. Warfield: The creep deformation of symmetrically loaded shells, J. Aero/Space Sci. 28 (1961), 507-508.

[45] A.E. Gemma: The steady creep of long pressurized cylinders, Journ. Aero/Space Sci. 29 (1962), 352-353.

[46] N.J. Hoff, W.E. Jahsman and W. Nachbar: A study of creep collapse of a long circular cylindrical shell under uniform external pressure, J. Aero/Space Sci. 26 (1959), 663-669.

[47] E.K.G. Odqvist: Applicability of the elastic analogy to creep problems of plates, membranes and beams, Creep in Structures, Berlin 1962, 137-160.

[48] H. Poritsky: Effect of creep on stresses in cylindrical shells, Creep in Structures, Berlin 1962, 229-244.

[49] C.R. Calladine: The steady creep of shells: a method of analysis, Nuclear Reactor Containment Buildings and Pressure Vessels, Proc. Symp. Glasgow 1960, London 1960, 411-431.

[50] C.R. Calladine: On the creep of a wrinkle, Creep in Structures, Berlin 1962, 245-271.

[51] C.R. Calladine: Upper and lower bound solutions for edge response of shells on steady creep, Proc. Symp. Non-Classica' Shell Problems (Warsaw 1963), Amsterdam 1964.

[52] V.I. Rozenblum: On the approximate creep equations, (in Russian), Izv. Akad. Nauk SSSR, Mekh. Mash. (1959) 5, 157-160.

[53] E.T. Onat and H. Yuksel: On the steady creep of shells, Proc. 3rd U.S. Natl. Congr. Appl. Mech. (Providence 1958), ASME, New York 1959, 625-630.

[54] Fr. B. de Veubeke: Creep Buckling. High Temperature Effects in Aircraft Structures, AGARDograph 28, New York 1958.

[55] S.A. Shesterikov: Creep buckling, (in Russian), Prikl. Mat. Mekh., 25 (1961), 754-755.

[56] A.P. Kuznetsov and L.M. Kurshin: Stability of circular cylindrical shells subjected to creep, (in Russian), Prikl. Mekh. Tech. Fiz. (1962), 3, 66-72.

[57] G. Gerard: Theory of creep buckling of perfect plates and shells, J. Aero/Space Sci. 29 (1962), 1087-1090.

[58] Z. Bychawski: Creep buckling of a cylindrical panel, Proc. World Conference on Shells, San Francisco 1962.

[59] V.I. Rozenblum: On a non-stationary creep of membrane shells, (in Russian), Prikl. Mekh. Techn. Fiz. (1960) 4, 82-84.

[60] F.A. Cozzarelli and S.A. Patel: Creep deformations in membrane shells, J. Frankl. Inst. 278 (1964), 45-61.

[61] E.A. Davis: Creep rupture tests for design of high-pressure steam equipment, J. Basic Engng. 2 (1960), 453-461.

[62] Tein Wah and R. Kirk: Creep collapse of long cylindrical shells under high temperature and external pressure, J. Aero/Space Sci. 28 (1961), 177-188, 208

[63] O.V. Sosin: Stationary anisotropic creep of discs, (in Russian), Prik. Mekh. Tekh. Fiz. (1963) 4, 128-131.

[64] A.A. Ilyushin: Finite relationship between forces and moments and their relation in the theory of shells, (in Russian), Prikl. Mat. Mekh. 9 (1945), 101-140.

[65] A.A. Ilyushin: an approximate theory of elastic-plastic deformations of an axisymmetrical shell, (in Russian), Prikl. Mat. Mekh. 8 (1944), 15-24.

[66] I.S. Tsurkov: Elastic-plastic equilibrium of rotationally symmetric shells subjected to small axially symmetric deformations, (in Russian), Izv. Akad. Nauk SSSR, OTN (1956) 11, 106-110.

[67] I.S. Tsurkov: Elastic-plastic equilibrium of shallow shells at small deformations, (in Russian), Izv. Akad. Nauk SSR, OTN (1957) 6, 139-142.

[68] I.S. Tsurkov: Elastic-plastic deformations at the clamped edge of a thin-walled cylinder, (in Russian), Inzh. sbornik (1961) 31, 93-100.

[69] R.A. Mezhlumyan: Bending and torsion of thin-walled shells beyond the elastic limit, (in Russian), Prikl. Mat. Mekh. 14 (1950), 253-264.

[70] R.A. Mezhlumyan: Boundary conditions of bending and torsion of thin-walled shells beyond the elastic limit, (in Russian), Prikl. Mat. Mekh. 14 (1950), 537-542.

[71] R.A. Mezhlumyan: Determination of ultimate load of a thin-walled structure taking account of the strainhardening material, (in Russian), Prikl. Mat. Mekh. 15 (1954), 174-182.

[72] R.A. Mezhlumyan: Approximate theory of elastic-plastic shells and its application to the analysis of structures, (in Russian), Inzh. sbornik 10 (1952), 35-70.

[73] V.M. Panferov: On the applicability of variational methods to problems of the theory of small elastic-plastic deformations, (in Russian), Prikl. Mat. Mekh. 16 (1952), 319-322.

[74] A.N. Ananina: An axisymmetrical deformation of a cylindrical shell subjected to elastic-plastic deformations, (in Russian), Inzh. Sbornik. 18 (1954), 157-160.

[75] P.G. Hodge: Displacements in an elastic cylindrical shell, J. Appl. Mech. 23 (1956), 73-79.

[76] V.S. Chernina: Elastic-plastic deformation of a welded non-homogeneous shell, (in Russian), Izv. Akad. Nauk SSSR, OTN, Mekh.i Mash. (1960) 1, 133-140.

[77] P. Klement: Theorie der elastisch-plastischen Zylinderschale, Österr.Ingenieur-Archiv,16 (1962) 199-211.

[78] A.A. Gvozdev: The determination of the value of the collapse – load for statically indeterminate systems undergoing plastic deformation, Inter. J. Mech. Sci. 1 (1960) 322-335 (Russian edition 1936).

[79] S.M. Feinberg: Principle of the limiting stress, (in Russian), Prikl. Mat. Mekh., 12 (1948), Izv. Akad. Nauk SSSR, OTN, Mekh. Mash. (1960) 4, 101-111.

[80] D.C. Drucker, H.I. Greenberg and W. Prager: The safety factor of an elastic-plastic body in plane strain, J. Appl. Mech., 18 (1951), 371-378.

[81] R. Hill: A note on estimating the yield point loads in a plastic-rigid body, Phil. Mag. 43 (7) (1952), 353-355.

[82] D.C. Drucker, W. Prager and H.I. Greenberg: Extended limit design theorems for continuous media, Quart. Appl. Math., 9 (1952) 381-389.

[83] A. Sawczuk and J. Rychlewski:On yield surfaces for plastic shells, Arch. Mech. Stos., 12 (1960) 29-53.

[84] M. Save: On yield conditions in generalized stresses, Quart. Appl. Math., 19 (1961) 259-267.

[85] V.V. Rozhdestvenski: Limit equilibrium of junctions of rotational shells, (in Russian), Nauchn. soobshcs. Akad. Stroi. i Arch., No. 1. Moscow 1957.

[86] M.I. Yerchov: Finite relationship between forces and moments of plastic deformations of shells, (in Russian), Stroit. Mekh. rasch. sooruzh, (1959), 3, 38-41.

[87] P.G. Hodge: The Mises yield condition for rotationally symmetric shells, Quart. Appl. Math., 18 (1961), 305-311.

[88] G.S. Shapiro: On yield surfaces for ideally plastic shells, Problems of Continuum Mechanics, Soc. Ind. Appl. Math. Philadelphia, 1961, 414-418.

[89] D.C. Drucker: Limit analysis of cylindrical shells under axially-symmetric loading, Proc. 1st Midwest Conf. Solid Mech. Urbana, 1953, 158-163.

[90] E.T. Onat and W. Prager: Limit analysis of shells of revolution, Proc. Ned. Akad. Wetensch. Ser. B. 57 (1954),534-548.

[91] P.G. Hodge: Rigid-plastic analysis of symmetrically loaded cylindrical shells, J. Appl. Mech., 21. (1954), 336-342.

[92] E.T. Onat: The plastic collapse of cylindrical shells under axially symmetrically loading, Quart. Appl. Math., 13 (1955), 68-72.

[93] W. Olszak and A. Sawczuk: Die Grenztragfähigkeit von zylindrischen Schalen bei verschiedenen Formen der Plastizitätsbedingung,Acta Techn. Hung. 26 (1959), 55-77.

[94] P.G. Hodge: Yield conditions for rotationally symmetric shells under axisymmetric loading, J. Appl. Mech. 27 (1960), 323-331.

[95] P.G. Hodge: A comparison of yield conditions in the theory of plastic shells, Problems in Continuum Mechanics, Soc. for Ind. and Appl. Mathem. Philadelphia, Pa, 1961, 165-177.

[96] P.G. Hodge and J. Panarelli: Interaction curves for circular cylindrical shells according to the Mises or Tresca yield criterion, J. Appl. Mech. 29 (1962), 375-380.

[97] P.G. Hodge: Limit analysis of rotationally symmetric plates and shells, Prentice-Hall, Englewood Cliffs N.J. 1963.

[98] Yu. N. Rabotnov: An approximate engineering theory of elastoplastic shells, (in Russian), Prikl. Mat. Mekh. 15 (1951), 167-174.

[99] P.G. Hodge: The linearization of plasticity problems by means of non-homogeneous materials, Proc. Symp. Nonhomog. Probl.,Warsaw 1958, Pergamon Press, London 1959, 147-156.

[100] V.I. Rozenblum: An approximate theory of equilibrium of plastic shells, (in Russian), Prikl. Mat. Mekh. 18 (1954), 289-302.

[101] D.C. Drucker and R.T. Shield: Limit analysis of symmetrically loaded thin shells of revolution, J. Appl. Mech., 26, 1959, 61-68.

[102] V.I. Rozenblum: On the yield conditions of thin shells, (in Russian), Prikl. Mat. Mekh. 1960, 364-366.

[103] W. Prager: On the plastic analysis of sandwich structures. Problems of Continuum Mechanics, Soc. Ind. Appl. Math. Philadelphia, 1961, 342-349.

[104] T. Nakamura: Plastic analysis of shells of revolution under axi-symmetric loads, Ph.D. Dissertation, Stanford Univ., 1961.

[105] P.G. Hodge: Piece-wise linear bounds on the yield-point load of shells, J. Mech. Phys. Solids, 11 (1963), 1-12.

[106] Yu. P. Listrova: On the bearing capacity of cylindrical shells under the condition of plasticity of the maximum reduced stress, (in Russian), Izv. Akad. Nauk SSSR, Mekh. Mash. (1963) 2, 173-176.

[107] A. Sawczuk and W. Olszak: A method of limit analysis of reinforced concrete tanks, Proc. Int. Coll. Simpl. Shell Calc. methods, (Brussels 1961), Amsterdam, 1962, 416-437.

[108] A. Sawczuk and J.A. König: Limit analysis of reinforced concrete cylindrical silos, (in Polish), Arch. Inż. Lądow., 8 (1962), 161-183.

[109] M.P. Nielsen: Yield conditions in the membrane state of reinforced concrete shells, Proc. Symp. Non-Classical Shell Problems, (Warsaw 1963), Amsterdam 1964.

[110] R. Sankaranarayanan and W. Olszak: The load carrying capacities of plates and shells, Proc. Symp. Non-Classical Shell Problems (Warsaw 1963), Amsterdam 1964.

[111] E.T. Onat: Plastic shells, Proc. Symp. Non-Classical Shell Problems, (Warsaw 1963), Amsterdam 1964.

[112] S.M. Feinberg: Plastic flow of a shallow shell for an axisymmetrical problem, (in Russian), Prikl. Math. Mekh. 21 (1957), 544-549.

[113] P.G. Hodge. The collapse load of a spherical cap, Proc. 4th Midwest Conf. Solid Mech., Austin Texas (1959), 1, 108-126.

[114] Z. Mróz and Xu-Bing-ye: The load carrying capacities of symmetrically loaded spherical shells, Arch. Mech. Stos. 15 (1963), 245-266.

[115] V.V. Rozhdestvenski: Bearing Capacity of Cylindrical Vessels with Conical and Spherical Heads, (in Russian), Moscow 1959.

[116] P.G. Hodge: Plastic analysis of circular conical shells, J. Appl. Mech., 27 (1960), 696-700.

[117] E.T. Onat: Plastic analysis of shallow conical shells, J. Engng. Mech. Div. Proc. Amer. Soc. Civ. Engrs. 86 (1960), 6, 1-12.

[118] P.G. Hodge and C. Lakshmikantham: Yield point loads of spherical caps with cut-outs, Proc. 4th U.S. Nat. Congr. Appl. Mech., (Berkeley, 1962) ASME, 1963, 951-954.

[119] P.G. Hodge and J. de Runtz: The carrying capacity of conical shells under concentrated and distributed loads, Proc. Symp. Non-Classical Shell Problems (Warsaw 1963), Amsterdam 1964.

[120] G. Eason and R.T. Shield: The influence of free ends on the load carrying capacity of cylindrical shells, Journ. Mech. Phys. of Solids, 4 (1955), 17-27.

[121] P.G. Hodge: Piecewise linear isotropic plasticity applied to a circular cylindrical shell with symmetrical radial loading, J. Franklin Inst. 263 (1957), 13-33.

[122] A.P. Rzhanitsyn: Plastic deformations of a tube under an axisymmetrical load, (in Russian), Izv. Nauk, SSSR, OTN (1958), 8, 60-65.

[123] R.T. Shield and D.C. Drucker: Limit strength of thin-walled pressure vessels with an ASME standard torispherical head, Proc. 3rd U.S. Nat. Congr. Appl. Mech. (Providence, 1958), ASME, New York 1959, 665-672.

[124] B. Paul and P.G. Hodge: Carrying capacity of elastic plastic shells under hydrostatic pressure, Proc. 3rd U.S. Nat. Congr. Appl. Mech. (Providence 1958) ASME, New York, 1959, 631-640.

[125] B. Paul: Carrying capacity of elastic-plastic shells with various end conditions under hydrostatic pressure, J. Appl. Mech. 26 (1959), 553-560.

[126] G. Eason: The load carrying capacity of cylindrical shells subjected to a ring of force, J. Mech. Phys. Solids, (1959), 169-181.

[127] A. Sawczuk and P.G. Hodge: Comparison of yield conditions for circular cylindrical shells, J. Franklin Inst., 269 (1960), 362-374.

[128] M.I. Yerchov: Determination of bearing capacity of a multi-span thin-walled pipeline, (in Russian), Vopr. teorii plastichnosti i prochnosti stroit. konstr. Trudy CNIISK, No. 4, Moscow 1961, 169-175.

[129] J. Paranelli and P.G. Hodge: Plastic analysis of cylindrical shells under pressure, axial load and torque, Proc. 8th Midw. Mech. Conf. (Cleveland 1963).

[130] Yu. P. Listrova and M.A. Rudis: Limit equilibrium of a toroidal shell, (in Russian), Mekh. i Mashinostroenie (1963), 3, 119-123.

[131] M.N. Fialkov: Limit analysis of simply supported circular shell roofs, J. Engng. Mech. Div. Proc. ASCE 84, 1958, Sept. 5706.

[132] C.A. Szmodits: A hyperbolic paraboloidal shell, (in Hungarian) Epitesi es Kozlekedestudomanyi Kozlemenyek 3 (1959), 1-2.

[133] A. Sawczuk: On experimental foundations of the limit analysis of reinforced concrete shells, Shell Research, North Holland, Amsterdam, 1961, 217-231.

[134] M. Janas: Limit analysis of a cylindrical shell, (in Polish), Arch. Inż. Lądow., 8 (1962), 365-374.

[135] M. Janas: Limit analysis of non-symmetric plastic shells by a generalized yield-line method, Proc. Symp. Non-Classical Shell Problems, (Warsaw 1963), Amsterdam 1964.

[136] T. Nakamura: Limit analysis of non-symmetric sandwich shells, Proc. Symp. Non-Classical Shell Problems, (Warsaw 1963), Amsterdam 1964.

[137] J. Rychlewski: Limit analysis of holicoidal shells, Proc. Symp. Non-Classical Shell Problems, (Warsaw 1963), Amsterdam 1964.

[138] N.V. Akhvledyani and V.N. Shaishmelashvili: On the limit analysis of shells, (in Russian), Soobsch. Akad. Nauk Gruz SSSR, 13 (1952).

[139] N.V. Akhvledyani and Shaishmelashvili: On the limit analysis of double-curvature shells, (in Russian), Trudy Inst. Stroit. Dela, Akad. Nauk Gruz SSR, Tbilisi 5, (1955), 61-71.

[140] G.I. Khazalya: Limit state analysis of shallow spherical shells, (in Russian), Soobch. Akad. Nauk Gruz SSR, 17 (1956), 815-822.

[141] J. Menyhard: Die statische Berechnung von zylindrischen Stahlbeton-Behältern auf Grund der Bruchteorie. Vorbericht des V. Kongr. Internat. Vereinigung f. Bruckenbau u. Hochbau, Lisboa 1956, 451-458.

[142] A.R. Rzhanitsyn: The design of plates and shells by the kinematical method of limit equilibrium, IX Cong. Appl. Mech. Brussels 1956, Actes, 6, 331-340.

[143] N.V. Akhvledyani: To the limit analysis of reinforced concrete rotational shells, Sobshch. Akad. Nauk. Gruz. SSR. 18 (1957), 209-210.

[144] A.P. Rzhanitsyn: Analysis of reinforced concrete shells using the method of limit equilibrium, (in Russian), Teoria rasch.i konstr. zhelezobet. konstr. Moscow, 1958, 155-175.

[145] A.R. Rzhanitsyn: Analysis of shells using the method of limit equilibrium, (in Russian), Issl. po vopr. teorii plast. i prochn. stroj. konstr. Moscow 1958, 7-35.

[146] A.M. Ovechkin: Equilibrium equations of reinforced concrete domes in the state of limit equilibrium, Nauchn. dokl. vyssh. shkoly stroit., (1958), 1, 35-46.

[147] A.R. Rzhanitsyn: Analysis of shallow shells using the method of limit equilibrium, Stroi. Mekh. Rasch. Sooruzh. (1959), 1, 5-11.

[148] A.R. Rzhanitsyn: Shallow and corrugated shells, (in Russian), Nauchn. Soobsch. Akad. Stroi.i Arkh SSSR, No. 14, Moscow 1960.

[149] A.M. Ovechkin: Calculation of reinforced concrete axisymmetric structural shells, Moscow 1961.

[150] S. Kaliszky: Unterschung einer Kegelstumpfschale aus Stahlbeton auf Grund des Traglastverfahrens, Acta Tech. Hung., 34, 1961, 159-175.

[151] N.V. Akhvledyani: To the analysis of bearing capacity of precast reinforced concrete domes, (in Russian), Stroi. Mekh. Rasch. Sooruzh. 3 (1961), 5, 15-17.

[152] N.V. Akhvledyani: On the ultimate load of shallow reinforced concrete shells of double curvature, (in Russian), Issled. Teorii Sooruzh. 11 (1962), 253-259.

[153] G.K. Haydukov: Limit equilibrium design of shallow shell panels, Proc. Sump. Non-Classical Shell Problems (Warsaw 1963), Amsterdam 1964.

[154] S. Kaliszky: Limit analysis of reinforced concrete truncated-cone shell, Proc. Symp. Non-Classical Shell Problems (Warsaw 1963), Amsterdam 1964.

[155] K.W. Johansen: Critical notes on calculation and design of cylindrical shells, Final Rep. 3rd Congr. IABSE, Liège 1948, 601-606.

[156] G. de Kazinczy: The limit design of shells, Final Rep. 3rd Congr. IABSE, Liège 1948.

[157] A.L.L. Baker: A plastic design theory for reinforced and prestressed concrete shell roofs, Mag. Concrete Research, 4, 1950, 27-34.

[158] A.L.L. Baker: Ultimate strength theory for short reinforced-concrete cylindrical shell roofs. Mag. Congr. Research, 10, 1952, 3-8.

[159] P.B. Morice: Research on concrete shell structures, Proc. 1st Symp. Shell Roof Constr. London 1952, Cem. Concr. Assoc. London 1954, 99-113.

[160] A.C. van Riel, W.J. Beranek and A.L. Bouma: Tests on shell roof models of reinforced concrete mortar, Proc. 2nd Symp. Shell Roof Constr. Oslo 1957, Teknisk Ukeblad, 1958, 315-324.

[161] A. Enami: Some experiments and the mechanism conditions of reinforced concrete prismatic folded plate structures, Proc. Symp. Non-Classical Shell Problems (Warsaw 1963), Amsterdam 1964.

[162] B.I. Bachrach, R.H. Lance: Plastic analysis of fiber reinforced shells of revolution, J. Appl. Mech., 41, 1974, 974-978.

[163] A. Biron: Limit analysis of cylindrical shells with longitudinal rib reinforcement, Int. J. Solids Structures, 6, 1970, 893-908.

[164] A. Biron, A. Sawczuk: Plastic analysis of rib-reinforced cylindrical shells, J. Appl. Mech., 34, 1967, 37-42.

[165] C.R. Calladine: On the derivation of yield conditions for shells, J. Appl. Mech., 39, 1972, 852-853.

[166] M.I. Erkhov: Problems of strength of ideally plastic shells, (in Russian), in: Stroit. Konstr., 4 CNIISK, Moscow 1969, 74-164.

[167] G.V. Ivanov: Approximation of the interaction between the membrane forces and moments at the Mises yield condition, (in Russian), Mekh. Tverd. Tela, 1968, No. 6, 74-75.

[168] G.V. Ivanov: Elastic-plastic yielding of shells at the Mises criterion, (in Russian), Mekh. Tverd. Tela, No. 3, 1969, 85-90.

[169] D.D. Ivlev: On the theory of limit analysis of shells at piecewise linear yield criteria, (in Russian), Mekh. Mashinostr.. 1962, No. 6, 95-102.

[170] D.D. Ivlev, Yu. P. Listrova, Yu. V. Nemirovsky: On the theory of limit analysis of layered plates and shells, (in Russian), Mekh. Mash., No. 4, 1964, 77-86.

[171] L.N. Kachanov: Problems of elastic-plastic theory of plates and shells, (in Russian), Proc. 6th All-Union Conf. Shells and Plates (Baku 1966), Nauka, Moscow 1966, 954-959.

[172] P.V. McLaughlin, Jr.: Plastic behavior of two layer sandwich structures, J. Appl. Mech., 40, 1972, 257-262.

[173] M. Sh. Mikeladze: Introduction to Technical Theory of Ideally Plastic Thin Shells, (in Russian), Metsnereba, Tbilisi 1969.

[174] M.S. Mikhailishin, Yu. V. Nemirovsky, O.N. Shablii: On the limit analysis of bimetalic shells of revolution, (in Russian), Zh. Prikl. Mekh. Tekh. Fiz. 1974, No. 2, 139-151.

[175] Z. Mróz, Xu Bin-ye: The load carrying capacities of symmetrically loaded spherical shells, Arch. Mech. Stos., 15, 1963, 245-266.

[176] Yu. V. Nemirovsky: Limit equilibrium of cylindrical waffer shells, (in Russian), Mekh. Tverd. Tela, 1967, No. 3, 52-59.

[177] Yu. V. Nemirovsky: Limit analysis of stiffened axisymmetric shells, Int. J. Solids Structures, 5, 1969, 1037-1058.

[178] M. Robinson: A comparison of yield surfaces for thin shells, Int. J. Mech. Sci., 15, 1971, 345-354.

[179] M. Robinson: The effect of transverse shear stresses on the yield surface for thin shells, Int. J. Solids Structures, 9, 1973, 810-818.

[180] R. Sankaranarayanan: Yield loci for nonhomogeneous shells of revolution, (in Polish), Rozpr. Inż. 14, 1966, 231-240.

[181] A. Sawczuk: Yield surfaces, Tech. Note 1, University of Waterloo, Waterloo 1971.

[182] A. Biron, P.G. Hodge Jr.: Limit analysis of rotationally symmetric shells under central boss loadings by a numerical method, J. Appl. Mech., 34, 1967, 644-650.

[183] A.A. Gvozdev, A.M. Protsenko: Prospects of application of the theory of limit analysis of shells, (in Russian), Proc. 7th All-Union Conf. Shells and Plates (Dniepropetrovsk 1969), Nauka, Moscow 1970, 736-748.

[184] M. Hamada, H. Nakanishi: A numerical method for the limit analysis of general shells of revolution, Bull. JSME, 14, 1971.

[185] P.F. Kiprichouk, O.H. Shablii: Evaluation of the collapse load of shallow rotationally symmetric shells, Prikl. Mekh. 6, 1970, No. 1, 34-42.

[186] Yu. P. Listrova, V.H. Potarov, M.A. Rudis: Limit equilibrium of shells of revolution made of sign sensitive materials, (in Russian), Mekh. Tverd. Tela, 1969, No. 1, 141-145.

[187] M.I. Reitman: Analysis of the equations of the theory of perfectly plastic shells, Arch. Mech. Stos., 19, 1967, 595-601.

[188] V.I. Rozenblyum: On the complete system of equations for plastic equilibrium of thin shells, (in Russian), Mekh. Tverd. Tela, 1966, No. 3, 127-132.

[189] J. Rychlewski: On the general theory of ideally plastic shells, (in Russian), in: Proc. 6th All-Union Conf. Shells Plates, (Baku 1966), Nauka, Moscow 1966, 873-880.

[190] J. Rychlewski, G.S. Shapiro: Ideal-plastic plates and shells, (in Russian), in: Proc. 6th All-Union Conf. Shells Plates, (Baku 1966), Nauka, Moscow 1966, 987-995.

[191] A. Sawczuk: On plastic analysis of shells, Proc. IUTAM, Thin Shells (Thibisi 1978), North-Holland, Amsterdam 1979.

[192] M. Sayir: Kollapsbelastung von rotations-symmetrischen Schalen, ZAMP, 17, 1966, 353-360.

[193] M. Sayir: Die rotations-symmetrische dünne Zylinderschale aus ideal-plastischem Material, ZAMP, 19, 1968, 185-219, 447-472.

[194] J. Schroeder, A.N. Sherbourne: A general theorem for thin shells in classical plasticity, J. Math. Phys., 47, 1968, 85-108.

[195] J. Schroeder, A.N. Sherbourne: Bounds for unsymmetrical yield point loads of thin shells, J. Math. Phys., 47, 1968, 249-261.

[196] J.G. Gerdeen, D.N. Hutula: Plastic collapse of ASME ellipsoidal head pressure vessels, J. Eng. Industry, 1970, 797-804.

[197] B. Górecki: Load carrying capacity of elliptic shells, (in Polish), Bull. WAT, 22, 1973, 31-43.

[198] H.M. Haydl, A.N. Sherbourne: Plastic analysis of shallow spherical shells, J. Appl. Mech., 40, 1974, 593-598.

[199] H.S. Ho, Limit pressures of rigidly clamped circular cylindrical shells, J. Eng. Mech. Div., EM 4, 100, 1974, 757-772.

[200] R.H. Lance, Chen-Hsiung Lee: The yield point load of conical shell, Int. J. Mech. Sci., 11, 1969, 129-143.

[201] J.A. Lellep: Limit equilibrium of shallow shells of sign sensive materials, (in Russian), Prikl. Mekh.,8, 1972, No. 2, 47-58.

[202] H.F. Muensterer, F.P.J. Rimrott: Elastic-plastic response of a sandwich cylinder subjected to internal pressure, J. Strain Anal., 6, 1971, 273-278.

[203] Nagarasan, E.P. Popov: Plastic and visco-plastic analysis of axisymmetric shells, Int. J. Solids Structures, 11, 1975, 1-19.

[204] H. Nakanishi, M. Suzuki, M. Hamada: A lower bound limit analysis of shells of revolution, J. Appl. Mech., 42, 1975, 494-495.

[205] N. Perrone: An experimental verification of limit analysis of short cylindrical shells, J. Appl. Mech., 36, 1964, 362-364.

[206] M. Save, Ch. Massonnet: Plastic analysis and design of plates, shells and disks, North-Holland, Amsterdam 1972.

[207] J. Schroeder, A.N. Sherbourne: Unsymmetrical yield point loads of spherical domes, J. Eng. Mech. Div. Proc. ASCE EM3, 1968, 823-839.

[208] O.N. Shablii, M.S. Mikhalishin: Limit equilibrium of cylindrical rigid-plastic shells of sign sensitive materials, (in Russian), Probl. Protchnosti, 1973, No. 8, 23-29.

[209] A.N. Sherbourne, H.M. Haydl: Limit loads of edge restrained shallow shells, Int. J. Solids Structures, 10, 1974, 873-881.

[210] V.E. Terrovere: Load carrying capacity and weight optimization of ellipsoidal shells, (in Russian), Prikl. Mekh., 8, 1972, No. 2, 68-72.

[211] D.C. Drucker and R.T. Shield: Design for minimum weight, Proc. 9th Int. Congr. Appl. Mech., Brussels 1956, Actes, 5, 212-222.

[212] D.C. Drucker and R.T. Shield: Bounds on minimum weight, Quart. Appl. Math., 15, 1957, 269-281.

[213] R.T. Shield: Optimum design methods for structures, Proc. 2nd Symp. Naval Struc. Mech. (Providence 1960), Oxford 1960, 580-591.

[214] Z. Mróz: On a problem of minimum weight design, Quart. Appl. Math. 19, 1961, 127-135.

[215] E.T. Onat and W. Prager: Limits of economy of materials in shells, De Ingenieur, 67, 1955, 10, 46-49.

[216] W. Freiberger: Minimum weight design of cylindrical shells, J. Appl. Mech., 23, 1956, 576-580.

[217] W. Freiberger: On the minimum weight design problem for cylindrical sandwich shells, J. Aero. Sci. 24, 1957, 847-848.

[218] R.T. Shield: On the optimum design of shells, J. Appl. Mech., 27, 1960, 316-322.

[219] M. Sh. Mikeladze: Analysis of weight and strength of rigid plastic shells, (in Russian), Arch. Mech. Stos. 11 (1959), 17-31.

[220] N. Sh. Mikeladze: On uniform-strength plastic shells, (in Russian), Soobshch. Akad. Gruz SSR, 25 (1960), 391-398.

[221] Z. Mróz: Optimal design of reinforced concrete shells, Proc. Symp. Non-Classical Shell Problems, (Warsaw 1963), Amsterdam 1964.

[222] W. Olszak and A. Sawczuk: Some problems of limit analysis and limit design of non-homogeneous axially symmetric shells, Proc. Symp. Concr. Shell Roof Constr. Oslo 1956, Teknisk Ukeblad, 1957, 249-256.

[223] H. Ziegler: Kuppeln gleicher Festigkeit, Ing. Archiv, 26, 1958, 378-382.

[224] W. Isler: Eine Kuppel gleicher Festigkeit, ZAMP, 10, 1959, 576-578.

[225] E. Melan: Theorie statisch unbestimmter Systeme aus ideal-plastischem Baustoff. Sitz.-Ber. Österr. Akad. Wiss. (2a), 145, 1936, p. 195-218.

[226] E. Melan: Zur Plastizität des räumlichen Kontinuums, Ing.-Arch., 9, 1938, p. 116-126.

[227] W.T. Koiter: A new general theorem of shake-down of elastic-plastic structures, Proc. Koninkl. Nederl. Akad. Wet. (B), 59, 1956, p. 24-34.

[228] W. Prager, Bauschinger adaptation of rigid-workhardening trusses, Mech. Res. Com. 1, 1974, p. 253-256.

[229] B. Halphen: Accommodation and adaptation, 1979, (in press).

[230] W.T. Koiter: Over de stabiliteit van het elastisch evenwicht, Thesis Delft (Amsterdam, 1945); two translations into English (Washington, 1967; Palo Alto, Cal. 1970).

[231] W.T. Koiter: On the nonlinear theory of thin elastic shells, Proc. Kon. Ned. Wet., B69, (1966), 1-54.

[232] W.T. Koiter: Foundations and basic equations of shell theory. A survey of recent progress, in Theory of Thin Shells (Copenhagen 1967), Springer Berlin Heidelberg 1969, 93-105.

[233] A.L. Goldenwejzer: Theory of thin elastic shells (in Russian), Gostekhizdat, Moscow 1953.

[234] P.M. Naghdi: Foundations of elastic shell theory, in: Progress in Solid Mechanics, 4, North Holland, Amsterdam 1963, 1-90.

[235] I.H. Donnell: Beams, plates and shells, McGraw Hill, New York 1975.

[236] M.K. Duszek: Equations of the theory of large deflections of plastic shells, (in Polish), Rozpr. Inż. 20, 1972, 389-407.

[237] M.K. Duszek: A systematic study of kinematics of shells at large strains and displacements, Bull. Acad. Pol. Sci., Cl IV, 26, 1978, 39-47.

[238] M.K. Duszek, A. Sawczuk: On fundamental relations of plastic shell theory, (in Polish), Rozpr. Inż., 18, 1970, 717-731.

[239] M. Kleiber: General theory of elastic-plastic shells, (in Polish), Shell Structures (Cracow 1974), PWN, Warsaw 1979 (in press).

[240] A. Sawczuk: Lagrangian formulation of large deflection theories of plastic shells, Bull. Acad. Pol. Sci., Cl. IV, 19, 1971, 247-252.

[241] A. Sawczuk: Problems of the theory of moderately large deflections of rigid-plastic shells, (in Polish), Mech. Teor. Stos., 9, 1971, 335-354.

[242] O.N. Shablii, M.S. Mikhalishin: Limit equilibrium of cylindrical rigid-plastic shells of sign sensitive materials, (in Russian), Probl. Protchnosti, 1973, No. 8, 23-29.

[243] Z. Waszczyszyn: Calculation of finite deflections of elastic-plastic plates and rotationally symmetric shells (in Polish), Technological University Cracow, Rep. No. 5, Cracow 1970.

[244] M.K. Duszek: Plastic analysis of cylindrical sandwich shells accounting for geometry changes, (in Polish), Rozpr. Inż. 15, 1967, 653-663.

[245] M.K. Duszek: Plastic analysis of cylindrical shells subjected to large deflections, Arch. Mech. Stos., 18, 1969, 599-614.

[246] M.K. Duszek: Plastic analysis of shallow spherical shells at moderately large deflections, in: Theory of Thin Shells, (Copenhagen 1967), Springer, Berlin 1969, 374-388.

[247] M.K. Duszek, M. Mitov: Plastic analysis of shells in the large deflection range, (in Polish), Rozpr. Inż., 1978.

[248] M.K. Duszek, A. Sawczuk: Load-deflection relations for rigid-plastic cylindrical shells beyond the incipient collapse load, Int. J. Mech. Sci., 12, 1970, 839-848.

[249] K.A. Koba, O.N. Shablii: Large deflections of rigid plastic shallow shells in hinged, restrained edge, Mekh. Tverd. Tela No. 5, 1973, 173-179.

[250] U.P. Lepik: Large deflection of rigid-plastic cylindrical shells under axial tension and external pressure, Nucl. Eng. Des., 4, 1966, 29-38.

[251] I.G. Teregulov: Large deflections of elastic plastic shallow, rigidly clamped shells, (in Russian), in: Proc. 7th All-Union Conf. Shells Plates (Dniepropetrovsk 1969), Nauka, Moscow 1970, 578-581.

[252] M.K. Duszek: Stability analysis of rigid, perfectly-plastic structures at the yield point load, Buckling of Structures, (Cambridge, Mass. 1974), Springer, Berlin 1976, 106-116.

[253] M. Kleiber: Kinematics and deformation processes in materials subjected to finite elastic-plastic strain, Int. J. Eng. Sci., 13, 1975, 513-525.

[254] M. Kleiber: Lagrangean and Eulerian finite element formulation for large strain elasto-plasticity, Bull. Acad. Pol. Sci., Cl. IV, 23, 1975, 117-126.

[255] R.H. Lance, J.F. Soechting: A displacement bounding principle in finite plasticity, Int. J. Solids Structures, 6, 1970, 1103-1118.

[256] G. Maier, D.C. Drucker: Effects of geometry changes on essential features of inelastic materials, Journ. Eng. Mech. Div. Proc. ASCE 1973, 819-834.

[257] M. Kleiber: Large deformations of elastic-plastic solids, Theory and numerical analysis of structures, (in Polish), IPPT Rep. 13/78, Warsaw 1978.

[258] M.I. Erkhov: Approximate theory of dynamic loading of plastic shells in: Proc. 7th All-Union Conf. Shells and Plates (Dniepropetrovsk 1969), Nauka, Moscow 1970, 230-234.

[259] M. Jones: A literature review on the dynamic plastic response of structures, Shock Vibr. Digest, 8, 1975, 89-105.

[260] H. Stolarski: Assessement of large displacements of a rigid-plastic shell withholding a localized impact, Nucl. Eng. Design, 41, 1977, 327-334.

[261] H. Stolarski: Extremum principle for dynamics of shells in the large displacement range, in: Proc. IUTAM Thin Shell (Tbilissi 1978), North Holland, Amsterdam 1979.

[262] Z. Waszczyszyn: Finite elastic-plastic deflections of orthotropic shells of revolution, Exact equations and calculations, Bull. Acad. Pol. Sci., Cl. IV, 16, 1968, 461-470.

[263] T. Wierzbicki: Dynamics of plastic structures, (in Polish), Arkady, Warsaw 1979 (in press).

[264] M. Janas, J.A. König: Limit analysis of shells, (in Polish), Arkady, Warsaw 1966.

[265] M.T. Huber: Specific distortion energy as a measure of (in Polish), Crasop. Techn. Lwów, vol. 22, 1904, p. 81, Reprinted in Pisma (collected works), vol. 2, PWN, Warsaw 1956.

[265] R. von Mises: Mechanik der festen Körper im plastisch deformablen Zustand. Gött. Nachr., Math. Phys. Kl. 1913, p. 582-592.

[266] Ch. A. Coulomb: Essai sur une application de règle de Maximis et Minimis à quelques problèmes statiques relatifs à l'architecture, Mém. de math. et phys., vol. 7, 1773, p. 343.

[267] H. Tresca: Mémoire pur l'écoulement des corps solides, Mém. prés. par div. sav., vol. 18, 1868, p. 733-799.

[268] R. von Mises: Mechanik der plastischen Formänderung von Kristallen, Zeitschr. für Angew. Math. und Mech., vol. 8, 1928, p. 161-185.

[269] R. Hill, Mathematical Theory of Plasticity, Oxford University Press, 1950.

[270] W. Olszak and W. Urbanowski: The generalized distortion energy in the theory of anisotropic bodies, Bull. Acad. Pol. Sci., Sér. Sci. Tech., 5, 1957, p. 29-37.

[271] W. Olszak and W. Urbanowski: The tensor of moduli of plasticity, Bull. Acad. Pol. Sci., Sér. Sci. Tech., 5, 1957, p. 39-45.

[272] W. Olszak and Urbanowski: The flow function and the yield condition for nonhomogeneous anistropic bodies, Bull. Acad. Pol. Sci., Sér. Sci. Tech., 4, 1957.

[273] M. Sh. Mikeladze: On the ultimate load of initially anisotropic shells, (in Russian), Dokl. Akad. Nauk SSSR, 98 (1954), 921-923.

[274] M. Sh. Mikeladze: On the plastic flow of anisotropic shells, (in Russian), Izv. Akad. Nauk. SSSR, OTN (1955), 8, 67-80.

[275] D. Niepostyn: The limit analysis of an orthotropic circular cylinder, Arch. Mech. Stos., 8, 1956, 565-580.

[276] M. Sh. Mikeladze: Rigid-plastic analysis of anisotropic plates and shells, IX Congr. Int. Mec. Appliquée (Brussels 1956), Actes 8, 1957.

[277] M. Sh. Mikeladze: General theory of anisotropic rigid plastic shells, (in Russian), Izv. Akad. Nauk. SSSR, OTN (1957), 1, 85-94.

[278] M. Sh. Mikeladze: Elasto-plastic equilibrium of anistropic shells, (in Russia ı), Soobshch. Akad. Nauk Gruz. SSR, 20 (1958), 13-20.

[279] A. Sawczuk: Piecewise linear theory of anistropic plasticity and its application to limit analysis problems, Arch. Mech. Stos., 11, 1959, 541-557.

[280] A. Sawczuk: Yield condition for anistropic shells, Bull. Acad. Polon. Sci., Cl. IV, 8, 1960, 213-277.

[281] Z. Mróz: The load carrying capacity of orthotropic shells, Arch. Mech. Stos. 12, 1960, 85-107.

[282] A. Sawczuk: On the theory of anisotropic plates and shells, Arch. Mech. Stos. 13, 1961, 355-366.

[283] Yu. Nemirovski and Yu. Rabotnov: Limit equilibrium of rib-reinforced cylindrical shells, (in Russian), Mekh. i Mashinostroenie (1963), 3, 83-94.

[284] W. Olszak and W. Urbanowski: The plastic potential and the generalized distortion energy in the theory of non-homogeneous anistropic elasto-plastic bodies, Arch. Mech. Stos., 8, 1956, 85-110.

[285] W. Olszak and A. Sawczuk: Théorie de la capacité portante des constructions non-homogènes et orthtropes, Ann. Inst. Techn. Bat. Trav. Publ., 13, 1960, 517-535.

[286] Yu. R. Lepik: To the ultimate load of non-homogeneous plates and shells, (in Russian), Mekh. i Mashinostroienie (1963), 4, 167-171.

[287] W.E. Jahsman and R.F. Hartung and J.E. Edwards: Plastic analysis of an axisymmetrically loaded shell of revolution with meridionally varying limit stress, Proc. Symp. Non-Classical Shell Problems, (Warsaw 1963), Amsterdam 1964.

[288] K. Hohenenser and W. Prager: Uber die Ansätze der Mechanik isotroper Kontinua, Zeitschr. f. Angew. Math. u. Mech., vol. 12, 1932, p. 216.

[289] W.W. Sokolowskij: Propagation of elastic-visco-plastic waves in rods (in Russian), Pokl. AN SSSR, 60, 5, 1948, p. 775-778.

[290] P. Perzyna: The constitutive equations for rate-sensitive plastic materials, Quart. Appl. Math., 20, 1963, p. 321.

[291] M. Reiner and K. Weissenberg: Rheology Leaflet, No. 10, 1939, p. 12-30.

[292] W. Olszak: Les critères de transition en elasto-visco-plasticité, Bull. Acad. Pol. Sci., Sér. Sci. Techn., 1, 1966.

[293] A. Yu. Ishlinski: General theory of plasticity with linear hardening, (in Russian), Ukr. Matemat. Zhurn. 6, (1954), 314-325.

[294] W. Prager: A new method of analysing stress and strain in work-hardening plastic solids, J. Appl. Mech., 23, 1956, 493-496.

[295] P. Perzyna: The constitutive equations for work-hardening and rate- sensitive plastic materials, Proc. Vibr. Probl., 4, 1963, p. 281.

[296] Z. Mróz: On the description of anisotropic work-hardening, J. Mech. Phys. Solids, vol. 15, 1967, p. 163-175.

[297] Z. Mróz: An attempt to describe the behaviour of metals under cyclic loads using a more general work-hardening model, Acta Mech., vol. 7, 1969, p. 199-212.

[298] A. Phillips and H. Moon: An experimental investigation concerning yield surfaces and loading surfaces, Rep. Yale Univ., Dept. of Engrg. and Appl. Sci., 1976, Cf. Acta Mechanics.

[299] A. Phillips and M. Ricciutti: Fundamental experiments in plasticity and creep of aluminium, Extension of previous results, Int. of Solids and Structure, vol. 12, 1976, p. 159-171.

[300] P.G. Hodge and F.Romano: Deformations of an elastic-plastic cylindrical shell with linear strain-hardening, J. Mech. Phys. Solids, 4, 1956, 145-161.

[301] N. Perrone and P.G. Hodge: On strain hardened circular cylindrical shells, J. Appl. Mech., 27, 1960, 489-495.

[302] E.T. Onat and S. Yamanturk: On thermally stressed elastic plastic shells, J. Appl. Mech., 29, 1962, 108-114.

[303] W. Prager: Linearization in visco-plasticity, Oester. Ingenieur-Archiv., 15, 1961, 155-157.

[304] R.E. Ball and S.L. Lee: Limit analysis of cylindrical shells, J. Eng. Mech. Div., Proc. ASCE, 89, No. EM3, 1963, 73-96.

[305] W. Prager: Nonisothermal plastic deformation, Proc. Konikl. Akad. Wet. (B) 61, 1958, p. 176-182 (Amsterdam).

[306] W. Olszak and P. Perzyna: The constitutive equations of the flow theory for a non-stationary yield condition, Proc. 11th Congr. Appl. Mech. Munich 1964, Springer 1966, p. 545-553.

[307] F. Campus: Plastification de l'acier doux en flexion plane simple, Bull. Cl. Sci., Scr. 5, vol. 49, 4, 1963.

[308] J.A. König and W. Olszak: The yield criterion in the general case of nonhomogeneous (nonuniform) stress and deformation fields, Proc. Symp. "Topics in Applied Continuum Mechanics", Springer Wien-New York 1974.

VARIATIONAL PRINCIPLES AND METHODS FOR VISCOELASTIC SHALLOW SHELLS

by

J. Brilla

Comenius University, Bratislava

VARIATIONAL PRINCIPLES AND METHODS FOR VISCOELASTIC SHALLOW SHELLS

J. Brilla

Abstract. Using the Laplace transform the generalized potential energy of viscoelastic shallow shells has been introduced and generalized and convolutional variational principles for quasistatic and dynamic problems for viscoelastic shallow shells have been formulated.

These variational principles form a good basis for generalized and convolutional variational methods and for a generalization of the finite element method for the solution of viscoelastic shallow shells.

Finally the convolutional variational principle for stability of viscoelastic shallow shells and their buckling and post-buckling behaviour are discussed.

1. INTRODUCTION

Many problems of mathematical physics defined in the form of differential equations can equally well be formulated as variational problems, that is, as problems of finding an extreme point for some real-valued functional over a certain set of admissible functions. The variational formulation of physical laws provides powerful tools for studying problems of mathematical physics. Variational principles have been widely employed in many branches of mechanics. However, the importance of these principles has been emphasized by the development of a variational method as effective as the finite element method.

Convolutional variational principles and their Laplace transforms have a similar importance in the formulation and solution of time dependent problems[1-5]. These variational principles form a good basis for convolutional and generalized variational methods and for a generalization of the finite element method for time dependent problems[6-9].

We shall deal with variational principles and methods for viscoelastic shallow shells, with a generalization of the finite element method for viscoelastic shallow shells and with the analysis of buckling and post-buckling problems of viscoelastic shallow shells.

2. CONSTITUTIVE EQUATIONS

When dealing with the infinitesimal theory or the large deflection theory of viscoelastic shallow shells, as it is common in the case of elastic structures, we shall employ linear constitutive equations.

The constitutive equation of an arbitrary linear viscoelastic material can be written in the form

$$\sigma^{ij} = H^{ijkl} \varepsilon_{kl} , \tag{2.1}$$

where H^{ijkl} represents a linear tensor operator. For non--polar materials this operator is symmetric. In addition, according to the second law of thermodynamics it is positive definite. The operator H^{ijkl} can assume a differential, integral or integro-differential form. Summation over repeated indices is implied. In the case of a differential form the constitutive equation can be written as [10]

$$K_{(r)} \sigma^{ij} = K^{ijkl}_{(s)} \varepsilon_{kl} , \tag{2.2}$$

or

$$L^{(s)} \varepsilon_{ij} = L^{(r)}_{ijkl} \sigma^{kl} , \tag{2.3}$$

where

$$K_{(r)} = \prod_{n=1}^{r} \left(\frac{d}{dt} + \kappa_n \right), \quad L^{(s)} = \prod_{n=1}^{s} \left(\frac{d}{dt} + \lambda_n \right) \tag{2.4}$$

are scalar operators and

$$K^{ijkl}_{(s)} = \sum_{n=1}^{s} \kappa^{ijkl}_{n} \frac{d^n}{dt^n}, \quad L^{(r)}_{ijkl} = \sum_{n=0}^{r} L^{n}_{ijkl} \frac{d^n}{dt^n} \qquad (2.5)$$

are tensor operators, $\kappa_n \geq 0$ are inverse relaxation times, $\lambda_n \geq 0$ are inverse retardation times, $N^{(o)} = 1$ and $s = r$ or $s = r+1$. In the case of non-polar materials tensor operators cannot be on both sides of (2.2-3).

In the case of a material of integral type the constitutive equation can be written in the form

$$\sigma^{ij} = \int_0^t G^{ijkl}(t - \tau) \frac{d\,\varepsilon_{kl}}{d\tau}\, d\tau\,, \qquad (2.6)$$

or

$$\varepsilon_{ij} = \int_0^t J_{ijkl}(t - \tau) \frac{d\,\sigma^{kl}}{d\tau}\, d\tau\,, \qquad (2.7)$$

where $G^{ijkl}(t-\tau)$ is a tensor of relaxation functions and $J_{ijkl}(t-\tau)$ a tensor of creep functions.

In the case of aging materials the coefficients in (2.2-3) are functions of t and tensors G^{ijkl}, J_{ijkl} are not of convolutional type.

3. GENERALIZED VARIATIONAL THEOREMS AND METHODS

We consider the governing differential equations of an anisotropic viscoelastic shallow shell in terms of displacements

$$K_{ijkl}(D)\left[\frac{h^3}{12}w,_{ijkl} + h(u_{i,j} + b_{ij}w)b_{kl}\right] = K(D)q \ ,$$
$$K_{ijkl}(D)(u_{k,jl} + b_{kl}w,_j) = 0 \ , \quad (i=1,2) \tag{3.1}$$

for materials of differential type and

$$\int_0^t G^{ijkl}(t-\tau)\frac{\partial}{\partial\tau}\left[\frac{h^3}{12}w,_{ijkl} + h(u_{i,j} + b_{ij}w)b_{kl}\right]d\tau = q \ ,$$
$$\int_0^t G^{ijkl}(t-\tau)\frac{\partial}{\partial\tau}(u_{k,jl} + b_{kl}w,_j)d\tau = 0 \ , \quad (i=1,2) \tag{3.2}$$

for materials of integral type, w being the displacement of the middle surface of the shell in x_3 direction (positive outward), u_i - the displacement in x_i direction, h - the thickness of the shell, b_{ij} - tensor of the curvature of the middle surface and $q(x,t)$ - the transverse loading of the shell,positive in the outward direction. Latin indices here and in the following assume the range of integers 1,2 . Subscripts preceded by a comma indicate differentiation with respect to corresponding Cartesian spatial coordinates and $D^n = \partial^n/\partial t^n$.

For real materials we have

$$K_{ijkl}(D)\varepsilon_{ij}\,\varepsilon_{kl} \geq 0 \ , \quad G_{ijkl}(o)\varepsilon_{ij}\,\varepsilon_{kl} \geq 0 \tag{3.3}$$

and the equality holds only for $\varepsilon_{ij} = 0$.

We shall consider the following boundary conditions

$$w = \frac{\partial w}{\partial n} = 0 \ , \quad u_1 = u_2 = 0 \ , \qquad \text{on } \partial\Omega \tag{3.4}$$

or

$$w = K_{ijkl}\, w_{,ij}\, \nu_k^n\, \nu_\ell^n = 0\ ,\ u_1 = u_2 = 0, \quad \text{on } \partial\Omega, \qquad (3.5)$$

where Ω is the domain of the definition of (3.1), $\partial\Omega$, its boundary and n denotes the outward normal to $\partial\Omega$ and ν_k^n the direction cosines of this normal.

From the physical point of view in the case (3.1) it is convenient to consider the initial conditions in the form

$$\left.\frac{\partial^\nu w}{\partial t^\nu}\right|_{t=0} = \left.\frac{\partial^\nu u_i}{\partial t^\nu}\right|_{t=0} = 0\ , \quad (\nu = 0,1,\ldots,r-1). \qquad (3.6)$$

Initial values at $t = 0^+$ can be different from zero and are to be obtained from the solution.

Applying the generalized Laplace transform[11] to (3.1) and (3.2) we arrive at

$$\begin{aligned} &\tilde{K}_{ijkl}(p)\left[\frac{h^3}{12}\,\tilde{w}_{,ijkl} + h(\tilde{u}_{i,j} + b_{ij}\tilde{w})b_{kl}\right] = \tilde{K}(p)\tilde{q}, \\ &\tilde{K}_{ijkl}(p)\,(\tilde{u}_{k,jl} + b_{kl}\,\tilde{w}_{,j}) = 0\ , \quad (i=1,2) \end{aligned} \qquad (3.7)$$

for materials of differential type and

$$\begin{aligned} &p\,\tilde{G}_{ijkl}(p)\left[\frac{h^3}{12}\,\tilde{w}_{,ijkl} + h(\tilde{u}_{i,j} + b_{ij}\tilde{w})b_{kl}\right] = \tilde{q}\ , \\ &p\,\tilde{G}_{ijkl}(p)\,(\tilde{u}_{k,jl} + b_{kl}\,\tilde{w}_{,j}) = 0\ , \quad (i=1,2)\ , \end{aligned} \qquad (3.8)$$

where tildas denote Laplace transforms and p is the parameter of transformation.

Similarly we obtain the Laplace transforms of boundary conditions. Thus we have arrived, in the sense of a Laplace transform, at the associated elastic problems for anisotropic viscoelastic shallow shells. This problem is parametrically dependent on the transform parameter p .

For this associated elastic problems we can, in an obvious way, derive the functionals of the generalized potential energy.

We introduce the Laplace transforms of stresses and strains. Then for real positive values of the transform parameter p, the generalized potential energy of a viscoelastic shallow shell can be written in the form

$$\tilde{\Pi} = \int_V \frac{1}{2} \tilde{\sigma}^{ij} \tilde{\varepsilon}_{ij} \, dV - \int_\Omega \tilde{q} \, \tilde{w} \, d\Omega \quad . \tag{3.9}$$

According to the assumptions of the shallow shell theory we have

$$\tilde{\varepsilon}_{ij} = \frac{1}{2}(\tilde{u}_{i,j} + \tilde{u}_{j,i}) + b_{ij} \, \tilde{w} - x_3 \, \tilde{w}_{,ij} \quad . \tag{3.10}$$

Denoting by

$$\tilde{\varepsilon}^{(o)}_{ij} = \frac{1}{2}(\tilde{u}_{i,j} + \tilde{u}_{j,i}) + b_{ij}\tilde{w} \tag{3.11}$$

the Laplace transform of middle surface strains, and integrating (3.9) with respect to x_3 through the thickness of the shell $\left[- {}^h/_2 \, , \, {}^h/_2\right]$, we obtain

$$\tilde{\Pi} = \int_\Omega \left[\frac{1}{2} (- \tilde{M}_{ij} \, \tilde{w}_{,ij} + h\tilde{\sigma}_{ij} \, \tilde{\varepsilon}^{(o)}_{ij}) - \tilde{q}\tilde{w} \right] d\Omega \quad , \tag{3.12}$$

where

$$\tilde{M}_{ij} = \int_{h/2}^{h/2} x_3 \, \tilde{\sigma}_{ij} \, dx_3 \quad . \tag{3.13}$$

Applying (2.2) and (3.13) we arrive at

$$\tilde{K}\,\tilde{M}_{ij} = \frac{h^3}{12}\,\tilde{K}_{ijkl}\,\tilde{w}_{,kl} \quad . \tag{3.14}$$

Then the generalized potential energy of viscoelastic shallow shells can be written in the form

$$2\tilde{K}\,\tilde{\Pi}(\tilde{w},\tilde{u}) = \int_{\Omega} \Big[\frac{h^3}{12}\,\tilde{K}_{ijkl}\,\tilde{w}_{,ij}\,\tilde{w}_{,kl} + \\ + h\,\tilde{K}_{ijkl}(\tilde{u}_{i,j} + b_{ij}\tilde{w})(\tilde{u}_{k,l} + b_{kl}\tilde{w}) - 2\tilde{K}\,\tilde{q}\,\tilde{w}\Big] d\Omega \quad . \tag{3.15}$$

Then it is easy to prove the generalized variational principle of minimum of the generalized potential energy for viscoelastic shallow shells :

The solution $\tilde{w}$, $\tilde{u}_1$, $\tilde{u}_2$ of the system (3.7) minimizes, for real positive values of the transform parameter p, the functional of the generalized potential energy (3.15) and inversely the functions $\tilde{w}$, $\tilde{u}_1$, $\tilde{u}_2$, which minimize (3.15), give the solution of (3.7). The same also is true for Boltzmann's materials. Then it is necessary to consider equations (3.8) instead of (3.7) and in the functional (3.15) to replace $\tilde{K}_{ijkl}$ by $p\,\tilde{G}_{ijkl}$ and to put $\tilde{K} = 1$.

When we want to formulate the principle of minimum of the generalized potential energy for viscoelastic shallow shells in terms of the stress function and of the transverse displacement, we have to modify (3.12) by making use of Lagrange's method of multipliers. Then (3.12) constrained by the condition (3.11) assumes the form

$$\tilde{\Pi} = \int_{\Omega} \{\frac{1}{2}\left[- \tilde{M}_{ij}\, \tilde{w}_{,ij} + h\, \tilde{\sigma}_{ij}\, \tilde{\varepsilon}^{(o)}_{ij}\right] - \tag{3.16}$$

$$- h\, \tilde{\sigma}_{ij}\left[\tilde{\varepsilon}^{(o)}_{ij} - \frac{1}{2}(\tilde{u}_{i,j} + \tilde{u}_{j,i}) - \tilde{b}_{ij}\tilde{w}\right] - \tilde{q}\, \tilde{w}\}\, d\Omega \ ,$$

where $h\, \tilde{\sigma}_{ij}$ is the Lagrange multiplier.

Introducing the stress function by

$$\sigma_{ij} = \epsilon_{ik}\, \epsilon_{jl}\, F_{,kl} \quad , \tag{3.17}$$

where ε_{ij} is the alternating tensor and making use of the equilibrium equation we arrive at

$$2\, \tilde{K}\, \tilde{L}\, \tilde{\Pi}(\tilde{w},\tilde{F}) = \int_{\Omega} \{\frac{h^3}{12}\, \tilde{L}\, \tilde{K}_{ijkl}\, \tilde{w}_{,ij}\, \tilde{w}_{,kl} -$$

$$- h\, \tilde{K}\, \tilde{L}_{ijkl}\, \epsilon_{im}\, \epsilon_{jn}\, \epsilon_{kr}\, \epsilon_{ls}\, \tilde{F}_{,mn}\, \tilde{F}_{,rs} \tag{3.18}$$

$$+ 2\, \tilde{K}\, \tilde{L}\, h\, \epsilon_{ik}\, \epsilon_{jl}\, \tilde{F}_{,ij}\, b_{kl}\tilde{w} - 2\, \tilde{K}\, \tilde{L}\, \tilde{q}\, \tilde{w}\}d\Omega \ ,$$

where

$$\left(\frac{\tilde{K}_{ijkl}}{\tilde{K}} \right)^{-1} = \frac{\tilde{L}_{ijkl}}{\tilde{L}} \ . \tag{3.19}$$

Then a familiar variational procedure provides the Laplace transform of the governing equations of viscoelastic shallow shells

$$\tilde{K}_{ijkl}\, \tilde{w}_{,ijkl} = \tilde{K}(\tilde{q} - \epsilon_{ik}\, \epsilon_{jl}\, b_{ij}\, h\tilde{F}_{,kl}) \ ,$$
$$\epsilon_{im}\, \epsilon_{jn}\, \epsilon_{kr}\, \epsilon_{ls}\, \tilde{L}_{ijkl}\, \tilde{F}_{,mnrs} = \tilde{L}\, \epsilon_{ik}\, \epsilon_{jl}\, b_{ij}\tilde{w}_{,kl} \ . \tag{3.20}$$

Thus we can formulate the generalized variational principles of minimum of the generalized potential energy for viscoelastic shallow shells in terms of the transverse displacement and the stress function. We have :

The solution $\tilde{w},\tilde{F}$ of the system (3.20) minimizes, for real positive values of the transform parameter p, the functional of the generalized potential energy (3.18) and, inversely, the functions $\tilde{w},\tilde{F}$, which minimize (3.18), are the solution of (3.20) .

The same is true also for Boltzmann's materials. Then it is necessary to replace in (3.18) and (3.20) $\tilde{K}_{ijkl}$ by $p\,\tilde{G}_{ijkl}$, and $\tilde{L}_{ijkl}$ by $p\,\tilde{J}_{ijkl}$, and to put $\tilde{K} = \tilde{L} = 1$.

Following the same procedure we can also derive other variational principles.

As an inversion of a general viscoelastic operator can be complicated for practical purposes, it is more convenient to make use of the principle of minimum of the generalized potential energy in terms of displacements.

4. GENERALIZED VARIATIONAL METHODS

On the basis of the derived principles of minimum of the generalized potential energy for viscoelastic shallow shells it is possible to generalize variational methods for the solution of viscoelastic shallow shells in the classical form[5] as well as in the form of finite element method[6]. We shall deal with a general formulation of the finite element method and make use of the minimum principle in terms of displacement.

Let $W \times U_1 \times U_2$ be vector subspace of the Sobolev space $W_{2,2} \times W_{2,1} \times W_{2,1}$ where W, U_1, U_2 are the spaces of functions which fulfil Laplace transforms of boundary conditions for $\tilde{w}, \tilde{u}_1, \tilde{u}_2$, respectively. Let $\Phi_\alpha(\alpha = 1,2,\dots,k)$, $\Psi_\alpha(\alpha = 1,2,\dots,m)$, $\kappa_\alpha(\alpha = 1,2,\dots,n)$ be the base functions in W_k, U_{1m}, U_{2n}, which are finite dimensional subspaces of W, U_1, U_2, respectively. Then the approximate value of $\tilde{w}$, $\tilde{u}_1, \tilde{u}_2$ can be sought in the form

$$\tilde{w}_k = \sum_{\alpha=1}^{k} \tilde{a}_\alpha(p)\, \Phi_\alpha ,$$

$$\tilde{u}_{1m} = \sum_{\alpha=k+1}^{k+m} \tilde{a}_\alpha(p)\, \Psi_{\alpha-k} , \tag{4.1}$$

$$\tilde{u}_{2n} = \sum_{\alpha=k+m+1}^{k+m+n} \tilde{a}_\alpha(p)\, \kappa_{\alpha-k-m} .$$

In this case we can put $\Psi_\alpha = \kappa_\alpha$ and $m = n$. However we shall use the general formulation.

Inserting (4.1) into (3.15) we transform the functional of the generalized potential energy into a function of $\tilde{a}_\alpha(p)$. We get

$$2\tilde{K}\tilde{\Pi}(\tilde{w}_k, \tilde{u}_{1m}, \tilde{u}_{2n}) = 2\tilde{K}\tilde{\Pi}(\tilde{a}_\alpha) = \tag{4.2}$$

$$= \sum_{\alpha=1}^{k}\sum_{\beta=1}^{k} \tilde{a}_\alpha \tilde{a}_\beta (\tilde{K}_{oo\alpha\beta} + \tilde{K}_{bb\alpha\beta}) + \sum_{\alpha=k+1}^{k+m}\sum_{\beta=k+1}^{k+m} \tilde{a}_\alpha \tilde{a}_\beta \tilde{K}_{11\alpha\beta} +$$

$$+2\sum_{\alpha=k+1}^{k+m}\sum_{\beta=k+m+1}^{k+m+n}\tilde{a}_\alpha\,\tilde{a}_\beta\,\tilde{K}_{12\alpha\beta}+\sum_{\alpha=k+m+1}^{k+m+n}\sum_{\beta=k+m+1}^{k+m+n}\tilde{a}_\alpha\tilde{a}_\beta\tilde{K}_{22\alpha\beta}+$$

$$+2\sum_{\alpha=1}^{k}\sum_{\beta=k+1}^{k+m}\tilde{a}_\alpha\,\tilde{a}_\beta\,\tilde{K}_{1b\alpha\beta}+2\sum_{\alpha=1}^{k}\sum_{\beta=k+m+1}^{k+m+n}\tilde{a}_\alpha\,\tilde{a}_\beta\,\tilde{K}_{2b\alpha\beta}-$$

$$-2\sum_{\alpha=1}^{k}\tilde{K}\,\tilde{a}_\alpha\,\tilde{q}_\alpha\ ,$$

where we have denoted

$$\begin{aligned}
\tilde{K}_{oo\alpha\beta} &= \frac{h^3}{12}\int_\Omega \tilde{K}_{ijkl}\,\Phi_{\alpha,ij}\,\Phi_{\beta,kl}\,d\Omega\ ,\\
\tilde{K}_{bb\alpha\beta} &= h\int_\Omega \tilde{K}_{ijkl}\,b_{ij}\,b_{kl}\,\Phi_\alpha\,\Phi_\beta\,d\Omega\ ,\\
\tilde{K}_{11\alpha\beta} &= h\int_\Omega \tilde{K}_{1j1l}\,\Psi_{\alpha-k,j}\,\Psi_{\beta-k,l}\,d\Omega\ ,\\
\tilde{K}_{22\alpha\beta} &= h\int_\Omega \tilde{K}_{2j2l}\,\kappa_{\alpha-k-m,j}\,\kappa_{\beta-k-m,l}\,d\Omega\ ,\\
\tilde{K}_{1b\alpha\beta} &= h\int_\Omega \tilde{K}_{1jkl}\,b_{kl}\,\Phi_\alpha\,\Psi_{\beta-k,j}\,d\Omega\ ,\\
\tilde{K}_{2b\alpha\beta} &= h\int_\Omega \tilde{K}_{2jkl}\,b_{kl}\,\Phi_\alpha\,\kappa_{\beta-k-m,j}\,d\Omega\ ,\\
\tilde{q}_\alpha &= \int_\Omega \tilde{q}\,\Phi_\alpha\,d\Omega\ ,\\
\tilde{K}_{12\alpha\beta} &= h\int_\Omega \tilde{K}_{1j2l}\,\Psi_{\alpha-k,j}\,\kappa_{\beta-k-m,l}\,d\Omega\ .
\end{aligned}\tag{4.3}$$

Then the minimum of $\tilde{\Pi}(\tilde{a}_\alpha)$ is determined by the equations

$$\frac{\partial\,\tilde{\Pi}(a_\alpha)}{\partial\tilde{a}_\alpha}=0\ ,\qquad (\alpha=1,2,\dots,k+m+n)\ ,\tag{4.4}$$

which gives

$$\frac{\partial \tilde{\Pi}}{\partial \tilde{a}_\beta} = \sum_{\alpha=1}^{k} \tilde{a}_\alpha \left(\tilde{K}_{oo\alpha\beta} + \tilde{K}_{bb\alpha\beta}\right) + \sum_{\alpha=k+1}^{k+m} \tilde{a}_\alpha \tilde{K}_{1b\alpha\beta} + \sum_{\alpha=k+m+1}^{k+m+n} \tilde{a}_\alpha \tilde{K}_{2b\alpha\beta} - \tilde{K}\,\tilde{q}_\beta = 0 \tag{4.5}$$

for $\beta \le k$,

$$\frac{\partial \tilde{\Pi}}{\partial \tilde{a}_\alpha} = \sum_{\alpha=1}^{k} \tilde{a}_\alpha \tilde{K}_{1b\alpha\beta} + \sum_{\alpha=k+1}^{k+m} \tilde{a}_\alpha \tilde{K}_{11\alpha\beta} + \sum_{\alpha=k+m+1}^{m+k+n} \tilde{a}_\alpha \tilde{K}_{12\alpha\beta} = 0 \tag{4.6}$$

for $k < \beta \le k+m$ and

$$\frac{\partial \tilde{\Pi}}{\partial \tilde{a}_\alpha} = \sum_{\alpha=1}^{k} \tilde{a}_\alpha \tilde{K}_{2b\alpha\beta} + \sum_{\alpha=k+1}^{k+m} \tilde{a}_\alpha \tilde{K}_{12\alpha\beta} + \sum_{\alpha=k+m+1}^{k+m+n} \tilde{a}_\alpha \tilde{K}_{22\alpha\beta} = 0 \tag{4.7}$$

for $k+m < \beta \le k+m+n$.

Equations (4.5-7) completely determine the coefficients $\tilde{a}_\alpha$ of the approximate solution. The whole system can be written in the form

$$\tilde{K}_{\alpha\beta}\,\tilde{a}_\beta = \tilde{K}\,\tilde{q}_\alpha \quad . \tag{4.8}$$

The solution of this system is given by the formula

$$\tilde{a}_\alpha\,(p) = \sum_{\beta=1}^{k} \frac{\tilde{K}\,\tilde{F}_{\beta\alpha}\,\tilde{q}_\beta}{|\tilde{K}_{\alpha\beta}|} \quad . \tag{4.9}$$

As $\tilde{K}_{ijkl}$ are polynomials in p of degree r

$$\tilde{K}_{\alpha\beta} = \sum_{\nu=0}^{r} K_{\alpha\beta}^{(\nu)}\,p^\nu \tag{4.10}$$

is a p - matrix, $|\tilde{K}_{\alpha\beta}|$ - the determinat is a polynomial in p of the degree $r(k+m+n)$ and $\tilde{F}_{\alpha\beta}$ - the adjoint matrix is a polynomial in p of the degree $r(k+m+n-1)$. As $\tilde{F}_{\beta\alpha}/|\tilde{K}_{\alpha\beta}|$ is a rational function of the transform parameter p, the inverse transform can be achieved by the method of decomposition into partial fractions.

We denote by $-p_i$ $(i = 1,2,\ldots,s)$ the distinct roots of the determinantal equation

$$\Delta(p) = |\tilde{K}_{\alpha\beta}| = 0 \tag{4.11}$$

with multiplicity α_i, where

$$\sum_{i=1}^{s} \alpha_i = (k+m+n)r \tag{4.12}$$

and r being the order of the operator K_{ijkl}.

From the physical properties of the problem we can assume that the matrix $K_{\alpha\beta}$ is a simple λ-matrix; then it holds[12]

$$\frac{\tilde{F}_{\alpha\beta}}{\Delta(p)} = \sum_{i=1}^{s} \frac{A_{\alpha\beta}(p_i)}{p + p_i} , \qquad (p \neq -p_i) , \tag{4.13}$$

where

$$A_{\alpha\beta}(p_i) = \frac{\alpha_i \tilde{F}_{\alpha\beta}^{(\alpha_i-1)}(-p_i)}{\Delta^{(\alpha_i)}(-p_i)} . \tag{4.14}$$

Hence we have

$$\tilde{a}_\alpha = \sum_{\beta=1}^{k} \sum_{i=1}^{s} \frac{A_{\alpha\beta}(p_i)}{p + p_i} \tilde{K} \tilde{q}_\beta . \tag{4.15}$$

Now applying the theorem on convolutional product we arrive at

$$a_\alpha(t) = \sum_{i=1}^{s} \sum_{\beta=1}^{k} \int_o^t A_{\alpha\beta}(p_i)(K\, q, \Phi_\beta)\, e^{-p_i(t-\tau)}\, d\tau \, , \qquad (4.16)$$

where

$$(K\, q, \Phi_\beta) = \int_\Omega K\, q\, \Phi_\beta\, d\Omega \quad . \qquad (4.17)$$

Finally, the approximate solution assumes the form

$$w_k = \sum_{\alpha=1}^{k} \sum_{\beta=1}^{k} \sum_{i=1}^{k} \Phi_\alpha \int_o^t A_{\alpha\beta}(p_i)(K\, q, \Phi_\beta)\, e^{-p_i(t-\tau)}\, d\tau \, ,$$

$$(4.18)$$

$$u_{1m} = \sum_{\alpha=k+1}^{k+m} \sum_{\beta=1}^{k} \sum_{i=1}^{s} \Psi_{\alpha-k} \int_o^t A_{\alpha\beta}(p_i)(K\, q, \Phi_\beta) e^{-p_i(t-\tau)}\, d\tau \, ,$$

$$u_{2m} = \sum_{\alpha=k+m+1}^{k+m+n} \sum_{\beta=1}^{k} \sum_{i=1}^{s} \kappa_{\alpha-k-m} \int_o^t A_{\alpha\beta}(p_i)(K\, q, \Phi_\beta) e^{-p_i(t-\tau)}\, d\tau \, .$$

In the case of a shell of Voigt material it is easy to prove that p_i are real and positive. From the positive definiteness of the viscoelastic operator it follows that also in a general case the roots of the determinantal equation are negative and thus p_i are positive. Further from the exact solution for plates, it is obvious, that in the case of the general anisotropy or orthotropy the roots of the determinantal equation are simple[13].

We have shown that in the case of quasistatic problems we can solve the associated elastic problems for discrete values of the transform parameter p and in combination with a convenient approximate discrete inverse transform we obtain the solution of the considered viscoelastic problems, without the determination of the roots of the determinantal equation.[7-8]

5. DYNAMIC PROBLEMS OF VISCOELASTIC SHELLS

When considering the associated elastic problems for viscoelastic shells, we can formulate the principle of minimum of generalized potential energy and apply generalized variational methods also to dynamic problems of viscoelastic shells.

Consider the governing equations for dynamic problems of viscoelastic shells

$$K_{ijkl}(D)\left[\frac{h^3}{12} w_{,ijkl} + h(u_{i,j} + b_{ij} w) b_{kl}\right] +$$

$$+ \rho K(D) \frac{\partial^2 w}{\partial t^2} = K(D) q \quad , \tag{5.1}$$

$$K_{ijkl}(D)(u_{k,jl} + b_{kl} w_{,j}) = 0 \quad , \quad (i = 1,2)$$

for materials of differential type with the boundary conditions (3.4) or (3.5) and initial conditions

$$\frac{\partial^n w}{\partial t^n} = w_n(x) \quad , \quad (n = 0,1,\ldots,r+1) \quad , \tag{5.2}$$

$$\frac{\partial^k u_i}{\partial t^k} = u_{ik} \quad , \quad (k = 0,1,\ldots,s-1) \quad .$$

Simultaneously we shall consider the integrodifferential equations

$$\int_0^t G_{ijkl}(t-\tau) \frac{\partial}{\partial \tau} \left[\frac{h^3}{12} w_{,ijkl} + h(u_{i,j} + b_{ij}w)b_{kl}\right] d\tau +$$

$$+ \rho \frac{\partial^2 w}{\partial t^2} = q \quad , \tag{5.3}$$

$$\int_0^t G_{ijkl}(t-\tau) \frac{\partial}{\partial \tau} (u_{k,jl} + b_{kl} w_{,j}) d\tau = 0 \quad , \quad (i=1,2)$$

with boundary conditions (3.4) or (3.5) and initial conditions

$$\frac{\partial^n w}{\partial t^n} = w_n \quad , \quad u_i(x,0) = u_{io} \quad , \quad (n = 0,1) \quad , \tag{5.4}$$

for Boltzmann's materials.

Applying ordinary Laplace transform to (5.1) and (5.3) we arrive at

$$\tilde{K}_{ijkl} \left[\frac{h^3}{12} \tilde{w}_{,ijkl} + h(\tilde{u}_{i,j} + b_{ij}\tilde{w})b_{kl}\right] + \rho \tilde{K} p^2 \tilde{w} = \tilde{K} \tilde{q} +$$

$$\sum_{\nu=0}^{s} K^{\nu}_{ijkl} \sum_{\rho=0}^{\nu-1} \left[\frac{h^3}{12} w_{\rho,ijkl} + h(u_{i\rho,j} + b_{ij}w_\rho)b_{kl}\right] p^{\nu-\rho-1} +$$

$$+ \sum_{\mu=0}^{r} \rho K_\mu \sum_{\rho=0}^{\mu+1} p^{\mu-\rho+1} w_\rho - \sum_{\mu=0}^{r} K_\mu \sum_{\rho=0}^{\mu-1} p^{\mu-\rho-1} q_\rho = \tilde{f}_o \quad , \tag{5.5}$$

$$\tilde{K}_{ijkl}(\tilde{u}_{k,jl} + b_{kl}\tilde{w}_{,j}) = \sum_{\nu=1}^{s} K^{\nu}_{ijkl} \sum_{\rho=0}^{\nu-1} (u_{k\rho,jl} + b_{kl} w_{\rho,j}) \cdot p^{\nu-\rho-1} = \tilde{f}_i$$

where $q_\rho = q^{(\rho)}(0)$ and

$$p\,\tilde{G}_{ijkl}(p)\left[\frac{h^3}{12}\tilde{w}_{,ijkl} + h(\tilde{u}_{i,j} + b_{ij}\tilde{w})b_{kl}\right] + \rho\, p^2\,\tilde{w} = \tilde{q} + \tilde{G}_{ijkl}(p)\left[\frac{h^3}{12} w_{o,ijkl} + h(u_{oi,j} + b_{ij} w_o)b_{kl}\right] + \rho(p\,w_o + w_1) = \tilde{f}_o\,, \tag{5.6}$$

$$p\,\tilde{G}_{ijkl}(p)(\tilde{u}_{k,jl} + b_{kl}\tilde{w}_{,j}) = \tilde{G}_{ijkl}(p)(u_{ok,jl} + b_{kl}w_{o,j}) = \tilde{f}_i\,.$$

Then the corresponding functional of the generalized potential energy assumes the form

$$2\tilde{K}\,\tilde{\Pi}(\tilde{w},\tilde{u}) = \int_\Omega \{\tilde{K}_{ijkl}\left[\frac{h^3}{12}\tilde{w}_{,ij}\tilde{w}_{,kl} + h(\tilde{u}_{i,j} + b_{ij}\tilde{w})(\tilde{u}_{k,l} + b_{kl}\tilde{w})\right] + \rho\tilde{K}\,p^2\tilde{w}^2 - 2(\tilde{f}_o\tilde{w} + \tilde{f}_i\,\tilde{u}_i + \tilde{f}_2\tilde{u}_2)\}d\Omega \tag{5.7}$$

or

$$2\tilde{K}\,\tilde{\Pi}(\tilde{w},\tilde{u}) = \int_\Omega \{p\,\tilde{G}_{ijkl}(p)\left[\frac{h^3}{12}\tilde{w}_{,ij}\tilde{w}_{,kl} + h(\tilde{u}_{i,j} + b_{ij}\tilde{w})(\tilde{u}_{k,l} + b_{kl}\tilde{w})\right] + \rho p^2\tilde{w}^2 - 2(\tilde{f}_o\tilde{w} + \tilde{f}_1\tilde{u}_1 + \tilde{f}_2\tilde{u}_2)\}d\Omega\,. \tag{5.8}$$

Thus we can formulate the generalized variational principles of minimum of the generalized potential energy for dynamic problems of viscoelstic shallow shells.

The solutions $\tilde{w}$, $\tilde{u}_1$, $\tilde{u}_2$ of the systems (5.1),(5.3) for q less then critical minimizes, for real positives values of the transform parameter p, the functionals of the generalized potential energy (5.7), (5.8), respectively, and inversely the functions $\tilde{w}$, $\tilde{u}_1$, $\tilde{u}_2$, which minimizes (5.7), (5.8) give the solutions of (5.1), (5.3), respectively.

The approximate solution of the problem can be obtained by a generalized variational method similar as in the quasi-static case.[14]

The roots of the determinantal equation are approximate values of the generalized eigenvalues of the problem. The real parts of eigenvalues then determine the damping factors and the imaginary parts determine the frequencies.[14]

6. CONVOLUTIONAL VARIATIONAL PRINCIPLES AND METHODS

Applying the theorem on the Laplace transform of the convolution to the functional (3.15) one obtains

$$
\begin{aligned}
2\, K\left(\frac{\partial}{\partial t}\right)\Pi(w,u) = \int_0^t \int_\Omega \Big\{ & \frac{h^3}{12}\, w_{,ij}(t-\tau)\, K_{ijkl}\left(\frac{\partial}{\partial \tau}\right) w_{,kl}(\tau) + \\
& + h\Big[u_{i,j}(t-\tau) + b_{ij}\, w(t-\tau)\Big] K_{ijkl}\left(\frac{\partial}{\partial \tau}\right)\Big[u_{k,l}(\tau) + \\
& + b_{kl}\, w(\tau)\Big] - 2\, w(t-\tau)\, K\left(\frac{\partial}{\partial \tau}\right) q(\tau) \Big\}\, d\Omega\, d\tau .
\end{aligned}
\qquad (6.1)
$$

Now we can formulate the convolutional variational principle for viscoelastic shallow shells.

The first variational $\delta_1 K \Pi$ of the functional of the convolutional potential energy of a viscoelastic shallow shell (6.1) vanishes (i.e., the functional is stationary) if and only if differential equations (3.1), boundary conditions (3.4) or (3.5) and initial conditions (3.6) are satisfied and inversely the functions w, u_1, u_2 which make the functional (6.1) stationary give the solution of the governing equations of a shallow viscoelastic shell (3.1).

Convolutional variational principles for linear viscoelastic bodies have been derived in a different way by Gurtin[1]. The present derivation shows their close connection with the governing equations of viscoelastic shallow shells. Generalized and convolutional principles do not have a physical meaning.

Similarly as in the case of generalized variational principles, we can make use of convolutional variational principles for an approximate solution of differential equations of viscoelastic shallow shells.

Choosing the base functions as in the case of the generalized finite element method, the approximate solution w, u_1, u_2 can be sought in the form

$$w_k(t) = \sum_{\alpha=1}^{k} a_\alpha(t)\,\Phi_\alpha \quad ,$$

$$u_{1m}(t) = \sum_{\alpha=k+1}^{k+m} a_\alpha(t)\,\Psi_{\alpha-k} \quad , \tag{6.2}$$

$$u_{2n}(t) = \sum_{\alpha=k+m+1}^{k+m+n} a_\alpha(t)\kappa_{\alpha-k-m} \cdot$$

Inserting (6.2) into (6.1) we arrive at

$$2\,K\,\Pi_{kmn}(w,\,u_1,\,u_2) = 2\,K\,\Pi_{kmn}(a_\alpha(t)) =$$

$$= \sum_{\nu=0}^{s} \{ \sum_{\alpha=1}^{k} \sum_{\beta=1}^{k} (K^\nu_{oo\alpha\beta} + K^\nu_{bb\alpha\beta}) \int_0^t a_\alpha^{(\nu)}(\tau)a_\beta(t-\tau)d\tau +$$

$$+ \sum_{\alpha=k+1}^{k+m} \sum_{\beta=k+1}^{k+m} K^\nu_{11\alpha\beta} \int_0^t a_\alpha^{(\nu)}(\tau)a_\beta(t-\tau)d\tau +$$

$$+ 2 \sum_{\alpha=k+1}^{k+m} \sum_{\beta=k+m+1}^{k+m+n} K^\nu_{12\alpha\beta} \int_0^t a_\alpha^{(\nu)}(\tau)a_\beta(t-\tau)d\tau +$$

$$+ \sum_{\alpha=k+m+1}^{k+m+n} \sum_{\beta=k+m+1}^{k+m+n} K^\nu_{22\alpha\beta} \int_0^t a_\alpha^{(\nu)}(\tau)a_\beta(t-\tau)d\tau + \qquad (6.3)$$

$$+ 2 \sum_{\alpha=1}^{k} \sum_{\beta=k+1}^{k+m} K^\nu_{1b\alpha\beta} \int_0^t a_\alpha^{(\nu)}(\tau)a_\beta(t-\tau)d\tau +$$

$$+ 2 \sum_{\alpha=1}^{k} \sum_{\beta=k+m+1}^{k+m+n} K_{2b\alpha\beta} \int_0^t a_\alpha^{\nu}(\tau)a_\beta(t-\tau)d\tau -$$

$$- 2 \sum_{\alpha=1}^{k} K^\nu \int_0^t q^{(\nu)}(\tau)\,a_\alpha(t-\tau)d\tau ,$$

where $K^\nu_{KL\alpha\beta}$ are given by formulae (4.3) after inserting K^ν_{ijkl} for $\tilde{K}_{ijkl}$, K^ν_{ijkl} beeing the coefficient at $\dfrac{\partial^\nu}{\partial t^\nu}$.

By the usual variational procedure we arrive at

$$\sum_{\nu=0}^{s} \{ \sum_{\alpha=1}^{k} (K^\nu_{oo\alpha\beta} + K^\nu_{bb\alpha\beta}) a_\alpha^{(\nu)}(t) +$$

$$+\sum_{\alpha=k+1}^{k+m} K^{\nu}_{1b\alpha\beta}\, a^{(\nu)}_{\alpha}(t) + \sum_{\alpha=k+m+1}^{k+m+n} K^{\nu}_{2b\alpha\beta}\, a^{(\nu)}_{\alpha}(t)\} - \tag{6.4}$$

$$- K(D) q_{\beta} = 0$$

for $\beta \leq k$,

$$\sum_{\nu=0}^{s} \{ \sum_{\alpha=1}^{k} K^{\nu}_{1b\alpha\beta}\, a^{(\nu)}_{\beta}(t) + \sum_{\alpha=k+1}^{k+m} K^{\nu}_{11\alpha\beta}\, a^{(\nu)}_{\alpha}(t) + \sum_{\alpha=k+m+1}^{k+m+n} K^{\nu}_{12\alpha\beta}\, a^{(\nu)}_{\alpha}(t)\} = 0 \tag{6.5}$$

for $k < \beta \leq k+m$ and

$$\sum_{\nu=0}^{s} \{ \sum_{\alpha=1}^{k} K^{\nu}_{2b\alpha\beta}\, a^{(\nu)}_{\alpha}(t) + \sum_{\alpha=k+1}^{k+m} K^{\nu}_{12\alpha\beta}\, a^{(\nu)}_{\alpha}(t) + \sum_{\alpha=k+m+1}^{k+m+n} K^{\nu}_{22\alpha\beta}\, a^{(\nu)}_{\alpha}(t)\} = 0 \tag{6.6}$$

for $k+m < \beta \leq k+m+n$.

Equations (6.4-6) constitute a full system of ordinary differential equations for determination of $a_{\alpha}(t)$.

The solution has to fulfil also the initial conditions

$$w^{(\rho)}_{\alpha}(o) = \int_{\Omega} w_{\rho}\, \Phi_{\alpha}\, d\Omega ,$$

$$u^{(\rho)}_{1\alpha}(o) = \int_{\Omega} u_{1\rho}\, \Psi_{\alpha}\, d\Omega , \tag{6.7}$$

$$u^{(\rho)}_{2\alpha}(o) = \int_{\Omega} u_{2\rho}\, \kappa_{\alpha}\, d\Omega ,$$

Thus we have arrived at a Cauchy problem for a system of

ordinary differential equation, which has a unique solution. Inserting this solution into (6.1) we get the approximate solution of the problem considered.

Similarly applying the theorem on Laplace transform of the convolution to the functional of the generalized potential energy for Boltzmann's material we obtain

$$2\,\Pi(w,u) = \int_{\Omega} \{ \frac{h^3}{12} \int_0^t \int_0^{t-\tau} G_{ijkl}\,(t-\tau-\sigma)\; \times$$

$$\times \frac{\partial}{\partial\sigma} w_{,ij}\,(\sigma) w_{,kl}(\tau) d\sigma\, d\tau \; +$$

$$+ h \int_0^t \int_0^{t-\tau} G_{ijkl}(t-\tau-\sigma)\, \frac{\partial}{\partial\sigma}\, (u_{i,j}(\sigma) + b_{ij}\, w(\sigma))\; u_{k,l}(\tau) \; +$$

$$+ b_{kl} w(\tau)\; d\sigma\, d\tau \; - \; 2 \int_0^t q(t-\tau) w(\tau) d\tau \}\; d\Omega \quad .$$

This functional is stationary at the point of the solution of (3.2). The solution of the problem can be sought in the form (6.1). Then by the usual variation procedure we obtain a sufficient system of integral equations of Volterra type for the determination of unknown functions $a_\alpha(t)$.

7. CONVOLUTIONAL VARIATIONAL PRINCIPLES OF STABILITY OF VISCOELASTIC SHALLOW SHELLS

We shall consider large deflections of viscoelastic shallow shells. The governing equations of large deflection theory of shallow viscoelastic shells are

$$\frac{h^3}{12} H_{ijk\ell} w_{,ijk\ell} = q + \epsilon_{ik} \epsilon_{j\ell} hF_{,k\ell}(w_{,ij} - b_{ij}) \; ,$$
$$C_{ijk\ell} F_{,ijk\ell} = \epsilon_{ik} \epsilon_{j\ell}(b_{ij} w_{,k\ell} - \frac{1}{2} w_{,ij} w_{,k\ell}) \; , \qquad (7.1)$$

where

$$C_{ijk\ell} = \epsilon_{im} \epsilon_{jn} \epsilon_{kr} \epsilon_{\ell s} B_{mnrs} \qquad (7.2)$$

and

$$(B_{mnrs})^{-1} = H_{ijk\ell} \; . \qquad (7.3)$$

We shall deal with the linearized stability theory. Then we put

$$- N_{ij} = - \lambda N^{c}_{ij} = \lambda \epsilon_{ik} \epsilon_{j\ell} h\overset{o}{F}_{,k\ell} \qquad (7.4)$$

at the nonlinear term in $(7.1)_1$ and neglect the nonlinear term in $(7.2)_2$. Thus we get

$$\frac{h^3}{12} H_{ijk\ell} w_{,ijk\ell} = q - \epsilon_{ik} \epsilon_{j\ell} hF_{,k\ell} b_{ij} - \lambda \overset{o}{N}_{ij} w_{,ij} \; , \qquad (7.5)$$

$$C_{ijk\ell} F_{,ijk\ell} = \epsilon_{ik} \epsilon_{j\ell} b_{ij} w_{,k\ell} \; . \qquad (7.6)$$

We shall consider the boundary conditions

$$w = w_{,n} = 0 \; , \qquad (7.7)$$

or

$$w = 0 \; , \quad H_{ijk\ell} w_{,ij} \nu^n_k \nu^n_\ell = 0 \qquad (7.8)$$

and

$$F_{,nn} = f_1(s) \; , \quad F_{,ns} = f_2(s) \; , \qquad (7.9)$$

Simultaneously we consider homogeneous initial conditions.

Assuming that N^o_{ij} is time independent we apply the Laplace transform to (7.5-6).

Thus

$$\frac{h^3}{12}\tilde{H}_{ijk\ell}(p)\tilde{w}_{,ijk\ell} = \tilde{q} - \epsilon_{ik}\epsilon_{j\ell}b_{ij}h\tilde{F}_{,k\ell} - \lambda N^o_{ij}\tilde{w}_{,ij} ,$$

$$\tilde{C}_{ijk\ell}(p)\tilde{F}_{,ijk\ell} = \epsilon_{ik}\epsilon_{j\ell}b_{ij}\tilde{w}_{,k\ell} . \qquad (7.10)$$

Thus we have arrived to an associated elastic problem and can find the corresponding generalized potential energy.

When deriving the functional for the potential energy of the associated elastic shell, we consider the same scheme of linearization and we arrive at

$$2\tilde{\Pi}(\tilde{w},\tilde{F}) = \int_\Omega \{ \tilde{H}_{ijk\ell}\tilde{w}_{,ij}\tilde{w}_{,k\ell} - h\tilde{C}_{ijk\ell}\tilde{F}_{,ij}\tilde{F}_{,k\ell} + $$
$$+ 2\epsilon_{ik}\epsilon_{j\ell}b_{ij}h\tilde{F}_{,k\ell}\tilde{w} - \lambda N^o_{ij}\tilde{w}_{,i}\tilde{w}_{,j} - 2\tilde{q}\tilde{w}\}d\Omega . \qquad (7.11)$$

We shall deal with stability problems of shallow shells loaded at the boundary. Thus we put $q = 0$ and rewrite (7.11) in the form

$$2\tilde{\Pi}(\tilde{w},\tilde{F}) = \int_\Omega \{\frac{h^3}{12}\tilde{H}_{ijk\ell}\tilde{w}_{,ij}\tilde{w}_{,k\ell} -$$
$$- h\tilde{F}_{,k\ell}(\tilde{C}_{ijk\ell}\tilde{F}_{,ij} - \epsilon_{ik}\epsilon_{j\ell}b_{ij}\tilde{w}) + \qquad (7.12)$$
$$+ \epsilon_{ik}\epsilon_{j\ell}b_{ij}h\tilde{F}_{,k\ell}\tilde{w} - \lambda N^o_{ij}\tilde{w}_{,i}\tilde{w}_{,j}\}d\Omega ,$$

where the second term expresses the condition given by $(7.10)_2$, and $\tilde{F}_{,k\ell}$ is the Lagrange multiplier.

Thus we have the following stability principle:

The solution $\tilde{w}$, $\tilde{F}$ of the system of Laplace transforms of differential equations of viscoelastic shallow shells (7.10) for buckling problems minimizes the functional of the generalized potential energy of the shell (7.12).

Now applying the theorem on the inverse Laplace transform of a convolutional product we get

$$2\Pi(w,F) = \int_0^t \int_\Omega \{ \frac{h^3}{12} w,_{ij}(t-\tau) H_{ijk\ell}\, w,_{k\ell}(\tau) -$$
$$- hF,_{k\ell}(t-\tau)[C_{ijk\ell}\, F,_{ij}(\tau) - \epsilon_{ik}\,\epsilon_{j\ell}\, b_{ij}\, w(\tau)] \qquad (7.13)$$
$$+ \epsilon_{ik}\,\epsilon_{j\ell}\, b_{ij}\, hF,_{k\ell}(t-\tau) w(\tau) - \lambda N^o_{ij}\, w,_i(t-\tau) w,_k(\tau)\} d\Omega\, d\tau .$$

We then have the following principle:

The solution w, F of the system of differential equations of viscoelastic shallow shells for buckling problems makes the functional of the convolutional potential energy of the shell stationary.

As we shall show in an example, in the case of viscoelastic shallow shells we have finite and infinite critical times and corresponding critical values of the loading.

In order to analyse these critical values we confine ourselves to constitutive equations of the differential type.

Then we have

$$\frac{h^3}{12} K_{ijk\ell}(D) w,_{ijk\ell} = -\epsilon_{ik}\,\epsilon_{j\ell}\, b_{ij}\, h\, K(D) F,_{k\ell} -$$
$$- \lambda N^o_{ij}\, K(D) w,_{ij} , \qquad (7.14)$$

$$\epsilon_{im}\,\epsilon_{jn}\,\epsilon_{kr}\,\epsilon_{\ell s}\, L_{mnrs}(D) F,_{ijk\ell} = \epsilon_{ik}\,\epsilon_{j\ell}\, b_{ij}\, L(D) w,_{k\ell}$$

and the generalized potential energy assumes the form

$$\tilde{K}(p)\tilde{L}(p)\,\tilde{\Pi}(\tilde{w},\tilde{F}) = \int_{\Omega}\{\frac{h^3}{12}\tilde{L}(p)\tilde{K}_{ijk\ell}(p)\tilde{w}_{,ij}\,\tilde{w}_{,k\ell} -$$

$$- h\tilde{F}_{,k\ell}\Big[\tilde{K}(p)\epsilon_{im}\,\epsilon_{jn}\,\epsilon_{kr}\,\epsilon_{\ell s}\,L_{mnrs}(p)\,\tilde{F}_{,ij} - \tag{7.15}$$

$$- \epsilon_{ik}\,\epsilon_{j\ell}\,b_{ij}\,\tilde{K}(p)\tilde{L}(p)\tilde{w}\Big] + \epsilon_{ik}\,\epsilon_{j\ell}\,h\tilde{K}(p)\tilde{L}(p)\tilde{F}_{,k\ell}\,b_{ij}\,\tilde{w} -$$

$$- \lambda\, N^o_{ij}\,\tilde{K}(p)\tilde{L}(p)\tilde{w}_{,i}\,\tilde{w}_{,j}\}d\Omega\,.$$

Assuming that the operators of the constitutive equations are of the same order and applying the Tauber theorem on limit values of the Laplace transform we arrive at

$$\lim_{p\to 0}\sum_{\alpha=0}^{r} p^{\alpha}K^{\alpha}_{ijk\ell}\,p\,\tilde{w}_{,ijk\ell} = -\lim_{p\to 0}\sum_{\alpha=0}^{r} p^{\alpha}K_{\alpha}\big(\epsilon_{ik}\epsilon_{j\ell}b_{ij}\,h\,p\,F_{,k\ell}$$

$$+ \lambda\,N^{(o)}_{ij}\,p\,\tilde{w}_{,ij}\big)\quad, \tag{7.16}$$

$$\lim_{p\to 0}\sum_{\alpha=0}^{s} p^{\alpha}\,\epsilon_{im}\,\epsilon_{jn}\,\epsilon_{kr}\,\epsilon_{\ell s}\,L^{\alpha}_{mnrs}\,p\,\tilde{F}_{,ijk\ell} =$$

$$\lim_{p\to 0}\sum_{\alpha=0}^{s} p^{\alpha}L_{\alpha}\,\epsilon_{ik}\,\epsilon_{j\ell}\,b_{ij}\,p\,\tilde{w}_{,k\ell}\quad,$$

what according to the Tauber theorem gives for $t=\infty$

$$K^o_{ijk\ell}\,w_{,ijk\ell}(\infty) = -\epsilon_{ik}\,\epsilon_{j\ell}\,b_{ij}\,K^o h\,F_{,k\ell}(\infty) -$$

$$- \lambda N^o_{ij}\,K^o w_{,ij}(\infty)\quad, \tag{7.17}$$

$$\epsilon_{im}\,\epsilon_{jn}\,\epsilon_{kr}\,\epsilon_{\ell s}\,L^o_{mnrs}\,F_{,ijk\ell}(\infty) = \epsilon_{ik}\,\epsilon_{j\ell}\,b_{ij}L^o w_{,k\ell}(\infty)\,.$$

Similarly we obtain for $t=0$

$$K^r_{ijk\ell}\, w_{,ijk\ell}(0) = -\epsilon_{ik}\,\epsilon_{j\ell}\, b_{ij}\, K^r\, h\, F_{,k\ell}(0) -$$
$$-\lambda N^o_{ij}\, K^r\, w_{,ij}(0) \quad , \qquad (7.18)$$

$$\epsilon_{im}\,\epsilon_{jn}\,\epsilon_{kr}\,\epsilon_{\ell s}\, L^s_{mnrs}\, F_{,ijk\ell}(0) = \epsilon_{ik}\,\epsilon_{j\ell}\, b_{ij} L^s w_{,k\ell}(0) .$$

Critical values for $t = 0$ and $t = \infty$ can be determined directly from (7.17), (7.18), respectively.

We shall consider the stability of an orthotropic viscoelastic cylinder uniformly compressed in the axial direction. Then buckling is symmetrical with respect to the axis of the cylinder and may occur at a certain value of the compressive stress.

We consider a circular cylindrical shell with radius R and thickness h . x_1 denotes the axial coordinate, x_2 the coordinate in the circumferential direction and x_3 in the direction of the outer normal to the surface of the shell.

Then

$$b_{11} = 0 \ , \quad b_{22} = \frac{1}{R} \qquad (7.19)$$

and

$$N_{11} = h\sigma \quad , \quad N_{12} = N_{22} = 0 \quad , \qquad (7.20)$$

The differential equations of the shell assume the form

$$\frac{h}{12} H_{1111}\, w_{,1111} + \frac{1}{R} h\, F_{,11} + h\sigma\, w_{,11} = 0 \quad ,$$
$$C_{1111}\, F_{,1111} - \frac{1}{R} w_{,11} = 0 \ . \qquad (7.21)$$

We assume w in the form

$$w = a(t) \sin \lambda x \quad , \tag{7.22}$$

where $\lambda = m\pi/\ell$, ℓ - beeing the length of the cylinder. Then we get from (7.20),

$$C_{1111} \; F_{,1111} = \frac{1}{R\lambda^2} a(t) \sin \lambda x \quad . \tag{7.23}$$

Inserting (7.22) and (7.23) into (7.15) we arrive at

$$\tilde{\Pi} = \pi R \ell \{ \frac{h^3}{12} \tilde{H}_{1111} \lambda^4 + \frac{h}{R^2 \tilde{C}_{1111}} - h\sigma \lambda^2 \} \tilde{a}^2(p) . \tag{7.24}$$

and thus

$$\delta_1 \tilde{\Pi} = 2\pi R \ell \{ \frac{h^3}{12} \tilde{H}_{1111} \lambda^4 + \frac{1}{R^2 \tilde{C}_{1111}} - h\sigma \lambda^2 \} \tilde{a}(p) = 0 \; . \tag{7.25}$$

In the special case of a shell of Zener material with a homogeneous spectrum, the constitutive equations of which are

$$\left(\frac{\partial}{\partial t} + k_1\right) \sigma_{ij} = \left(k \frac{\partial}{\partial t} + k_1\right) E_{ijk\ell} \; \varepsilon_{k\ell} \quad ,$$

$$\left(k \frac{\partial}{\partial t} + k_1\right) \varepsilon_{ij} = \left(\frac{\partial}{\partial t} + k_1\right) B_{ijk\ell} \; \sigma_{k\ell} \tag{7.26}$$

with $E^{-1} = B$ and $C_{1111} = B_{2222}$.

Inserting (7.26) into the inverse transform of (7.25) we arrive at

$$\{ \frac{h^2}{12} E_{1111} \lambda^4 \left(k \frac{\partial}{\partial t} + k_1\right) + \frac{1}{R^2 B_{2222}} \left(k \frac{\partial}{\partial t} + k_1\right) - \sigma \lambda \left(\frac{\partial}{\partial t} + k_1\right) \} a(t) = 0 \tag{7.27}$$

and then for non-zero initial condition we have

$$a(t) = C \exp\{- \frac{\left(\frac{h^2}{12} E_{1111} B_{2222} \lambda^4 + \frac{1}{R^2} - \sigma \lambda^2 B_{2222}\right) k_1}{\left(\frac{h^2}{12} E_{1111} B_{2222} \lambda^4 + \frac{1}{R^2}\right) k - \sigma \lambda^2 B_{2222}} t\} . \tag{7.28}$$

The stability of the cylindrical shell depends on the sign of the exponent. For negative values of the exponent the deflection decreases as time goes on and the shell is stable. For positive values of the exponent the deflection increases with time and the shell becomes unstable for $t=\infty$. The exponent is negative for small values of σ and the numerator changes its sign at

$$\sigma_{cr}^{(1)} = \frac{h^2}{12} E_{1111} \lambda^2 + \frac{1}{R^2 B_{2222}} \frac{1}{\lambda^2} . \tag{7.29}$$

The deflection tends to infinity in a finite time if the denominator of the exponent tends to zero. Thus the critical value σ for a finite (zero) critical time is given by

$$\sigma_{cr}^{(2)} = k\sigma_{cr}^{(1)} . \tag{7.30}$$

As we see, the critical values depend on λ. Assuming that λ changes continuously we get the lowest critical values for

$$\lambda^2 = \frac{2\sqrt{3}}{Rh\sqrt{E_{1111} B_{2222}}} . \tag{7.31}$$

Hence

$$\sigma_{cr}^{(1)} = \frac{h}{\sqrt{3}\,R} \sqrt{\frac{E_{1111}}{B_{2222}}} , \tag{7.32}$$

which in the case of isotropy gives

$$\sigma_{cr}^{(1)} = \frac{Eh}{R} \frac{1}{\sqrt{3(1-\nu^2)}} , \quad \sigma_{cr}^{(2)} = k\, \sigma_{cr}^{(1)} . \tag{7.33}$$

It is obvious that for $\sigma = \sigma_{cr}^{(2)}$ the velocity of the deflection becomes infinite.

The critical value $\sigma_{cr}^{(1)}$ for the infinite critical time corresponds to the critical stress of the elastic element and $\sigma_{cr}^{(2)}$ corresponds to the critical stress of both springs k E .

Linearization can be applied only for the infinitesimal deformation and as the deformation increases above infinitesimal deformations, it is necessary to apply a non-linear analysis. As we can show using a nonlinear analysis, the deflection for the loading greater than $\sigma_{cr}^{(1)}$ tends for $t=\infty$ to a finite value corresponding to the postbuckling deformation of the shell.

In order to find limit values for $t=\infty$ we formally apply the Laplace transform to nonlinear equations for shallow shells of differential type materials. We get

$$\frac{h^3}{12} K_{ijk\ell}(p)\tilde{w}_{,ijk\ell} = \epsilon_{ik}\epsilon_{j\ell} Kh\left(\widetilde{w_{,ij} F_{,k\ell}} - b_{ij}\tilde{F}_{,k\ell}\right),$$

$$\epsilon_{im}\epsilon_{jn}\epsilon_{kr}\epsilon_{\ell s} L_{mnrs}(p)\tilde{F}_{,ijk\ell} = \tag{7.34}$$

$$= \epsilon_{ik}\epsilon_{j\ell} L\left(b_{ij}\tilde{w}_{,k\ell} - \frac{1}{2}\widetilde{w_{,ij} w_{,k\ell}}\right) .$$

Then applying Tauber's theorem for $t=\infty$, we arrive at

$$\frac{h^3}{12} K^o_{ijk\ell} w_{,ijk\ell} = \epsilon_{ik}\epsilon_{j\ell} K^o h\, F_{,k\ell}(w_{,ij} - b_{ij}) , \tag{7.35}$$

$$\epsilon_{im}\epsilon_{jn}\epsilon_{kr}\epsilon_{\ell s} L^o_{mnrs} F_{,ijk\ell} = \epsilon_{ik}\epsilon_{j\ell} w_{,k\ell} L^o\left(b_{ij} - \frac{1}{2} w_{,ij}\right),$$

where it holds

$$\left[\frac{K^{o}_{ijk\ell}}{K^{o}}\right]^{-1} = \frac{L^{o}_{ijk\ell}}{L_{o}} \quad . \tag{7.36}$$

Hence for $t = \infty$ the post-buckling deformation of a viscoelastic shallow shell reaches the magnitude of the post-buckling deformation of an elastic shell with the long-time moduli of elasticity,and is finite. This post-buckling deformation occurs for an infinitesimal initial perturbation and a loading greater than the critical loading with the infinite critical time.

Applying Tauber's theorem for $t = 0$, we get equations similar to (7.35) with $K^{o}_{ijk\ell}$, K^{o}, $L^{o}_{ijk\ell}$ and L^{o} replaced by $K^{r}_{ijk\ell}$, K^{r}, $L^{s}_{ijk\ell}$ and L^{s}, respectively.

Thus the instant buckling loads and post-buckling deformations of viscoelastic shells at $t = 0$, are equal to those loads and deformations of elastic shells with instant moduli. For loads less than the critical load with zero critical time,the deformation of the shell occurs only when initial perturbations are applied. For the load less than the critical one with infinite critical time,the shell is stable. For the load greater then the critical one with infinite critical time,the shell becomes unstable with finite buckles. For loads equal or greater than the critical load with zero critical time,the deformation the bifurcation occurs without initial perturbations. However, it can be shown that similarly as in the elastic case there are post-buckling loads smaller than the classical, initial buckling loads.[15]

REFERENCES

1. Gurtin, M.E., Variational principles in the linear theory of viscoelasticity, *Arch.Rat.Mech.Anal.*,13, 167, 1963.
2. Brilla, J., Convolutional variational principles in plane viscoelasticity, in *Proc.of the Int.Conf.on Variational Methods in Engineering*, Southsmpton University, 1972.
3. Brilla, J., Variational methods in mathematical theory of viscoelasticity, in *Proc.of Czechoslovak Conf. on Diff. Equations*, J.E. Purkyně University, Brno, 1973.
4. Brilla, J., Generalized variational methods in linear viscoelasticity, in *Mechanics of Visco-Elastic Media*, J.Hult, Ed., Springer, Berlin-Heidelberg-New York, 1975.
5. Brilla, J., Convolutional variational methods for viscoelastic shallow shells, in *Proc.of 2nd National Cong. on Theor. and Appl. Mech.*, Varna 8.-14.X.1973, Bulgarian Academy of Sciences, Sofia, 1975.
6. Brilla, J., Finite element analysis of a system of quasiparabolic partial differential equations, *Acta Univ. Carolinae, Mat. et Physics*, 1-2, 5, 1974.
7. Brilla, J., Némethy, A., Analysis of viscoelastic anisotropic plates by finite element method (in Slovak),

Stavebnícky časopis, 22, 3, 1974.

8. Brilla,J.,Lichardus,S.,Némethy,A., The generalization of finite element method for the solution of viscoelastic two-dimensional problems,in *Mechanics of Visco-Elastic Media*, J.Hult,Ed.,Springer, Berlin-Heidelberg-New York, 1975.
9. Brilla,, J., Convolutional variational principles and stability of viscoelastic plates,*Ing.Arch.*, 45,275, 1976.
10. Brilla, J., Linear viscoelastic bending analysis of anisotropic plates, in *Proc. of the XIth Int. Cong. of Appl. Mech., Munich 1964*, Berlin-Heidelberg-New York, Springer, 1966.
11. Carslaw, H.S., Jaeger, J.C., *Operational Methods in Applied Mathematics*, 2nd Edition, Oxford, University Press, 1953.
12. Lancaster, P., *Lambda matrices and vibrating systems*, Pergamon Press, New York-Oxford, 1966.
13. Brilla, J., Viscoelastic bending of anisotropic plates (in Slovak), *Stavebnícky časopis*, 17, 153, 1969.
14. Brilla, J., A generalization of finite element method to dynamic viscoelastic analysis, in *The Mathematics of Finite Elements and Applications II.*, J.R.Whiteman, Ed., Academic Press, London-New York, 1976.
15. Brilla, J., Stability of viscoelastic shallow shells, in *Proc. of the 3rd IUTAM Symposium on Shell Theory*,Tbilisi, August 22-26, 1978, in print.

THE STATIC-GEOMETRIC ANALOGY IN THE EQUATIONS OF THIN SHELL STRUCTURES

by

C.R. Calladine
University of Cambridge

THE STATIC-GEOMETRIC ANALOGY IN THE EQUATIONS OF THIN SHELL STRUCTURES

C.R. Calladine

Abstract

The 'static-geometric analogy' in thin shell structures is a formal correspondence between equilibrium equations on the one hand and geometric compatibility equations on the other. It is well known as a fact, but no satisfactory explanation of its basis has been given. The paper gives an explanation for the analogy, within the framework of shallow-shell theory. The explanation is facilitated by two innovations: (i) separation of the shell surface conceptually into separate stretching *(S)* and bending *(B)* surfaces; (ii) use of *change of Gaussian curvature* as a prime variable. Various limitations of the analogy are pointed out, and a scheme for numerical calculation which embodies the most useful features of the analogy is outlined.

Introduction

There is a curious formal analogy between the statical equilibrium equations and the geometric compatibility equations in the classical small-deflexion theory of thin shell structures which are free from surface tractions. It was pointed out by Goldenweiser (15, 16) and Lur'e (20, 21) who called it the 'static-geometric analogy' and it has been examined by Sanders (27), Novozhilov (25), Budiansky and Sanders (2) and Naghdi (24). Naghdi remarks that it is a kind of analogy peculiar to shell structures, and has no counterpart for solid continua within the context of classical elasticity. Long before the analogy had been stated explicitly one of its consequences was well known: that the same differential operators crop up in different guises in the equations of elastic shells. Examples of this can be found in Timoshenko and Woinowsky-Kreiger (32), Flügge (10, 11) and other texts, and in many papers.

It is odd that none of these writers who set out the analogy explicitly expresses (in print, at least) any curiosity about the *origins* of the analogy. Koiter (19), apparently alone among students of shell structures, seeks an underlying reason for the analogy, and gives an argument based on the principle of virtual work which involves complicated tensorial manipulations. We may conclude that no simple explanation of the origins of the analogy has hitherto been discovered; and that no explanation has been given for the curious restriction of the analogy to the somewhat impractical case of shells which are free from surface tractions.

This paper aims at explaining the basis of the analogy. To this end two innovations are introduced. The first is to split the shell, conceptually, into two coincident surfaces S and B, which carry separately the 'stretching' and 'bending' stresses, respectively. The second is to introduce *change of Gaussian curvature* as a prime variable in the problem. In consequence of these innovations we can not only extend the analogy to cover shells sustaining normal surface tractions, but also we can establish a simple rational connexion between the analogous equations.

In this paper the treatment is restricted to the 'shallow-shell' version of the shell equations which was originally given by Marguerre (22) and has been set out, with justification of the approximations involved, by several writers including Reissner (26), Sanders (28) and Flügge (10, 11); see also Donnell (9).

The reasoning behind this restriction is threefold. First, the main ideas which are essential for the proof are exposed more clearly for these simplified equations. Second, many practically important problems not only in simple shell theory but also in buckling (see, for example, Koiter (18)) lie within the scope of shallow-shell theory. For example, as Reissner (26) and Flügge and Elling (12) point out, in many problems of localized loading the region of interest is shallow even if the shell itself is not necessarily shallow.

Third, although the present methods allow the static-geometric analogy to be extended to the case of normal surface tractions, they also demonstrate, as we shall see, that a further extension to the case of tangential surface tractions makes the analogy unduly complicated. In these circumstances a treatment in terms of 'non-shallow' shell theory seems unwarranted. Indeed, we might be tempted to argue that these real limitations render the whole matter of the analogy a subject for academic interest only, and that effort should be rather devoted to the study of numerical schemes of computation. It is here, paradoxically, that the 'two-surface' idea shows its most promising potential, for it suggests two new variables which seem to be ideally suited to numerical work on shells of arbitrary configuration. We give a brief outline of such a numerical scheme at the end of the paper.

Gaussian Curvature and Change of Gaussian Curvature

In the same way that the curvature of a (smooth) plane curve may be defined locally as rate of change of angle with respect to arc length *($d\psi/ds$)* and shown to be equal to the reciprocal of the radius of curvature, the Gaussian curvature of a smooth surface may be defined as rate of change of solid angle with respect to surface area *($d\alpha/dA$)* and shown to be equal to the reciprocal of the product of the local principle radii, $R_1 R_2$. This second aspect of Gaussian curvature, and the concepts of principal curvatures, etc., are well known, and will not be described further here (see, for example, Hilbert and Cohn-Vossen (14)). But the first aspect is little known, and some remarks will useful. Hilbert and Cohn-Vossen (14) describe clearly the idea (due to Gauss (13)) that an arbitrary closed loop drawn on a surface is associated with a solid angle whose value is found by mapping (*via* parallel normals) onto a unit sphere. For present purposes it will be simpler to describe a related treatment (after Maxwell (23) and Caspar and Klug (6)) of a 'faceted' version of the surface consisting of small triangles (of a compact, i.e. non-elongated kind) whose vertices lie on the original surface. For such a polyhedron all of the Gaussian curvature resides in the vertices, and it may be shown (14, 23) that the solid angle subtended by any vertex is precisely equal to the 'angular defect' of the vertex, defined as 2π minus the sum of the included angles of the faces meeting at the vertex. In other words the angular defect is the 'missing' angle when a vertex is laid out flat, with a single cut and no stretching or tearing; see Fig. 1. Taking the area associated with a vertex as one-third of the sum of the areas of faces meeting at the vertex, we see that an approximation to the Gaussian curvature in the region of the vertex is given by the expression

$$\frac{\text{angular defect at a vertex}}{\text{area associated with the vertex}}.$$

This idea is directly useful in the calculation of 'Geodesic, triangulated surfaces. For the design of a nearly spherical surface or dome it is merely necessary to arrange for the angular defect at each vertex to be proportional to the associated area. For a mesh with triangles of approximately equal area a very simple scheme is to have equal angular defect at all vertices. Now the solid angle subtended by a complete enclosure is 4π, so that if there are v vertices the angular defect per vertex must be $720/v°$. (This result connects with Euler's theorem connecting the number of faces, edges and vertices of an arbitrary polyhedron: see Aleksandrov and Zalgaller (1)). Once a triangulation scheme has been decided upon (see Caspar and Klug (6) for an enumeration of these) it is easy to calculate the angles of all the triangles, and hence the sides. The same basic idea is equally helpful in the cutting of

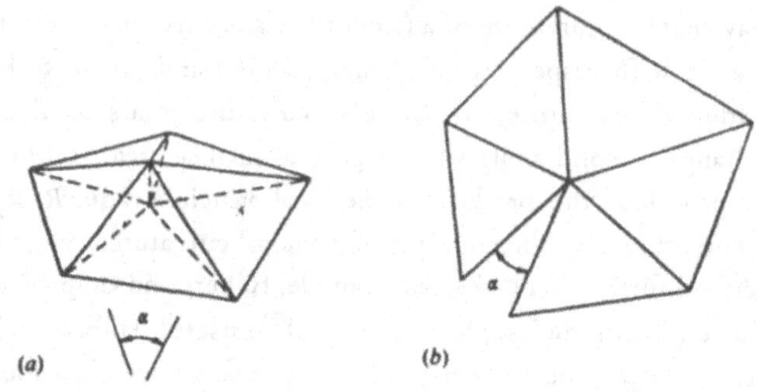

Fig. 1. *(a)* A vertex of a triangulated polyhedral version of a surface. The vertex subtends a solid angle α, shown schematically, which is precisely equal to the 'angular defect' α of the vertex defined in *(b)* which shows the same vertex flattened inextensionally after having been cut. Angle α is in radian measure.

plywood panels to make the casting surface for a doubly-curved shell roof.

This approach to Gaussian curvature brings out clearly some important features of the geometry of surfaces. The first is that inextensional deformation of a surface (in which there is neither stretching nor tearing) does not affect the value of Gaussian curvature. This follows immediately from the fact that in such a deformation the angular defects at the vertices do not change. Second, a surface with zero Gaussian curvature everywhere is *developable*, i.e. it can be deformed inextensionally into a plane. We are of course supposing in these examples that the surface has enough freedom at its edges to be unrestrained in this kind of deformation. The third feature (which will be useful later) is that two surfaces which coincide, i.e. fit each other, or are 'mutually applicable', have equal values of Gaussian curvature at every point; and that, further, equality of Gaussian curvature at corresponding points is a guarantee that the surfaces can be fitted to each other.

It is a simple exercise on the same lines (see Calladine (5)) to show that small strains in an initially flat sheet lying initially in the x,y plane impart Gaussian curvature K according to the relation

$$K = -\frac{\partial^2 \varepsilon_y}{\partial x^2} + \frac{\partial^2 \gamma_{xy}}{\partial x \partial y} - \frac{\partial^2 \varepsilon_x}{\partial y^2}, \tag{1}$$

where ε_x, ε_y and γ_{xy} are tensile and shear strains with respect to the x,y coordinate system. If the surface is constrained to be plane we have $K = 0$ and (1) reduces to the plane strain compatibility equation (e.g. (31)). Thus, any straining of a plane surface which does not satisfy the plane strain compatibility equation must impart some Gaussian curvature, and distort the surface out-of-plane. A simple physical illustration of this is provided by a plane triangulated network of bars connected by frictionless joints. Although some distributions of bar elongation leave the network planar, arbitrary patterns of bar elongation produce non-zero angular defects and thus impart Gaussian curvature to the mesh.

In the study of shell structures we shall be concerned with initially curved surfaces which are subsequently strained, thereby undergoing changes of Gaussian curvature. Throughout this paper we shall use the symbol K for the original Gaussian curvature of the unstrained surface, and the symbol g for the change in Gaussian curvature imparted by the strains.

In this paper we are concerned almost exclusively with *shallow* shells. Over the region of interest the surface makes such small angles with a particular tangent plane that the x,y coordinate system for the plane (with origin at the point of tangency) may be used for the surface without significant loss of accuracy. Within the conventional framework of 'shallow-shell' assumptions we obtain from (1) our required compatibility expression:

$$g = -\frac{\partial^2 \varepsilon_y}{\partial x^2} + \frac{\partial^2 \gamma_{xy}}{\partial x \partial y} - \frac{\partial^2 \varepsilon_x}{\partial y^2} . \tag{2}$$

The other, better known, aspect of Gaussian curvature expresses K in terms of principal radii of curvature:

$$K = \frac{1}{R_1 R_2} . \tag{3}$$

In general, when the surface suffers a small distortion the principal radii are altered and the change in value of $(1/R_i)$ is called k_i, or the change of curvature in principal direction i. Thus, by differentiating (3) as a product we obtain

$$g = \frac{k_2}{R_1} + \frac{k_1}{R_2} . \tag{4}$$

A general distortion of the surface will also involve a change of twist k_{12} of the surface: however, this is not involved in the expression for g as derived above.

In equations (2) and (4) we have two distinct formulae for change of Gaussian curvature of a surface; one is expressed in terms of surface strains and the other in terms of changes of curvature of lines in the surface. We shall need both formulae in the following development.

A Two-Surface Model of a Shell

Figure 2(a) shows the various stress resultants acting on an elementary piece of the shell, together with the applied normal surface traction (i.e. pressure) p. The diagram defines the positive senses of direct and in-plane shear stress resultants N_x, N_y, N_{xy} $(= N_{yx})$, out-of-plane shear stress resultants Q_x, Q_y and bending and twisting stress resultants

M_x, M_y, M_{xy} ($= M_{yx}$). In Fig. 2(*b*) and (*c*) these have been divided into two families; N and (Q,M). At the same time the shell has been divided into two physical entities, an *S*-surface and a *B*-surface. The *S*- surface (*b*) carries in-plane forces but is physically incapable of transmitting bending moment and transverse shear force. It is easy to envisage simple physical systems such as strings and thin sheets which can carry tensile loads but are incapable of withstanding significant shear force or bending moment. The *S*-surface, unlike these examples, is equally capable of withstanding compressive forces. The *B*-surface (*c*) on the other hand is capable of sustaining bending moment and shearing force, but it offers no resistance to in-plane forces. A mechanical analogue is more difficult to envisage than for the *S*-surface, except for a one-dimensional 'beam' version shown at (*d*). It is perhaps worth remarking that in 'small-deflexion' theory of plates, in which in-plane stretching effects are ignored, we are in effect using a '*B*-surface' model of the plate. This simple picture of the *B*-surface must be modified a little in the context of *non-shallow* shells. We shall take up this point at the end of the paper.

Later on we shall endow the *S*- and *B*-surfaces with stretching and bending stiffness respectively (or strength, in a plastic theory), but for the moment we are concerned with the physical idea of separation into a 'two-phase continuum'. A good example of the kind of separation we are considering is Terzaghi's (30) view of saturated soil as a 'soil skeleton' and 'pore-water', each with its own mechanical properties and in a sense occupying the same space simultaneously.

Now, whenever we make a *cut* in the course of structural analysis we must introduce a force variable to express the mutual reaction between the parts, and a displacement variable to describe the fact that the parts do not separate. In our two-surface model for a shell we express the force-interaction in terms of an interface stress or pressure p_B ; thus the portions of applied load p_S, p_B carried by the *S*- and *B*-surfaces respectively (see Fig. 2) are related to the applied loading p by the equilibrium equation

$$p = p_S + p_B \, . \tag{5}$$

The geometrical or kinematic aspect of the interaction is simply that the surfaces coincide not only in the original configuration but also in any subsequent distortion of the shell. This is achieved very simply by making the values of g_S and g_B for the surfaces coincide: the compatibility condition corresponding to (5) is thus

$$g_S = g_B \, . \tag{6}$$

Note that it is not necessary to define the displacement of the surfaces in order to express the compatibility condition.

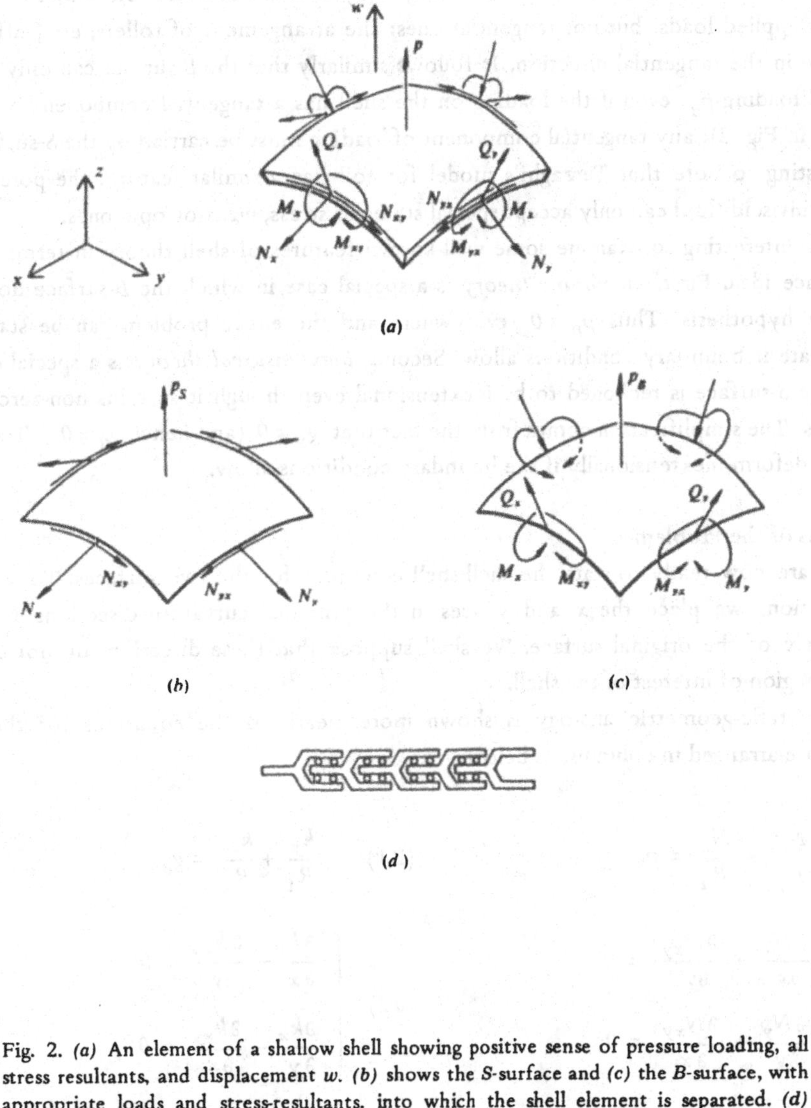

Fig. 2. *(a)* An element of a shallow shell showing positive sense of pressure loading, all stress resultants, and displacement w. *(b)* shows the S-surface and *(c)* the B-surface, with appropriate loads and stress-resultants, into which the shell element is separated. *(d)* shows a straight beam illustrating the physical characteristics of the B-surface. In *(a)* principal curvatures are in the local x, y-directions. Principal radii are counted positive if the corresponding centres of curvature are on the opposite side of the shell from the positive p and w directions.

In the foregoing discussion I have not mentioned an important point. It is clear from the 'beam' version of the B-surface shown in Fig. 2(d) that its constitution allows to accept normally applied loads, but not tangential ones: the arrangement of rollers, etc., offers no resistance in the tangential direction. It follows similarly that the B-surface can only accept a *normal* loading p_B even if the loading on the shell has a tangential component (which it does not in Fig. 2): any tangential component of loading must be carried by the S-surface. It is interesting to note that Terzaghi's model for soil has a similar feature: the pore water being an inviscid fluid can only accept special states of stress, *viz.* isotropic ones.

It is interesting to examine some well-known features of shell theory in terms of the two-surface idea. First, *membrane theory* is a special case in which the B-surface does not exist, by hypothesis. Thus $p_B = 0$ everywhere and the entire problem can be statically determinate if boundary conditions allow. Second, *inextensional theory* is a special case in which the S-surface is reckoned to be inextensional even though it sustains non-zero stress resultants. The simplification comes from the fact that $g_S = 0$ (and hence $g_B = 0$). The shell can only deform inextensionally if the boundary conditions allow.

Equations of the Problem

We are now ready to state the shell-shell equations for the two surfaces. To simplify the equations we place the x and y aces in the principal curvature directions 1 and 2 respectively of the original surface. We shall suppose that these directions do not change over the region of interest of the shell.

The static-geometric analogy is shown more clearly if the equations for the two surfaces are arranged in columns, as below.

(S1) $$\frac{N_x}{R_1} + \frac{N_y}{R_2} = p_S$$

(B1) $$\frac{k_y}{R_1} + \frac{k_x}{R_2} = g_B$$

(S2) $$\left\{ \begin{aligned} &\frac{\partial N_x}{\partial x} + \frac{\partial N_{xy}}{\partial y} = 0 \\ &\frac{\partial N_y}{\partial y} + \frac{\partial N_{xy}}{\partial x} = 0 \end{aligned} \right.$$

(B2) $$\left\{ \begin{aligned} &\frac{\partial k_y}{\partial x} - \frac{\partial k_{xy}}{\partial y} = 0 \\ &\frac{\partial k_x}{\partial y} - \frac{\partial k_{xy}}{\partial x} = 0 \end{aligned} \right.$$

(S3) $$\left\{ \begin{aligned} &N_y = \frac{\partial^2 \phi}{\partial x^2}, \quad N_{xy} = -\frac{\partial^2 \phi}{\partial x \partial y}, \\ &N_x = \frac{\partial^2 \phi}{\partial y^2} \end{aligned} \right.$$

(B3) $$\left\{ \begin{aligned} &k_x = -\frac{\partial^2 w}{\partial x^2}, \quad k_{xy} = -\frac{\partial^2 w}{\partial x \partial y}, \\ &k_y = -\frac{\partial^2 w}{\partial y^2} \end{aligned} \right.$$

(S4) $$\frac{\partial^2 \varepsilon_y}{\partial x^2} - \frac{\partial^2 \gamma_{xy}}{\partial x \partial y} + \frac{\partial^2 \varepsilon_x}{\partial y^2} = -g_S$$ (B4) $$\frac{\partial^2 M_x}{\partial x^2} + \frac{2\partial^2 M_{xy}}{\partial x \partial y} + \frac{\partial^2 M_y}{\partial y^2} = -p_B$$

(S5) $$\begin{cases} \varepsilon_x = (N_x - \nu N_y)/Et \\ \varepsilon_y = (N_y - \nu N_x)/Et \\ \gamma_{xy} = 2(1+\nu)N_{xy}/Et \end{cases}$$ (B5) $$\begin{cases} M_y = D(k_y + \nu k_x) \\ M_x = D(k_x + \nu k_y) \\ M_{xy} = D(1-\nu)k_{xy} \end{cases}$$

Consider first the S-surface equations. Equations (S1), (S2) are the three equilibrium equations for an element of the S-surface. Resolving forces out-of-plane we obtain (S1), and resolving in the x- and y-directions we obtain (S2). As we have no tangential traction, the RHS of (S2) are zero and these two equations are exactly the same as for plane stress. Accordingly, we can define an Airy stress function ϕ in (S3), which automatically satisfies (S2). Equation (S4) is the single compatibility equation, already given as (2). This completes the equilibrium and compatibility equations for the S- surface, but we also give, as (S5) the simple isotropic version of Hooke's law for the S-surface, for the sake of a subsequent example. E is Young's modulus, ν is Poisson's ratio and t is the (uniform) thickness of the shell.

Turning to the B-surface, let us start with the equilibrium equation (B4). This relates the applied pressure P_B to the bending and twisting moments on the element of Fig. 2(c): it is actually a compound of a force equilibrium equation and two moment equilibrium equations, obtained by eliminating Q_x and Q_y. There are no force-equilibrium equations in the x- and y-directions as there is no tangential loading on the B-surface, as explained already. (This is only strictly true for the shall-shell equations: see Flügge (11)). The decision to obtain a single equation by eliminating Q_x and Q_y produces an equation (B4) which is formally analogous to the S-surface compatibility equation(S4). The equation is just the same as for an element of a flat plate.

We proceed, as in plate theory, to express curvature changes k_x, k_y, k_{xy}, in terms of the normal component w of displacement in the compatibility equation (B3), which is formally analogous to the stress-function definition (S3). The positive sense of w is defined in Fig. 2. The positive sense of k_x, etc., corresponds to that for M_x, etc. Compatibility equations (B3) and are not particularly useful: they do, however, make an analogy with (S2). Finally (B1) is a compatibility condition already given, and it is obviously formally analogous to (S1). We also give as (B5) the version of Hooke's law for a uniform isotropic shell: bending stiffness D is defined by

(7) $$D = Et^3/12(1-\nu^2).$$

We list below the variables which constitute the analogy.

(8) $$\begin{array}{lll} N_x \leftrightarrow k_y & \varepsilon_x \leftrightarrow M_y & \phi \leftrightarrow -w \\ N_y \leftrightarrow k_x & \varepsilon_y \leftrightarrow M_x & g_S \leftrightarrow p_B \\ N_{xy} \leftrightarrow -k_{xy} & \gamma_{xy} \leftrightarrow -2M_{xy} & p_S \leftrightarrow g_B \end{array}$$

There are several notable features of the analogy, as follows.

(i) The minor untidiness (e.g. $\gamma_{xy} \leftrightarrow -2M_{xy}$ and various minus signs) is a consequence of arbitrary definitions and sign conventions, and is of no fundamental significance.

(ii) Force quantities on one side are analogous with strain quantities on the other side, and equilibrium equations on one side correspond to compatibility equations on the other.

(iii) Note that for subscripted quantities there is an exchange of subscripts: e.g. $M_x \leftrightarrow \varepsilon_y$, $N_x \leftrightarrow k_y$.

(iv) The notion of separating the shell into two distinct surfaces brings out the analogy between g and $p(g_S \leftrightarrow p_B;\ p_S \leftrightarrow g_B)$ to complement the well-known analogy between ϕ and w. In the usual treatment equations (S1) and (B4) are compounded by means of the overall equilibrium equation (5) and (S4) and (B1) are compounded by means of the overall compatibility equation (6). In this way the variable g disappears completely and the two compound equations are only analogous if $p = 0$, i.e. there is zero surface traction. The present scheme thus extends the analogy to pressure-loading cases.

(v) Only the normal component, w, of displacement enters explicitly the equations and the analysis. Tangential displacement components are not zero: they are simply not required (in the shallow-shell theory) in the relations between displacement and change of curvature (cf. the non-appearance of Q_x, Q_y in our equilibrium equation (B4)). It is the introduction of g as a variable which enormously simplifies the kinematic aspects of the problem.

Explanation of the Static-Geometric Analogy

Let us now investigate the reasons behind the analogy. We shall divide the analogous equations into three parts, as follows:

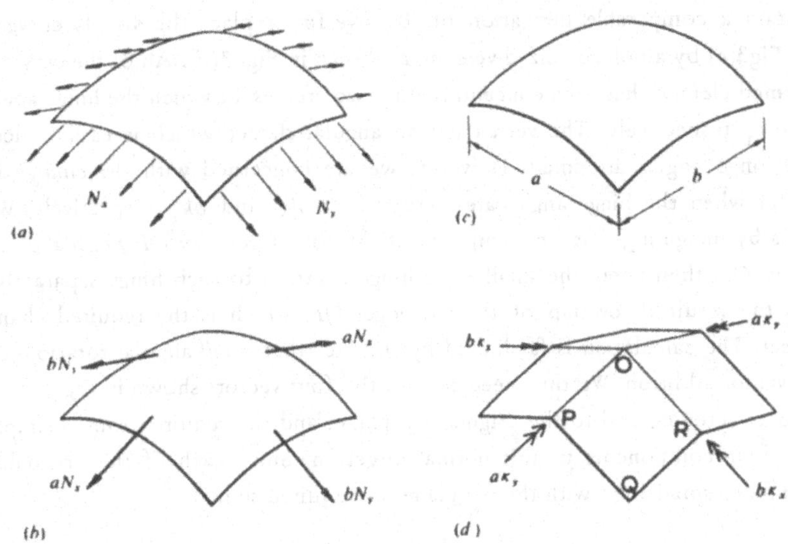

Fig. 3. *(a)*, *(b)* Diagrams relating to the derivation of equilibrium equation (S1). *(c)*, *(d)* Diagrams relating to the derivation of compatibility equation (B1); *(d)* shows a polygonalized version of the small element *(c)*, and the double-headed arrows represent additional folding corresponding to changes of curvature of the smooth element.

(i) (S 2), (S 3) ↔ (B 2), (B 3),
(ii) (S 1) ↔ (B 1),
(iii) (S 4) ↔ (B 4).

The first part (i) of the analogy is obvious from the fact that the relation between the Airy stress function ϕ and stress resultants N_x, etc., is just the same as that between displacement w and curvature changes k_y, etc. In the plane-stress equations and plate-bending equations important variables are related by second derivatives: in one case force quantities, in the other displacement quantities. Note that the subscript exchange noted in (iii) above enters at this point.

The connexion (ii) between (S1) and (B1) can be seen with the help of Fig. 3. In (*a*) we have a diagram showing the forces acting on the *S*-element. Since the element is not twisted in our chosen x, y coordinate system the stress resultants on the various edges are statically equivalent to single forces, which are inclined to the x,y plane as indicated in *(b)*.

Equilibrium of the forces normal to the x,y plane gives, for a sufficiently small element, the equilibrium equation (S1). All of this, of course, is well known.

To obtain a comparable derivation of (B1) we first replace the doubly curved shell element of Fig3(c) by a polygonalized version, as shown in Fig. 3(d). All of the curvature of the undeformed element has been concentrated in two creases, at which the hinge angles are a/R_2 and b/R_1 respectively. The vertex has an angular defect which is easily calculated when these hinge angles are small. However, we are concerned with the *change* in the angular defect when the hinge angles are increased by ak_y and bk_x respectively. We can calculate this by imagining that we clamp one of the four faces, say $OPQR$, make a slit on one edge, say OP, then apply the small extra hinge rotation to each hinge separately, and finally find the resulting overlap of the cut edges OP, which is the required change of angular defect. The calculation is facilitated by the fact that *small* angular rotations satisfy the laws of vector addition. We thus need to sum the four vectors shown in Fig. 3(d). Each pair sums to a vector normal to the original x,y plane, and the required sum is simply the sum of the four components in this normal direction. Since each of the original hinge directions makes a small angle with the x,y plane the required sum is

$$ab(k_x/R_2 + k_y/R_1).$$

Since each of the planes makes but a small angle with the x,y plane, this represents the change of angular defect to sufficient accuracy. The required result follows directly.

We see from this way of deriving equation (B1) that the key to the analogy is that both *forces* and *small angular rotations* are vectors.

The final stage (iii) of linking (S4) and (B4) is more complicated. Let us begin with a simpler problem concerning a beam, shown in Fig. 4(a). Here the problem is to find the equilibrium equation relating the bending moment $M(x)$ in the beam to the transverse loading p per unit length. The simplest plan is to use virtual work, taking a virtual transverse displacement, shown in Fig. 4(a), consisting of a shallow triangle of width $2h$ and height Δ. This gives an abrupt change in slope, or hinge rotation, at stations $x-h, x$ and $x+h$ equal to $\Delta/h, -2\Delta/h$ and Δ/h respectively; and the virtual work equations gives

$$(\Delta/h)(M_{x-h} - 2M_x + M_{x+h}) = -ph\Delta\,. \tag{9}$$

Here we have assumed that p is constant over the small interval $2h$ in x. In the limit as $h \to 0$ (and keeping $\Delta/h \ll 1$) we obtain

$$p = -d^2M/dx^2\,. \tag{10}$$

Note that by this choice of virtual displacement function we have avoided introducing shearing force to the equation.

We can use essentially the same method to derive equilibrium equation (B4) for a flat plate, as follows. Consider a plate in the x,y plane with a given distribution of bending and twisting moment. Take an arbitrary shallow pyramidal virtual normal displacement, as shown in Fig. 4(b), centred on the point in question. If the plan of the pyramid is given and the height is Δ, it is a matter of simple geometry to calculate the (small) rotation θ_i of each of the radial and peripheral hinges. Applying virtual work and assuming that p_B is constant over the area in question we find

$$p_S A \Delta / 3 = \sum_i M_i l_i \theta_i , \tag{11}$$

where A is the area of the base of the pyramid, l_i is the length of the hing i and M_i is the (mean) bending moment about the hinge.

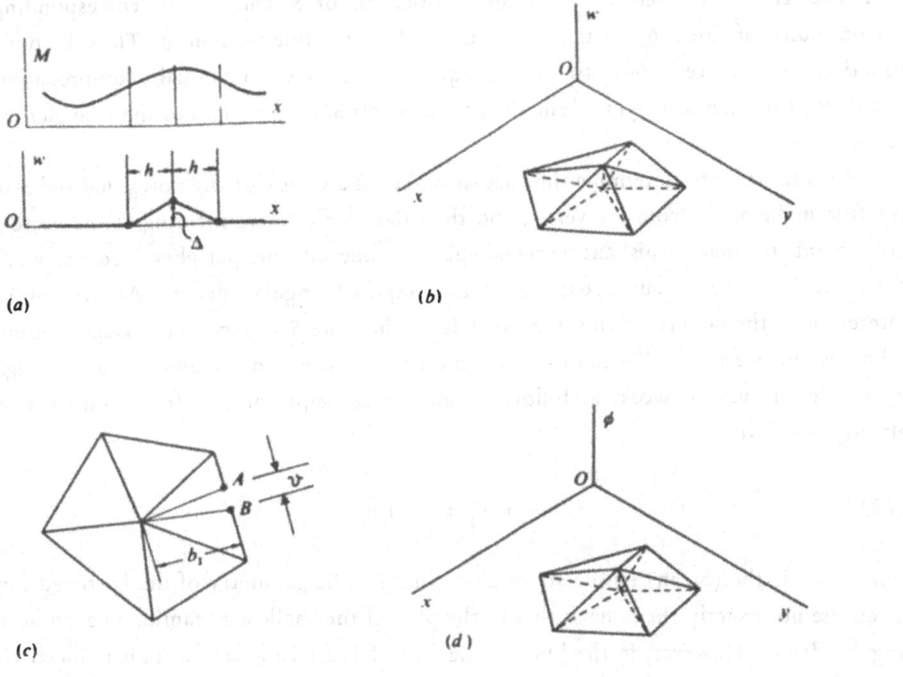

Fig. 4. Diagrams relating to the connexion between equations (S3) and (B3). *(a)* Derivation of equation (10) by virtual work. *(b)* Displacement function used for derivation of equation (B3) by virtual work. *(c)* Flattened version of a typical vertex. *(d)* Airy stress function used for derivation of equation (S3) by virtual work.

It is easy to show that if the pyramid is given a square base of side $2h$ aligned with the x, y axes (or indeed a rectangular base), application of (11) in the limit $h \to 0$ gives precisely equation (B4). It is of course necessary to express (by means of Mohr's circle or otherwise) the bending moment on the diagonal hinges in terms of M_x, M_y, M_{xy}; but this is straightforward.

We now use an analogous method to find the change in Gaussian curvature at a point on the *S*-surface in terms of the strains in the surface. Figure 4(c) shows a flattened view of a small part of the undeformed polygonalized *S*-surface, consisting of the triangles surrounding a particular vertex. When the *S*-surface is strained there will be a consequent change of angular defect, which we wish to calculate. Figure 4(d) shows a stress function ϕ which we shall employ to find a local self-equilibrating virtual force field, which in turn we shall use in a virtual work calculation. It consists of a single shallow pyramid of exactly the same form as that previously used (see Fig. 4(b)): its base matches the triangular faces which surround the vertex. From the definition of ϕ (equation (S3)) we see that all stress resultants are zero where ϕ is a linear function of x and y; but corresponding to a discontinuity of slope θ_i in the function ϕ there is a line tension θ_i. Thus the pyramidal function ϕ corresponds to a self-equilibrating set of radial compressions and circumferential tensions lying along the plan projection of the edges of the triangles meeting at the vertex.

Now in order to determine the angular defect at a vertex of the polygonalized *S*-surface we first make a cut from the vertex and then flatten the faces meeting at the vertex. It is convenient to make this cut perpendicular to one of the peripheral edges, which we designate $i = 1$. The cut opens up to the required angular defect. At present we are interested in the *change* of this angular defect when the *S*-surface in the vicinity is given an arbitrary surface strain. We can calculate the change, say v, in the dimension AB (Fig. 4(c)) by employing virtual work, as follows, using the self-equilibrating force system described already. We have

$$v \cdot \theta_1 = \sum_i \varepsilon_i l_i \theta_i \,, \tag{12}$$

where ε_i is the (mean) tensile strain along line i. The geometry of the flattened vertex is of course not exactly the same as that of the plan of the shallow pyramid, on account of the angular defect. However, in the limit as the size of the triangular facets is reduced to zero the difference becomes negligible. Now the change in angular defect is equal to v/b_1, where b_1 is the dimension defined in Fig. 4(c). The change in Gaussian curvature is equal to this divided by the associated area, $A/3$. Thus $v = g_B\, b_1 A/3$, and noting that $b_1\, \theta_1 = \Delta$ we can rearrange (12) to give

$$g_B A\Delta/3 = \sum_i \varepsilon_i l_i \theta_i . \tag{13}$$

This equation is identical in form to (11), with M_i replaced by ε_i and p_S replaced by g_B. It follows that, just as (11) may be specialized to give (B4) so (13) may be specialized to give (S4). This analysis thus establishes the formal link between the two equations. It pivots on the analogy between w and ϕ, and the use of virtual work.

Elastic Shallow-Shell Equations

So far we have been concerned entirely with analogies between the equations of equilibrium and compatibility for shallow shells. The two-surface idea continues to be advantageous when we specify the material properties for the shell. In order to illustrate this we now consider a shell of uniform thickness made from isotropic material, for which the appropriate forms of Hooke's law have already been given in equations (S5), (B5). Using equations (S3) and (S5) we can express the S-surface compatibility condition (S4) in terms of ϕ:

$$(1/Et)\,\nabla^4 \phi = g_S . \tag{14}$$

Similarly we can express the B-surface equilibrium equation in terms of w:

$$D\nabla^4 w = p_B . \tag{15}$$

Note that Poisson's ratio disappears in both sets of substitutions (notwithstanding differences of sign between (S5) and (B5)) leaving the biharmonic operator ∇^4 in both cases. In the special case of plane stress g_S is necessarily zero and we obtain the well-known governing equation $\nabla^4 \phi = 0$ for plane stress in an isotropic elastic material.

In the general case of a shallow shell we can proceed by using (S1) and (B1) together with (5) and (6) to obtain the well-known coupled equations of shallow-shell theory:

$$\frac{1}{Et}\nabla^4 \phi - \Gamma^2 w = 0 , \tag{16}$$

$$\Gamma^2 \phi + D\nabla^4 w = p , \tag{17}$$

where Γ^2 (a 'shell operator') is defined by

$$(18) \qquad \Gamma^2(\ldots) = (1/R_1)\,\partial^2(\ldots)/\partial y^2 + (1/R_2)\partial^2(\ldots)/\partial x^2.$$

Corresponding equations for non-isotropic shells can be developed once the form of Hooke's law has been given.

Boundary Conditions

We have been concerned in this paper with setting up the governing equations in a particular symmetric form, but it is appropriate to say something about boundary conditions. It is obvious from the way in which we have endowed the *S*- and *B*-surfaces with specific mechanical properties that some edge conditions will have to apply to the *S*-surface, and others to the *B*-surface. Where an edge loading is applied tangentially, it must be regarded as being carried by the *S*-surface, but a load normal to the surface will apply to the *B*-surface. Similarly, edge displacement conditions in the tangent plane apply to the *S*-surface, but normal constraints are transferred to the *B*-surface. In fact the well-known edge problems which occur in the classical theory of plates (concerning the separability of transverse shear force and twisting moment) seem to carry through to the *B*-surface.

When the surface is plane (so $\Gamma^2 = 0$), and also $p = 0$, equations (16) and (17) themselves are analogous, and are interchangeable if ϕ and w are suitably normalized. This gives a well-known analogy (32) between plane stress and the bending of plates with zero normal pressure, and the usefulness of the analogy is limited only (but in practice severely) by the zero-pressure restriction and by boundary conditions. Johnson (17) and Collins (8) have described a related analogy in plasticity between plane strain and the bending of plates, but again the absence of pressure loading from the plate side of the analogy limits usefulness.

Again, in shell theory there is an analogy between the stress function ϕ for a membrane solution (for which $p_B = 0$, as explained) when $p = 0$ and the displacement function w for an inextensional solution (for which $g_S = 0$, as explained). It is easy to furnish simple and interesting but not particularly useful examples; but as before it is boundary conditions which dominate.

General Surface Tractions and the Need for Numerical Procedures

Many practical shell structures are required to carry surface tractions. These loadings may be distributed or localized, and applied normally or tangentially. The equations developed in this paper are satisfactory for normal (pressure) loadings and indeed localized normal loads; these are carried straight through to the *B*-surface at the point of application, where they produce singularities exactly as for a flat plate; compare (7), (12) and (29).

Tangentially applied surface loads, on the other hand, present a problem. The crux of the matter is that when there are tangential surface tractions of intensity q_x and q_y in the x- and y-directions respectively, equations (S2) must be replaced by

$$\begin{cases} \dfrac{\partial N_x}{\partial x} + \dfrac{\partial N_{xy}}{\partial y} = q_x , \\ \\ \dfrac{\partial N_y}{\partial y} + \dfrac{\partial N_{xy}}{\partial x} = q_y . \end{cases} \tag{19}$$

This loading is analogous to body-force loading in plane stress, and it is easy to see (cf. Timoshenko and Goodier (31)) that the equations are satisfied by

$$N_x = \frac{\partial^2 \phi}{\partial y^2} + U; \quad N_y = \frac{\partial^2 \phi}{\partial x^2} + V; \quad N_{xy} = -\frac{\partial^2 \phi}{\partial x \partial y}, \tag{20}$$

where

$$\frac{\partial U}{\partial x} = q_x, \quad \frac{\partial V}{\partial y} = q_y .$$

So far, so good; but clearly the U, V will now have to be introduced to the compatibility equation (14). Also, in order to preserve the static-geometric analogy, (B3) must be modified to give

$$k_x = \frac{\partial^2 w}{\partial x^2} + \xi, \quad k_y = \frac{\partial^2 w}{\partial y^2} + \zeta, \tag{21}$$

where ξ, ζ are analogous to U, V. The functions ξ, ζ can be identified with initial curvatures which the shell has perhaps by virtue of initial through-thickness variations of temperature. The consequence of this can be carried through, but it is not difficult to see that there is room for confusion about the definitions of k_x, k_y. If we wish to tackle thermal stress problems, by far the simplest plan is to have

$$w = k_x = k_y = k_{xy} = 0$$

at an initial, isothermal configuration, and to introduce any thermal strains due to temperature (or analogous) variations over the surface and through the thickness by adding extra terms to Hooke's law, equations (S5) and (B5).

We conclude, then, that while the static-geometric analogy *may* be extended to include tangential surface loadings, the consequence is to complicate the essentially simple picture which we have described for the case of normal pressure loadings. In other words, attempts to generalize the analogy to include tangential loadings seem condemned by a law of strongly diminishing returns. We are thus driven to conclude that since tangential surface loadings are important in engineering practice, the static-geometric analogy is of academic interest at most.

Practical Applications

There are, however, some strong practical implications in the present work. The principal of these is that the pair of variables p_B and $g_B(=g_S)$ seem to be ideally suited to a wide range of numerical calculations. Provided we can find a satisfactory scheme (finite element or finite difference) for making the calculations discrete, we can envisage a number of ways of realizing a 'two-surface' computational scheme. The concept of a 'polygonalized' surface, which was useful in getting to grips with Gaussian curvatures, should lead naturally to suitable discrete systems. The *S*-surface is statically determinate for direct stress resultants in terms of the applied loading and the unknown interface pressure p_B, and in principle therefore g_S can be worked out in terms of p_B for a given constitutive law. The *B*-surface is kinematically determinate by virtue of the static geometry analogy, in the sense that once g_B is known the curvature changes can be worked out; hence bending moments and finally p_B can be found. Boundary conditions need to be accounted for properly at various stages, but we can see in principle the beginning of a viable computational scheme which yields a set of ordnary simultaneous equations in p_B for linear-elastic material, or iterative computations for various non-linear materials.

The above remarks apply strictly only to shallow-shell problems. But in a numerical scheme in which the shell is represented by a multi-faced polyhedron and the equations are set up for each vertex in local normal and tangential directions it is arguable that we are justified in using a basically shallow-shell scheme throughout. It is instructive to examine this proposition further, in order to see in what circumstances it is inadequate.

The shallow-shell equations involve a series of approximations concerning the application of essentially plane coordinates onto curves surfaces. But the crucial assumption is the neglect, in the present scheme, of the two equilibrium equations in the tangential directions. This general point is well known: see, for example, Reissner (26), Flügge (11). By inspection of Fig. 2(*c*) there are unbalanced forces per unit area of Q_x/R_1 and Q_x/R_2 in the *x*- and *y*-directions, respectively; and since the *B*-surface has zero direct stress resultants everywhere, equilibrium can only be preserved if they are transferred to the *S*-surface by interface shearing stresses. It is the neglect of these interface shearing stresses which is the critical simplifying step in the setting-up of shallow-shell theory. In order to obtain an estimate of the relative size of the interface shear and normal stresses, consider a simplified case in which $Q_y = 0$. Then, from Fig. 2(*c*) we find that the magnitude of the interface normal stress is dQ_x/dx, while that of the interface shear stress is Q_x/R_1. Following a line of argument which is well known in the theory of shells (see, e.g. (4), (11), (32)) we shall be justified in neglecting the interface shear stress provided the length of the zone in which Q_x varies rapidly is small in comparison with the local value of R_1. An example in which this is *not* the case is the 'torispherical' head of a thin cylindrical pressure vessel. In the toroidal region one of the principal radii is small, and may be of the same order as the width of the zone over which the shearing stress resultant varies rapidly (3). No one, of course, could regard this as a shallow shell. The same sort of difficulty arises in a more acute form in the case of junctions between shells. The lesson here is to make sure, in a numerical version of the calculations, that the equilibrium of the *B*-surface is satisfied properly. This presents no special difficulties, and indeed raises questions which must be faced anyway at the boundaries even of a genuinely shallow shell.

I am grateful to Dr. A. Klug for explaining the geometry of curved surfaces to me, and to Dr. J.A. Greenwood, Dr. P.G. Lowe, Dr. M. Pavlovic, Mr. D.J. Payne and Mr. A.C. Smith for discussing various points with me.

This article has been transcribed, with minor alterations and corrections, from the *Mathematical Proceedings of the Cambridge Philosophical Society*, vol. 82 (1977) 335-351, with the author's permission.

REFERENCES

(1) Aleksandrov, A.D. and Zalgaller, V.A. *Intrinsic geometry of surfaces* (transl. by J.J. Danskin) (Providence: American Mathematical Society, 1967).

(2) Budiansky, B. and Sanders, J.L. On the 'best' first-order linear shell theory. In *Progress in applied mechanics; the Prager anniversary volume*, pp. 129-140. (New York: Macmillan, 1963).

(3) Calladine, C.R. Creep in torispherical pressure vessel heads. In *Creep in structures: Proc. IUTAM Symposium, Gothenburg*, pp. 247-268. (Berlin: Springer-Verlag, 1972).

(4) Calladine C.R. Structural consequences of small imperfections in elastic thin shells of revolution. *Int. J. Solids Structures* 8(1972), 679-697.

(5) Calladine, C.R. Thin-walled elastic shells and analysed by a Rayleigh method. *Int. J. Solids Structures* 13(1977), 515-530.

(6) Caspar, D.L.D. and Klug, A. Physical principles in the construction of regular viruses. *Cold Spring Harbour Symposia on Quantitative Biology* 27 (1962), 1-24.

(7) Chernyshev, G.N. On the action of concentrated forces and moments on an elastic thin shell of arbitrary shape. *J. Appl. Math. Mech. (Prikl. Mat. Mekh.)* 27 (1963), 172-184.

(8) Collins, I.F. On an analogy between plane strain and plate bending solutions in rigid/perfect plasticity theory. *Int. J. Solids Structures* 7 (1971), 1057-1073.

(9) Donnell, L.H. Stability of thin-walled tubes under torsion. *National Advisory Committee for Aeronautics*, report 497 (Washington, 1933).

(10) Flügge, W. *Statik und Dynamik der Schalen*, 3rd ed., pp. 202-204 (Berlin: Springer-Verlag, 1962).

(11) Flügge, W. *Stresses in Shells*, 2nd ed., pp. 414-422 (Berling: Sprihger-Verlag, 1973).

(12) Flügge, W. and Elling, R.E. Singular solutions for shallow shells. *Int. J. Solids Structures* 8 (1972), 227-247.

(13) Gauss, K.F. General investigation of curved surfaces (in Latin) (transl. by J.C. Morehead, and A.M. Hiltebeitel), (Hewlett, New York: Raven Press, 1965).

(14) Hilbert, D. and Cohn-Vossen, S. *Geometry and the imagination*, pp. 193-204 (New York: Chelsea Publ. Co., 1952).

(15) Goldenweiser, A.L. Equations of the theory of thin shells (in Russian). *Prikl. Mat. Mekh.* 4 (1940), 35-42.

(16) Goldenweiser, A.L. *Theory of elastic thin shells* (transl. ed. A. Herrmann), pp. 92-96 (Oxford, Pergamon for A.S.M.E., 1961).

(17) Johnson, W. Upper bounds to the load for the transverse bending of flat rigid-perfectly plastic plates. Part 2. An analogy: slip-line fields for analysing the bending and torsion of plates. *Int. J. Mech. Sci.* 11 (1969), 913-938.

(18) Koiter, W.T. The effect of axisymmetric imperfections on the buckling of cylindrical shells under axial compression. *Proc. Kon. Ned. Acad. Wet.* (B) 66 (1963), 265-279.

(19) Koiter, W.T. On the nonlinear theory of thin elastic shells. *Proc. Kon. Ned. Acad. Wet.* (B) 69 (1966), 1-54.

(20) Lur'e, A.I. General theory of elastic shells (in Russian). *Prikl. Mat. Mekh. 4 (1940), 7-34.*

(21) Lur'e, A.I. On the static geometric analogue of shell theory. *In Problems of continuum mechanics; the Muskhelisvili anniversary volume*, pp. 267-274 (Philadelphia: Society for Industrial and Applied Mathematics, 1961).

(22) Marguerre, K. Zur Theorie der gekrümmten Platte grosser Formänderung. *Proc. 5th Intern. Cong. Appl. Mech., Cambridge, Mass.* IUTAM (1939), 93-101.

(23) Maxwell, J.C. On the transformation of surfaces by bending. *Trans. Cambridge Philos. Soc.* 9 (1856), 445-470.

(24) Naghdi, P.M. The theory of shells and plates. *Handbuck der Physik, Band VIa/2*, pp. 425-640 (esp. p. 613) (Berlin: Springer-Verlag, 1972).

(25) Novozhilov, V.V. *The theory of thin shells* (transl. P.G. Lowe, ed. J.R.M. Radok) (Groningen: P. Noordhoff, 1959).

(26) Reissner, E. Stresses and small displacements of shallow shells, II. *J. Math. Phys.* 25 (1946), 279-300.

(27) Sanders, J.L. An improved first-approximation theory for thin shells. *National Aeronautics and Space Administration Technical Report R24* (Washington, 1959).

(28) Sanders, J.L. Singular solutions to the shallow shell equations. *J. Appl. Mech.* 37 (1970), 361-364.

(29) Sanders, J.L. and Simmonds, J.G. Concentrated forces on shallow cylindrical shells. *J. Appl. Mech.* 37 (1970), 367-373.

(30) Terzaghi, K. *Theoretical soil mechanics*, pp. 11-15 (New York: Wiley, 1943).

(31) Timoshenko, S.P. and Goodier, J.N. *Theory of elasticity*, 2nd ed., pp. 24, 26-27 (New York: McGraw-Hill, 1951).

(32) Timoshenko, S.P. and Woinowsky-Krieger, S. *Theory of plates and shells*, 2nd ed. (New York: McGraw-Hill, 1959).

Printed in the United States
By Bookmasters